SUBSTRATE NOISE COUPLING IN MIXED-SIGNAL ASICs

Substrate Noise Coupling in Mixed-Signal ASICs

Edited by

Stéphane Donnay

IMEC, Leuven, Belgium

and

Georges Gielen

Katholieke Universiteit, Leuven, Belgium

KLUWER ACADEMIC PUBLISHERS

BOSTON / DORDRECHT / LONDON

A C.I.P. Catalogue record for this book is available from the Library of Congress.

ISBN 978-1-4419-5341-4 e-ISBN 978-0-306-48170-3

Published by Kluwer Academic Publishers,
P.O. Box 17, 3300 AA Dordrecht, The Netherlands.

Sold and distributed in North, Central and South America
by Kluwer Academic Publishers,
101 Philip Drive, Norwell, MA 02061, U.S.A.

In all other countries, sold and distributed
by Kluwer Academic Publishers,
P.O. Box 322, 3300 AH Dordrecht, The Netherlands.

Printed on acid-free paper

CONTENTS

Contributors xi

Foreword xix

Projects in the mixed-signal design cluster xxi

Introduction xxv
Georges G.E. Gielen, Stéphane Donnay
1. Context xxv
2. Book overview xxvii

Chapter 1: Technology impact on substrate noise 1
Francois J.R. Clément
1. Introduction 1
2. Substrate physics 4
   2.1 Resistive effect 4
   2.2 Capacitive effect 4
   2.3 Depletion regions 6
   2.4 Latch-up 7
3. Parasitic substrate effects 8
4. Wafer impact 11
   4.1 Lightly doped wafer 12
   4.2 Epitaxial wafer 14
5. Fabrication processes 17
   5.1 Surface implant 17
   5.2 Buried layers 18

6. Conclusions 19

Chapter 2: Substrate noise generation in complex digital systems 23
Stéphane Donnay, Marc van Heijningen, Mustafa Badaroglu
1. Introduction 23
2. Sources of substrate noise 24
3. Substrate modeling 25
4. How to measure substrate noise 26
5. First mixed-signal test chip with simple inverter chains 28
5.1 Time-domain substrate noise 30
5.2 Dominant noise coupling source analysis 31
5.3 Frequency domain substrate noise 33
6. Second test chip: a 86-Kgate digital filter bank 35
6.1 Measurement results 37
6.2 Substrate noise analysis 38
7. Conclusions 42

Chapter 3: Modeling and analysis of substrate noise coupling in mixed-signal ICs 47
Nishath Verghese, Wen Kung Chu and Jim McCanny
1. Introduction 47
2. Substrate noise analysis methodology 50
3. Modeling parasitics 51
3.1 Device/Well/Interconnect parasitics 52
3.2 Package parasitics 52
4. Substrate parasitics 53
5. Analysis of substrate noise 55
6. Analysis of impact of substrate noise 57
7. Substrate noise analysis data flow 58
8. A design example 59
9. Summary 63

Chapter 4: SPACE for substrate resistance extraction 65
N.P. van der Meijs
1. Introduction 65
2. Substrate analysis overview 68
2.1 Modeling 68
2.2 Extraction 70
3. The Boundary Element Method 72
3.1 Introduction 72
3.2 Discretization 74
3.3 Matrix inversion 75

3.4 Results ... 79
4. Parametric modeling method ... 80
4.1 Methodology ... 80
4.2 Implementation and results ... 85
4.3 Conclusion ... 86
5. Combined BEM/FEM Modeling ... 86
6. The SPACE Layout to Circuit Extractor ... 89
7. Conclusion ... 90

Chapter 5: Models and parameters for crosstalk simulation ... 93
Valentino Liberali
1. Introduction ... 93
2. Design methodology ... 95
2.1 Top-down design ... 95
2.2 Bottom-up verification ... 96
3. Modeling ... 97
4. Parameters ... 100
4.1 Package parasitics ... 100
4.2 On-chip parasitics: capacitances ... 100
4.3 On-chip parasitics: resistances ... 101
5. Simulation ... 106
6. Validation of the proposed approach ... 106
6.1 Comparison with simulations from the back-annotated netlist ... 106
6.2 Comparison with experimental measurements on a test chip ... 108
7. Conclusion ... 110

Chapter 6: High-level simulation of substrate noise generation in complex digital systems ... 113
Mustafa Badaroglu, Marc van Heijningen and Stéphane Donnay
1. Introduction ... 113
2. Library characterization ... 115
2.1 Substrate macro model characterization ... 115
2.2 Effect of load and input transition time ... 118
2.3 Gate-level VHDL library extension for monitoring the switching activities ... 120
3. Substrate noise simulation ... 121
3.1 Overview of substrate noise simulation ... 121
3.2 Chip-level substrate lumped network ... 121
3.3 Extraction of the noise sources ... 124
3.4 Substrate noise simulation ... 125

4. Experimental results ... 125
4.1 Four Bit Counter ... 125
4.2 Multiplier ... 127
4.3 Accuracy of SWAN in comparison with measurements for 86K gate digital ASIC ... 129
4.4 Speed-up of SWAN in comparison with SPICE simulations ... 130
5. Conclusions ... 132

Chapter 7: Modeling the impact of digital substrate noise on analog integrated circuits ... 135
Yann Zinzius, Georges Gielen, Willy Sansen
1. Introduction ... 135
2. Overview of substrate noise impact in analog circuits ... 138
3. Modeling the digital substrate noise impact on analog circuits ... 140
3.1 Principle of the modeling method ... 140
3.2 Description of the model extraction methodology ... 142
3.3 Illustration and validation of the modeling methodology ... 143
4. Measurements of the impact of digital substrate noise on analog designs ... 151
4.1 Digital substrate noise impact on a comparator ... 151
4.2 Experimental test chip and measurement setup ... 153
4.3 Comparator measurement results ... 155
4.4 Impact of substrate noise on an embedded analog-to-digital converter ... 157
5. Conclusions ... 159

Chapter 8: Measuring and modeling the effects of substrate noise on the LNA for a CMOS GPS receiver ... 161
Min Xu and Bruce A. Wooley
1. Introduction ... 161
2. General model of the effect of substrate noise on analog circuits ... 163
3. Substrate noise characterization ... 165
3.1 Substrate noise caused by a single digital transition ... 168
3.2 Substrate noise spectra distribution for the digital circuit emulator ... 171
4. Noise coupling into the LNA ... 174
4.1 LNA output spectrum ... 174
4.2 Noise coupling mechanism ... 175

4.3 Experimental verification 179
5. A statistical approach to substrate noise characterization for digital circuits 180
6. Conclusion 185

Chapter 9: A practical approach to modeling silicon-crosstalk in systems-on-silicon 189
Paul T.M. van Zeijl
1. Introduction 189
2. Problem statement 190
3. Limitations in state-of-the-art approaches to silicon-crosstalk 195
4. Our strategy 196
4.1 Modeling the digital circuitry 196
4.2 Modeling the analog circuitry 201
4.3 Substrate modeling for low-impedance substrate (0.35μ pure CMOS) and overall simulations 201
4.4 Substrate modeling (BiCMOS/RFCMOS substrates) 204
4.5 Overall simulations 205
5. Conclusions 207

Chapter 10: The reduction of switching noise using CMOS current steering logic 209
Maher Kayal, Richard Lara Saez and Marc Pastre
1. Introduction 209
2. Definitions 210
3. CSL inverter 210
3.1 Static Characteristics 211
3.2 Noise Margins 212
3.3 Dynamic Characteristics 213
3.4 Current Spikes 215
4. CSL NAND and NOR gates 216
5. FSCL inverter 217
5.1 Static Characteristics 218
5.2 Noise Margin 219
5.3 Dynamic Characteristics 220
5.4 Complex gates in FSCL 222
6. Experimental comparison between static logic and CSL ... 223
6.1 Switching noise sensing 223
6.2 Comparison between CSL and standard static logic in a mixed-mode application 225
7. Comparative evaluation of CSL, FSCL and

conventional static logic 227
7.1 Power consumption in CMOS conventional static logic 227
7.2 Power consumption in CMOS CSL and FSCL 228
7.3 Summary 229
8. CSL design and layout CAD tools 230
8.1 CSL libraries design CAD tool 230
8.2 CSL layout generator CAD tool 231
9. Conclusion 232

Chapter 11: Low-noise digital design techniques 233
Mustafa Badaroglu and Stéphane Donnay
1. Introduction 233
2. Reducing substrate noise generation 235
2.1 Supply current transfer function to the substrate 235
2.2 Shaping the supply current 236
2.3 Changing the supply current transfer function to the substrate 240
3. Clock tree with different latencies 242
3.1 Introduction 242
3.2 Clock region assignment 243
3.3 Folding of the supply current transients 244
3.4 Clock latency optimization 245
3.5 Experimental results 245
4. Measurements to evaluate the low-noise design techniques 248
4.1 Overview of the simulated reduction factors for the generated substrate noise 248
4.2 Time- and frequency-domain measurements 250
4.3 Effect of I/O cells 252
5. Conclusions 253

Chapter 12: How to deal with substrate bounce in analog circuits in epi-type CMOS technology 257
Bram Nauta and Gian Hoogzaad
1. Introduction 257
2. Substrate noise 258
3. Problems in analog 260
4. Strategy for analog 262
5. Examples 265
6. Conclusions 268

Chapter 13: Reducing substrate bounce in CMOS RF-circuitry 271
Domine M.W. Leenaerts
1. Introduction 271
2. Substrate bounce due to a sigma-delta modulator 274
3. Guard rings on a low-ohmic substrate 276
4. Guard rings on a high-ohmic substrate 281
5. Substrate bounce in an RF system 282
6. Concluding remarks 286

CONTRIBUTORS

Francois Clément received his Ph.D. in EE from the Swiss Federal Institute of Technology in Lausanne (EPFL), 1995. In 1996, he was visiting scholar at Stanford University. Currently, he is Layin Product Manager with Simplex Solutions (formerly Snake Technologies), France, and a research associate with the Swiss Federal Institute of Technology (EPFL). His research interests include integrated circuit fabrication and physical design.

Stéphane Donnay received the M.S. and Ph.D. degree in electrical engineering from the Katholieke Universiteit Leuven (K.U.Leuven), Belgium in 1990 and 1998 respectively. He was a research assistant in the ESAT-MICAS laboratory of the K.U.Leuven from 1990 until 1996, where he worked in the field of analog and RF modeling and design automation. In 1997 he joined IMEC, where he is now responsible for the Mixed-Signal and RF group. His current research interests are: integration of RF front-ends for digital telecommunication applications, in particular 5 GHz WLAN front-ends, chip-package co-design, modeling and simulation of substrate noise coupling in mixed-signal ICs and modeling and simulation of RF front-ends. He is responsible for many industrial projects in this area and has authored or co-authored more than 75 papers in books, journals and conference proceedings.

Marc van Heijningen was born in Laren, the Netherlands, in 1973. He received the M.S. degree in Electrical Engineering in 1998 from the Eindhoven University of Technology, the Netherlands. In 1997 he did his master thesis on 1/f noise modeling of low-power, low-voltage CMOS technologies at the Advanced Semiconductor Processing group of IMEC,

Leuven, Belgium. After that the worked in the Mixed-Signal and RF Applications group of IMEC, focusing on substrate noise coupling in mixed-signal integrated circuits. Currently he is working in the Radar Technology group of TNO Physics and Electronics Laboratory, The Netherlands, on millimeter wave MMIC design.

Mustafa Badaroglu received the B.S. degree in electrical engineering from Bilkent University, Ankara, Turkey, in 1995, and the M.S. degree in electrical engineering from Middle East Technical University, Ankara, Turkey, in 1998. He did his master thesis on design, verification and VLSI implementation of embedded microcontrollers. He is currently a researcher in the Mixed-Signal and RF Applications group of IMEC, Leuven, Belgium. He is also working towards the Ph.D. degree in electrical engineering at the Katholieke Universiteit Leuven (K.U.Leuven), Belgium. His Ph.D. promoters are Prof. Hugo De Man and Prof. Georges Gielen. At IMEC he is currently working on substrate noise coupling in mixed-signal ICs focusing on gate level characterization and reduction of substrate noise in integrated digital circuits. He has also worked on design and VLSI implementation of OFDM-based wireless LAN modems. His current research interests include deep submicron signal integrity analysis, design of low-noise digital circuits and supply distribution networks.

Nishath Verghese received the B. E degree in Electrical Engineering from Birla Institute of Technology and Science, India in 1990 and the M.S and Ph.D. degrees in Electrical and Computer Engineering from Carnegie Mellon University, Pittsburgh, PA in 1993 and 1995 respectively. He was a principal at EDA startups Apres Technologies and CadMOS Design Technology which was subsequently acquired by Cadence Design Systems. He is currently an Engineering Director with the CadMOS group at Cadence responsible for signal integrity analysis tools for mixed-signal ICs. He has several publications in the areas of noise analysis, layout verification and synthesis and is co-author of the book "Simulation Techniques and Solutions for Mixed-Signal Coupling in Integrated Circuits."

Wen Kung Chu received the B.S. degree in Electrical Engineering from National Tsing Hua University, Taiwan. He received a master's degree in Computer Engineering from the University of Southern California in 1990. He was with the mixed signal group of Cadence Design Systems, San Jose, since 1990. He joined EDA startup Apres Technologies in 1997 as principal engineer, which in October 1998 was acquired by CadMOS Design Technology, which in April 2001 was acquired by Cadence Design Systems.

He is currently a senior member of the consulting staff at Cadence, working on signal integrity analysis products for mixed-signal solutions.

Jim McCanny is currently Technical Marketing Director for signal integrity at Cadence Design Systems. He was Vice President of Business Development at CadMOS Design Technology which was acquired by Cadence in February 2001. Before CadMOS, Mr. McCanny served as the Executive Vice President of Sales and Marketing for Ultima Interconnect Technology (now Celestry) and was first Engineering Manager and then Major Account Technical Program Manager at EPIC Design Technology. He also spent ten years with Texas Instrument's Design Automation Department where he was a Member of Group Technical Staff responsible for development of static timing analysis and parameterized layout EDA tools. He received his B.Sc. in Computer Science and Mathematics from Manchester University, England in 1980.

Nick van der Meijs received the M.Sc. degree (cum laude) and the Ph.D. degree, both in electrical engineering, from Delft University of Technology, the Netherlands, in 1985 and 1992, respectively. Currently, he is an associate professor in Delft. His teaching responsibilities include VLSI design, electronic design automation and circuit theory. His research focuses on physical/electrical verification of integrated circuits, and he is one of the main authors of the SPACE layout to circuit extractor. He has also worked on various other topics, including module generation and frameworks. He is a recipient (1993) of the STW/NWO (Dutch Technology Foundation/ Netherlands Organization for Scientific Research) “Pionier” (Pioneer) research grant.

Valentino Liberali was born in Broni, Italy, in 1959. He received the degree in Electronic Engineering from the University of Pavia in 1986. In the same year he was granted a one-year scholarship from SGS (now STMicroelectronics). From 1987 to 1990 he was with Italian Nuclear Physics Institute (INFN) working on the development and characterisation of low-noise electronics for particle detectors. From 1990 to 2000 he was an Assistant Professor at the Department of Electronics of the University of Pavia. He is now an Associate Professor at the Department of Information Technologies of the University of Milan. His main research interests are the design of analog/digital interfaces and mixed-signal integrated circuits. Valentino Liberali is a member of AEI (Associazione Elettrotecnica ed Elettronica Italiana).

Yann A. Zinzius was born in Sarrebourg (France) in 1973. He received his D.E.A. in Microelectronic and Instrumentation from the Department of Physics from the Université Louis Pasteur, Strasbourg (France) in 1997. He is currently working toward a Ph.D. degree at the Katholieke Universiteit Leuven (K.U.Leuven). In 1997 he worked for the Laboratoire d'Electronique et de Physique des Systèmes Instrumentaux (L.E.PS.I), Strasbourg (France), were he was involved in the design of a Digital to Analog Converter for pixel sensors. From 1997 to 1998 he worked as a researcher in the group of microelectronics for instrumentation in the European Organization for Nuclear Research (CERN) in Geneva (Switzerland), were he was involved in the design of Analog to Digital Converters for particle physics experiments in radiation hard technology, DMILL. He is currently a researcher in the group ESAT-MICAS from the Katholieke Universiteit Leuven, where he is working on analysis and modeling of the digital noise impact on embedded analog circuits. His current research interests include the high – level analysis of the impact of substrate noise on embedded data converters.

Georges G. E. Gielen is a Professor in Electrical Engineering at the Katholieke Universiteit Leuven, Belgium. His research interests are in analog and mixed-signal design and design automation. He is responsible for many industrial projects in this area, he has published more than 200 papers in books, journals and conference proceedings, and is invited regularly for program committees of conferences and for presenting tutorials and invited speeches on analog and mixed-signal design and CAD. He is a Fellow of the IEEE, and a member of the Board of Governors of the IEEE Circuits and Systems Society. He was the 1997 Laureate of the Belgian National Academy of Science, Literature and Arts in the category of engineering sciences, and he received the 2000 Alcatel Award for innovation in telecommunications from the Belgian National Fund of Scientific Research.

Willy Sansen was born in Poperinge, Belgium on May 16, 1943. He received the M.Sc. degree in electrical engineering from the Katholieke Universiteit (K.U.) Leuven in 1967 and the Ph.D. degree in electronics from the University of California, Berkeley, in 1972. Since 1981 he has been full professor at the ESAT laboratory of the K.U.Leuven. During the period 1984-1990 he was the head of the Electrical Engineering Department. He was a visiting professor at Stanford University in 1978, at the Federal Technical University Lausanne in 1983, at the University of Pennsylvania, Philadelphia in 1985 and at the Technical University Ulm in 1994.

He has been involved in design automation and in numerous analogue integrated circuit designs for telecom, consumer electronics, medical applications and sensors. He has been supervisor of 39 Ph.D. theses in that field and has authored and co-authored more than 400 papers in international journals and conference proceedings and six books, among which the textbook (with K. Laker) on Design of Analog Integrated Circuits and Systems. Dr. Sansen is member of several editorial committees of journals, such as the IEEE Journal of Solid-State Circuits, Sensors and Actuators, and High Speed Electronics. He is a member of the executive and program committees of the IEEE International Solid State Circuits Conference, and program chair of the ISSCC'02 conference. He servers regularly on the program committees of conferences such as ESSCIRC, ASCTT, Eurosensors, and Transducers.

Min Xu received the B.S degree in Physics from the University of Science and Technology of China in 1994, and the M.S. and Ph.D. degrees in Electrical Engineering from Stanford University, Stanford, CA, in 1997 and 2001, respectively. She is now a senior design engineer at Big Bear Networks, Inc., Milpitas, CA, working on high-speed electronics for optical communications with data rate up to 40 Gb/s. Her research interests include substrate noise in mixed-signal circuits, system and integrated circuit design.

Bruce A. Wooley is the Robert L. and Audrey S. Hancock Professor of Engineering and the Chairman of the Department of Electrical Engineering at Stanford University. He received the B.S., M.S. and Ph.D. degrees in Electrical Engineering from the University of California, Berkeley in 1966, 1968 and 1970, respectively. From 1970 to 1984 he was a member of the research staff at Bell Laboratories in Holmdel, NJ, and he joined the faculty at Stanford in 1984. His research is in the field of integrated circuit design, where his interests include oversampling A/D and D/A conversion, low-power mixed-signal circuit design, circuit design techniques for video and image data acquisition, high-speed embedded memory, high-performance packaging and testing, noise in mixed-signal integrated circuits, and circuits for high-speed communications. Prof. Wooley is a Fellow of the IEEE and the Past President of the IEEE Solid-State Circuits Society. He has served as the Editor of the *IEEE Journal of Solid-State Circuits* and as the Chairman of both the International Solid-State Circuits Conference (ISSCC) and the Symposium on VLSI Circuits. He was awarded the University Medal by the University of California, Berkeley, and he was an IEEE Fortescue Fellow. He was also a recipient of the IEEE Third Millennium Medal. He has published more than 130 technical articles and is a co-author of *The Design of Low-Voltage, Low-Power Sigma-Delta Modulators*.

Paul T.M. van Zeijl received his Masters Degree from the Delft University of Technology, Delft, the Netherlands in 1985. He received his

Ph.D. in 1990 from the same University for the study towards a fully integrated FM Upconversion Receiver Front-end with fully-integrated on-chip SAW filters. He was awarded the "VEDER PRIJS" for his contribution to the succesfull integration of radio circuitry in 1991. From 1991 onwards he works for Ericsson. From 1991-1997 he worked on the standarization of DECT and on the realization of integrated circuits for DECT cordless phones and base-stations. From 1997 onward he works on the integration of radio and system-on-a-chip ASICs for paging and Bluetooth in deep-submicron CMOS technology. He is a Senior Specialist in radio ASIC integration since 2000. His interests are analog, radio and RF integrated circuits in bipolar and CMOS technologies. He has written several technical papers and hold several patents.

Maher Kayal was born in 1959 in Lebanon, received his Master and Ph.D. degree in electrical engineering from the Swiss Federal Institute of Technology (EPFL) in 1983 and 1989 respectively. He is now a Full professor in the electronics laboratories of the Swiss Federal Institute of Technology. His current research & development includes: Low Power Low Voltage Mixed-mode circuit design, sensors interface, signal processing and CAD tools for analog integrated circuits. He authored or co-authored over 50 scientific publications. He received in 1990 the Ascom award for the best work in telecommunication fields and in 1997 the best ASIC award at the European Design and Test Conference ED&TC.

Richard Lara Sàez was born in 1967 in Chile, received his Master in Electrical Engineering from the Universidade de São Paulo, SEL-EESC, Brazil in 1993 and Ph.D. degree from the Swiss Federal Institute of Technology (EPFL) in 1998. His current research & development includes: Mixed-mode circuit design and RF. Since 1999 he is with Motorola Brazilian design center.

Marc Pastre was born in 1977 in Switzerland, received his Master degree in Computer Science Engineering from the Swiss Federal Institute of Technology (EPFL) in 2000. He is currently working towards his Ph.D. in the electronics laboratories of the Swiss Federal Institute of Technology. His research interests include mixed-mode circuit design, digital trimming of analog circuits and CAD tools for analog and mixed-mode design.

Bram Nauta was born in Hengelo, The Netherlands, in 1964. In 1987 he received the M.Sc degree (cum laude) in Electrical Engineering from the University of Twente, Enschede, The Netherlands. In 1991 he received the Ph.D. degree from the same university on the subject of analog CMOS filters for very high frequencies. In 1991 he joined the Mixed-Signal Circuits and Systems Department of Philips Research, Eindhoven the Netherlands, where he worked on high speed AD converters. From 1994 he led a research group in the same department, working on "analog key modules". In 1998 he returned to the University of Twente, as full professor heading the IC Design group in the MESA+ Research Institute and department of Electrical Engineering. His current research interest is analog CMOS circuits for transceivers. Besides, he is also part-time consultant in industry and in 2001 he co-founded Chip Design Works. His Ph.D. thesis was published as a book: Analog CMOS Filters for Very High Frequencies, Kluwer, Boston, MA, 1993. He holds 9 patents in circuit design and he received the "Shell Study Tour Award" for his Ph.D. Work. From 1997-1999 he served as Associate Editor of IEEE Transactions on Circuits and Systems -II; Analog and Digital Signal Processing. In 1998 he served as Guest Editor, and from 2001 as Associate Editor for IEEE Journal of Solid-State Circuits.

Gian Hoogzaad was born in Blokker, The Netherlands, in 1972. He received the M.S. degree (with honors) in electrical engineering from the University of Twente, Enschede, The Netherlands, in 1996. His graduation research was on the subject of white and 1/f noise in CMOS ring oscillators. In the same year, he joined Philips Research Laboratories, Eindhoven, The Netherlands, as a Member of the Mixed-Signal Circuits and Systems group. He worked on different subjects like substrate noise sensitivity in analog circuits and high-speed, high-resolution analog-to-digital converter design for video and broad-band communication applications. In 2001, he joined Philips Semiconductors, Delft, The Netherlands, as a Member of the Delft Design Center of Business Line Standard Analog. Currently he is designing products in the field of power management and motor driver ICs.

Domine M. W. Leenaerts studied electrical engineering at Eindhoven University of Technology. He gained his Ph.D. in 1992. From 1992 until 1999 he was with this university as an associate professor of the micro-electronic circuit design group. His focus was on analog circuit design and circuit theory (e.g. vco, analog memory, pll, analog multipliers, ADC/DAC). In 1995, he has been a Visiting Scholar at the Department of Electrical Engineering and Computer Science of the University of California, Berkeley and visiting professor at the Electronic Research Laboratory of the same department. In 1997 he has been an invited professor at the Technical University of Lausanne (EPFL). Dr. Leenaerts is a principal scientist at Philips Research Laboratories, Eindhoven, where he is involved in RF integrated transceivers design. Dr. Leenaerts is IEEE Distinguished Lecturer and Associate Editor of the IEEE Transactions of Circuits and Systems –

part I. His research interests includes nonlinear dynamic system theory, ADC/DAC design and RF and microwave techniques. He has published over 100 papers in scientific and technical journals and conference proceedings. As co-author, he received the best paper award 1995 of the Int. Journal of Circuit Theory and Applications. He has written 3 books among which *Circuit Design for RF Transceivers*, Kluwer.

FOREWORD

This book is the first in a series of three dedicated to advanced topics in Mixed-Signal IC design methodologies. It is one of the results achieved by the Mixed-Signal Design Cluster, an initiative launched in 1998 as part of the TARDIS project, funded by the European Commission within the ESPRIT-IV Framework. This initiative aims to promote the development of new design and test methodologies for Mixed-Signal ICs, and to accelerate their adoption by industrial users.

As Microelectronics evolves, Mixed-Signal techniques are gaining a significant importance due to the wide spread of applications where an analog front-end is needed to drive a complex digital-processing subsystem. In this sense, Analog and Mixed-Signal circuits are recognized as a bottleneck for the market acceptance of Systems-On-Chip, because of the inherent difficulties involved in the design and test of these circuits. Specially, problems arising from the use of a common substrate for analog and digital components are a main limiting factor.

The Mixed-Signal Cluster has been formed by a group of 11 Research and Development projects, plus a specific action to promote the dissemination of design methodologies, techniques, and supporting tools developed within the Cluster projects. The whole action, ending in July 2002, has been assigned an overall budget of more than 8 million EURO.

The novelty of the TARDIS initiative is that in addition to the standard R&D work, the participating projects have a compromise to publicize the new methodological results obtained in the course of their work. A Cluster Coordinator, Instituto de Microelectrónica de Sevilla, in Sevilla (Spain) has the role to coordinate and promote actions to carry out effectively the dissemination work and foster cooperation between the participating projects. All public results from the dissemination action are available from the Cluster Web site (http://www.imse.cnm.es/esd-msd).

Mixed-Signal design is a critical part for many IC designs. The advantages brought by System-on-Chip will only be fully successful if techniques are developed that allow coexistence of high-perfomance analog functions sharing a common substrate with large blocks of digital functions. Interfaces between the analog and the digital world, materialized in data converters will always be present in any mixed-signal design, and he verification of those embedded analog functions, may be in many cases the factor limiting the production-test throughput. New technologies, like Silicon-on-Insulator (SOI), offer interesting possibilities for the design of mixed-signal ICs, but require the mastering of new design techniques. The work of projects in the Cluster has been focused on four main areas (Substrate Noise Coupling, Advanced Data Converters, Testability and Special Technologies).

This book addresses the specific problem of Substrate Noise Coupling in Mixed-Signal circuits and incorporates the results achieved by the Cluster projects with activity in that area complemented by contributions from external experts that have occasionally participated in activities organized by the Cluster.

We hope that the reader will find this book useful, and we would like to thank all partners of the MSD Cluster for contributing to the success of the initiative. Special thanks are given to all the authors and to the editors for their effort to make this book a reality.

José Luis Huertas, Juan Ramos-Martos; Sevilla, September 2002

PROJECTS IN THE MIXED-SIGNAL DESIGN CLUSTER:

ABACUS: Active Bus Adaptor and Controller for Remote Units

The objective of this project is the development of an integrated circuit for space applications, that implements the analog/digital interface between the spacecraft On-Board Data Handling (OBDH) bus, and the Remote Terminal Units (RTUs). The design will use 0.8μm SOI technology.

BANDIT: Embedding Analog-to-Digital Converters on Digital Telecom ASICs

The goal of BANDIT is to develop a general design methodology for embedding high-speed analog/digital converters (ADCs) on large digital telecom ASICs, with special attention to the problems caused by mixed-signal integration.

HIPADS: High-Performance Deep Sub-micron CMOS Analog-to-Digital Converters using Low-Noise Logic

The aim of this project is to develop different A/D Converters in deep sub-micron digital CMOS process, using a new Current Steering Logic (CSL) family approach that has the property of inducing a very low substrate noisc. The converters are intended to become integrated components of larger systems, and should be considered presently as products under specifications covering end-user applications.

MADBRIC: Mixed Analog-Digital Broadband IC for Internet Power-Line Data Synchronous Link

The project main objective is the development of prototype building blocks of a chipset for high-speed communications through the power lines, that will improve achievable data rates using state of the art mixed-signal integrated circuits and DSP techniques.

MIXMODEST: Mixed Mode in Deep Submicron Technology

The technical target of the MIXMODEST project is to develop design techniques that permit the implementation of mixed-signals systems in the most advanced 0.35μm and 0.25μm deep sub-micron digital CMOS technology.

OPTIMISTIC: Optimisation Methodologies in Mixed-Signal Testing of ICs

The OPTIMISTIC project, concerned with Optimisation Methodologies in Mixed-Signal Testing of ICs, aims at the development and introduction of advanced test generation in mixed-signal IC design. Building upon existing advanced tools for control and test systems, a new approach is to be developed that will allow the mixed-signal chip designer to take large responsibility in the generation of test as part of the design activity.

RAPID: Retargetability for Reusability of Application-Driven Quadrature D/A Interface Block Design

This project is concerned with the development of an advanced methodology for the design of a mixed-signal application-driven quadrature D/A interface sub-system, aiming at its reusability by a retargeting procedure with minimal changes to their structural sub-blocks.

SUBSAFE: Substrate Current Safe Smart Power IC Design Methodology

The overall technical objective of this project is to develop a design methodology that employs device and circuit simulation to assure IC digital functionality under current injection in the substrate produced by forward bias conditions in N-wells (i.e. during switching of power stages driving inductive loads). The design methodology will change from the current largely empirical approach to Computer-Aided Design guided critical parameter evaluation, validated by a relatively small number of measurements.

SYSCONV: Systematic Top-Down Design and System Modeling of Oversampling Converters

This project develops a system-level model for oversampling delta-sigma converters suitable for use in mixed-signal system simulations and verifications. It addresses the development of a model of the entire converter as a block on its own, that can then be used in efficient mixed-signal system simulations where the converter is only a block in the overall system

TERMIS: High-Temperature / High-Voltage Mixed Signal SOI ASICs for Aerospace Applications

The project addresses the development of a fully integrated high-voltage driver IC for two different electromagnetic micro-motors which are dedicated for satellite applications. Each circuit, in die form, will be packaged in the corresponding micro-motor. The systems must operate at 200°C under a 30V power supply and must survive space irradiation.

VDP: Video Decoder Platform

This project develops a prototype video decoder platform. The result will be an IC that captures video signals and decodes the information for use in, for instance digital TV, set top boxes, and PC video capture. It will exploit innovative architectures trading signal to noise ratio versus accuracy, decoding both analog and digital video sources.

INTRODUCTION

Georges G.E. Gielen
ESAT-MICAS, Katholieke Universiteit Leuven, Belgium

Stéphane Donnay
IMEC-DESICS, Leuven, Belgium

1. CONTEXT

Driven by cost-constrained applications such as telecommunications, computing and consumer/multimedia and facilitated by the continuing miniaturization in the CMOS ULSI technology, the microelectronics IC market is characterized by an ever-increasing level of integration complexity. Today complete systems, that previously occupied one or more boards, are integrated on a few chips or even on one single multi-million transistor chip – a so called System-on-Chip (SoC). Examples are single-chip cameras or new generations of integrated telecommunication systems that include analog, digital and eventually radio-frequency (RF) sections on one single chip. Although most functions in such integrated systems are implemented with digital or digital signal processing (DSP) circuitry, the analog circuits needed at the interface between the electronic system and the continuous-valued outside world are also being integrated on the same die for reasons of cost and performance. Modern System-on-Chip designs are therefore more and more mixed-signal, and this will even be more prevalent if we move towards the intelligent homes, the mobile road/air offices and the wireless workplaces of the future.

Unfortunately, the integration of both analog/RF and digital circuits on the same die not only offers many benefits, but also creates some technical difficulties, especially in ultra-deep-submicron CMOS technologies. Since the analog circuits exploit the low-level physics of the fabrication process, they remain difficult and costly to design, but they are also vulnerable to any kind of noise or crosstalk signals. The higher levels of integration (moving towards 100 million transistors per chip clocked at ever higher frequencies) make the mixed-signal signal integrity problem increasingly challenging. One of the most important problems is the parasitic supply and substrate noise coupling, caused by the fast switching of the digital circuitry that then propagates to the sensitive analog circuitry via the common substrate.

The continuously on-going increase in speed and complexity of the digital circuitry on mixed-signal integrated systems also means an increase of the amount of digital switching noise generated by this circuitry. This noise is coupled into the substrate, which is shared with the sensitive analog circuits. Also the supply and substrate connection networks play a role here, since the inductances of the bondwires create ringing and this may even be a very significant contributor to the substrate noise. At the same time, the performance and precision levels required from the analog circuits will also increase as dictated by today's applications such as emerging communication systems (e.g. WLAN). This goes together with an increase of the sensitivity or the susceptibility of the analog circuits to digital substrate noise. It is therefore important to be able to predict the impact of digital switching noise on the analog circuit performance at the design stage of the integrated system, before the chip is taped out for fabrication. Methodologies and tools for substrate noise analysis and simulation at all stages of the design flow (before and after layout) are therefore needed to anticipate this problem.

There are three aspects to such a substrate noise analysis and simulation methodology for mixed-signal integrated systems. The first is the modeling of the digital switching noise injected in the substrate. Note that this depends on the activity level (the amount of switching) of the digital gates, and therefore depends on the signal patterns. As a result the injected noise is both non-stationary time-varying as well as frequency-dependent. The second part of the analysis methodology is the analysis of the transmission of the noise from the source (the digital circuitry) to the reception point (the analog circuitry embedded in the same substrate). This requires a modeling of the substrate, which can be considered as a kind of resistive/capacitive mesh. For CMOS technologies with high-ohmic substrates the resistive nature of the substrate has to be fully taken into account, while for low-ohmic substrates the bulk can be considered as one equipotential node leaving only the epi layer as a resistive layer. Finally, the third part of the analysis methodology is the modeling of the impact of substrate noise on the analog side. The analog circuitry is not a single noise reception point but has many noise sensing nodes that all have a different sensitivity to the noise. This analysis therefore can become quite complex and time consuming for large analog circuitry such as entire front-ends. Hence it may be needed to introduce higher-level (behavioral or macro) modeling for the analog circuits in order to make this analysis tractable.

In addition, besides an analysis methodology, also design guidelines and techniques to reduce or avoid the substrate noise problem need to be developed. This requires measures to both quiet the talker (the digital circuitry), to increase the transmission impedance between talker and listener, and to desensitize the listener (the analog circuitry). Some of the measures can be executed at technology level, others at circuit design level or at the layout level.

The purpose of this book is to provide an overview of very recent research results in the field of substrate noise analysis. Much of the reported work has been established as part of the Mixed-Signal Initiative of the European Union. It is a representative sampling of the current state of the art in this area of substrate noise analysis and reduction techniques. This volume complements other similar volumes that focus on analog and RF circuit design techniques. The book consists of 13 contributions that will briefly be introduced next.

2. BOOK OVERVIEW

The first five chapters describe **techniques for modeling the substrate**, in relation to the technology used, and present some substrate noise measurements on experimental ASICs.

Chapter 1 presents an overview of technology impacts on substrate noise. The electromagnetic substrate behavior of integrated circuits (IC's) is reviewed and the significant parasitic phenomena are presented. The technology impact is examined from three complementary points of view. The respective influence of the lightly-doped and epitaxial wafers is detailed. Fabrication process steps changing the substrate characteristics are addressed for CMOS and bipolar technologies. Die attachment is considered as a means to reduce substrate parasitics.

Chapter 2 presents a SPICE-level modeling technique for the analysis of substrate noise generation by digital circuits on low-ohmic substrates. Two experimental ASICs in a low-ohmic epi-type CMOS technology are presented. The ASICs contain digital noise generating circuits and analog substrate noise sensor amplifiers that can measure the substrate voltage directly. The first ASIC contains some simple digital test structures and is used to verify the SPICE substrate models. The second ASIC contains a 86-Kgate digital multirate filterbank for telecom applications. The measurements on both ASICs provide valuable insight into the mechanisms of substrate noise generation in digital circuits.

Chapter 3 presents techniques for the modeling and analysis of substrate noise coupling in mixed-signal integrated circuits. The physical phenomena responsible for the creation of the undesired signals as well as the transmission mechanisms and media are described. Modeling and analysis techniques to quantify the noise coupling phenomena are presented. A computer-aided design methodology based on the modeling approaches and developed for the analysis and design of noise coupling in mixed-signal integrated circuits is described and illustrated for some practical designs.

Chapter 4 then presents a general introduction to the problem of substrate-resistance extraction and gives an overview of three extraction techniques: a boundary-element method (BEM), an empirical parametric method, and a combination of a BEM with a finite-element method (FEM). All three methods exhibit a different but useful accuracy/performance trade-off and suit different situations in the design flow. It will also be shown how to produce reduced-order equivalent circuits (rather than full detailed models that mandate a-posteriori model-order reduction techniques to be useful) and how this can actually reduce extraction time and memory. The methods are implemented in the SPACE layout-to-circuit extractor that is a comprehensive tool for transforming a layout into a netlist with all relevant parasitics, including the substrate resistances.

The above methods start from a completed layout. It would however also be interesting to predict the problem at early stages of the design. *Chapter 5* therefore describes a simplified model for the analysis of crosstalk effects in deep-submicron CMOS technologies. It is described how the substrate bias resistance value can be obtained either from technology parameters or by experimental measurements on a test structure, and crosstalk effects can then be easily estimated through a SPICE-level simulation. The proposed approach is validated by means of a test chip.

The next four chapters **describe techniques to model the digital substrate noise injection as well as the analog substrate noise reception.** This is illustrated with several design examples.

Chapter 6 describes a methodology for the high-level simulation of substrate noise generation in complex digital systems. Existing approaches usually extract the model of the substrate from layout information and then simulate the extracted transistor-level netlist with this substrate model using a transistor-level simulator like SPICE. For large digital circuits the substrate simulation is however not feasible with a transistor-level simulator. A high-level methodology is therefore presented that simulates the substrate noise generation in EPI substrates by taking the noise coupling from the switching gates and also from the supply rails into account. Experimental results show good accuracy results while maintaining a speedup of three orders of magnitude with respect to SPICE simulations. The approach is also applied to an 86K digital ASIC and compared to measurements.

Chapter 7 on the other hand describes a methodology to model the impact of digital substrate noise on analog integrated circuits embedded in mixed-signal integrated systems. A high-level substrate sensitivity modeling methodology is presented that allows simulating the impact in acceptable CPU times. Measurements are also presented on an embedded comparator, that show the important impact of the digital noise on this design. The measurement results are used to predict the impact on the performance of an embedded analog-to-digital converter.

Chapter 8 presents measurements and modeling results from another design example. A general model for the effects of substrate noise on analog circuits is described, and the fundamental coupling mechanisms are revealed using a frequency-domain approach. Measurement results of the substrate noise induced by an experimental digital circuit emulator for different operating conditions are described, as well as an analysis of both the time-domain and frequency-domain characteristics of this noise. Next, the measured effects of substrate noise on the performance of an LNA for a CMOS GPS receiver are presented and explained using the developed models. Finally, a statistical approach is outlined to a generalized modeling of the substrate noise generated in digital circuits.

Finally, *chapter 9* will demonstrate a simple approach in modeling crosstalk on silicon. By splitting the problem into three parts (the digital interference caused by the digital circuitry or source, the transfer of interference in the substrate, and the (undesired) reception of the interference by the analog part) and modeling these three parts in a simple, yet effective manner, simulations for the complete system can easily be done. A comparison of measured data and simulation results shows the effectiveness of the approach for a low-ohmic substrate. A second application, a single-chip Bluetooth ASIC, demonstrates the approach in a system-on-chip.

The final four chapters then present **techniques to reduce the effect of switching noise** in embedded systems.

Chapter 10 explains the reduction of switching noise using CMOS current-steering logic. The main advantage of the current-steering technique is the small amount of noise generated during state commutations of logic gates. However, it presents a steady state consumption, which is considered as a limitation for low power applications when compared to the conventional static logic.

Since in most cases ringing of the power supply is the major source of substrate noise generation, techniques targeting at shaping the supply current and its transfer function to the substrate can reduce substrate noise generation significantly. *Chapter 11* describes such reduction techniques, which modify the supply current and its transfer function, and therefore which reduce the substrate noise. To demonstrate the techniques, a mixed-signal ASIC is fabricated in a 0.35μm CMOS epi process. The test chip contains one reference design and two digital low-noise designs with the same basic architecture. Measurements show more than a factor of 2 on average in RMS noise reduction with penalties of 3% in area and 4% in power for the low-noise design employing a supply-current waveform shaping technique based on a clock tree with latencies. The second low-noise design employing separate substrate bias for both n and p-wells, dual-supply, and on-chip decoupling, achieves more than a factor of two reduction in RMS noise, with however a 70% increase in area but with a 5% decrease in power consumption

Chapter 12 describes how to deal with substrate bounce in analog circuits in epi-type CMOS technology. Although measures are known to reduce substrate noise, the noise will never be completely eliminated since this requires larger chip area or exotic packages and thus higher cost. Analog circuits on digital ICs simply have to be resistant to substrate noise. A general strategy is given which can be summarized as: the supply of the analog circuits must be referred to the substrate and the analog signals must be referred to a clean analog ground. Furthermore several design constraints are given to minimize the effect of substrate noise on analog. Two bandgap circuits are discussed and it is shown that apparently minor design issues, such as the connection of an n-well of a PMOS differential pair, can have large impact on the substrate sensitivity of this circuit. This has been verified by measurements.

Finally, *chapter 13* describes techniques to reduce substrate bounce in CMOS RF-circuitry. The use of guard rings as a mean to reduce the effects of substrate bounce in a mixed-signal IC are commented. Measurements are reported on lightly and heavily doped substrates in several CMOS technologies. Furthermore, the problems of substrate bounce in RF applications where the substrate bounce is caused by digital circuitry, are described.

The editors wish the reader much pleasure in exploring the different chapters in this book, and in adopting the presented techniques in his/her daily practice to reduce the impact of supply and substrate noise couplings in analog, RF and mixed-signal integrated circuits and systems-on-chip, enabling in this way more and more powerful and reliable designs that will make our lives easier and more comfortable in the years to come.

The editors would like to thank all chapter authors who contributed to this book. We also thank the Mixed-Signal Cluster Coordinator CNM-Sevilla for giving us the opportunity to assemble this book.

Stéphane Donnay, Georges Gielen, September 2002

Chapter 1

TECHNOLOGY IMPACT ON SUBSTRATE NOISE

Francois J.R. Clément
EPFL, Lausanne, Switzerland

Abstract: The electromagnetic substrate behavior of integrated circuits (IC) is reviewed and the significant parasitic phenomena are presented. The technology impact is examined from three complementary points of view. The respective influence of the lightly doped and epitaxial wafers is detailed. Fabrication process steps changing the substrate characteristics are addressed for CMOS and bipolar technologies. Die attachment is considered as a means to reduce substrate parasitics.

1. INTRODUCTION

With the increasing complexity of mixed digital-analog designs, and with the decreasing feature size of current technologies, taking into account parasitic coupling through the substrate has become a key issue in reducing time-to-market of new circuits [1, 2]. Understanding the wide variety of parasitic effects represents a major concern. Additionally, the increasing impact of IC fabrication technology on parasitic behaviors is especially challenging in terms of circuit design [7-13] and modeling [14-26].

The fabrication of an integrated circuit entails a long and costly process. Current processes use planar technology [3, 5]. In this method, an initial bulk material is altered through successive steps to create electrical devices. The bulk material is either lightly doped (i.e. doping concentration around 10^{15} cm^{-3}) or epitaxial (i.e. a lightly-doped layer built on top of heavily doped material with a concentration of 10^{19} cm^{-3}), according to the type of wafer introduced in the fabrication facility. The main process steps include growth, oxidation, deposition, implantation and diffusion. A set of masks is used to determine which areas of the circuit are concerned by each process step.

Figure 1-1 illustrates how the original substrate is modified uniformly or selectively, using masks.

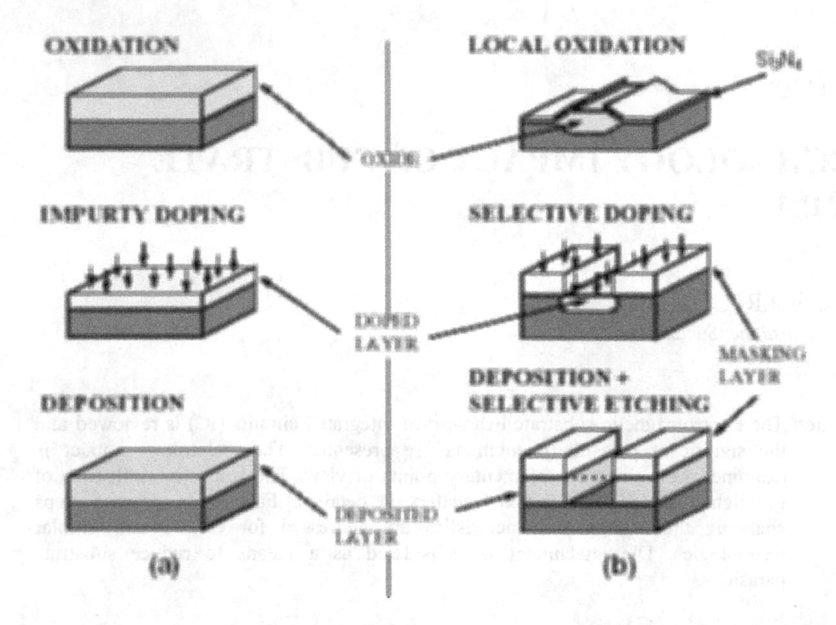

Figure 1-1. Schematic representation of forming layers in silicon with planar technology: (a) uniform, (b) selective.

The masks are computed from graphic geometries known as the "layout". Hence, as shown in *Figure 1-2*, the vertical structure of an integrated circuit at a particular surface point (x, y) is completely determined by the process and by the layout.

Today's technologies have become very complex in order to reduce size and to preserve characteristics of integrated devices. There exist many different processes from the most common to the more specific that are suitable for particular applications such as high-speed or high-voltage applications.

The number of fabrication steps varies from 100 for the simplest technologies to 400 for the most complex. The corresponding number of masks varies between 10 and 40. Hence, for a specific technology, the vertical doping profile of the substrate is completely determined by the combination of the masks. The substrate characteristics will change significantly with the wafer type and the fabrication process, as well as the die-attachment technique.

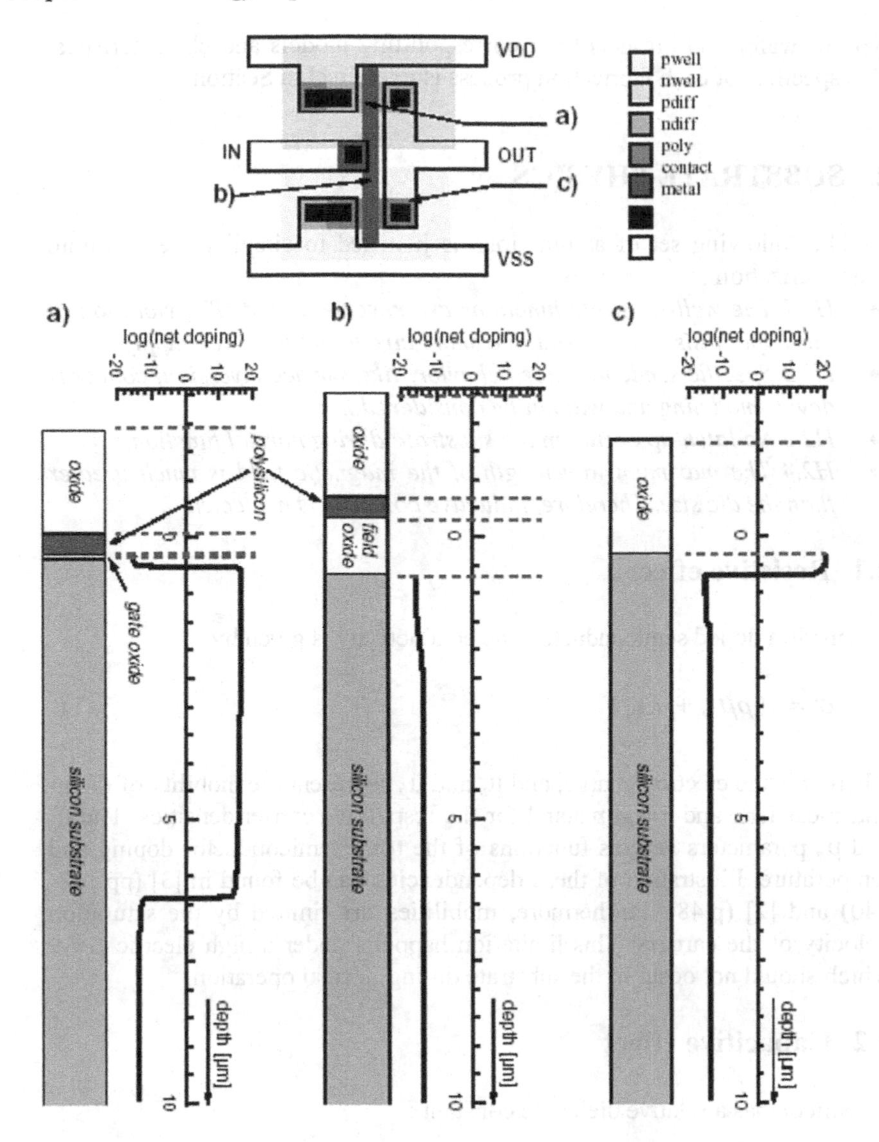

Figure 1-2. Layout of CMOS inverter and corresponding IC vertical structures for a standard process.

Parasitic currents will flow differently according to the substrate structure and the die attachment. The substrate physics, addressed in Section 2, determine simplification assumptions and relevant substrate characteristics from semiconductor physics. The significant substrate behaviors and their corresponding models are summarized in Section 3. Section 4 reviews each type of wafer and presents their corresponding models and characteristics. The specifics of each fabrication process is examined in Section 5.

2. SUBSTRATE PHYSICS

The following set of assumptions is justified to simplify the substrate characterization :

- *H2.1 The well-substrate junctions are reverse biased. The violation of this hypothesis would result in a short-circuit of the power supply.*
- *H2.2 Specific semiconductor behavior, like surface inversion, concerns device modeling and will not be considered here.*
- *H2.3 No latch-up occurs in the substrate during normal function.*
- *H2.4 The maximum wavelength of the magnetic field is much greater than the die size. Therefore, inductive coupling is neglected.*

2.1 Resistive effect

Inside a doped semiconductor, the conductivity is given by:

$$\sigma = q(p\mu_p + n\mu_n) \tag{1}$$

where q is the electron charge, and μ_n and μ_p represent the mobility of the n- and p-carriers, and n and p stand for the respective carrier densities. The μ_n and μ_p parameters vary as functions of the total semiconductor doping and temperature. Illustration of these dependencies can be found in [3] (pp.138-140) and [2] (p.48). Furthermore, mobilities are limited by the saturation velocity of the carriers. This limitation happens under a high electric field, which should not occur in the substrate during normal operation.

2.2 Capacitive effect

Silicon has a relative dielectric constant :

$$\varepsilon_{r_{Si}} = 11.7 \tag{2}$$

which leads to the absolute dielectric constant :

$$\varepsilon_{Si} = \varepsilon_{r_{Si}} \varepsilon_o = 1.035 \left[\frac{pF}{cm}\right] \tag{3}$$

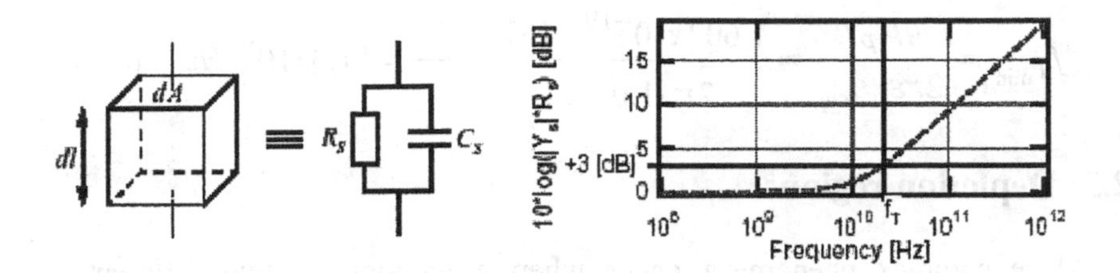

Figure 1-3. RC model for a piece of homogeneous substrate.

Capacitive and resistive effects occur throughout the substrate. However, for substrates doped homogeneously, the capacitance of the substrate can be neglected in the operating frequency range. In the frequency domain the equivalent admittance *Ys* for a piece of substrate, such as that illustrated in *Figure 1-3*, is given by :

$$Y_S = \frac{1 + sR_S C}{R_S} = \frac{1 + j\omega T_S}{R_S} \tag{4}$$

Therefore, equations (1) and (3) lead to a substrate time constant, T_s in equation (5), that no longer relates to the piece dimensions.

$$T_S = R_S C_S = \frac{\rho_S dl}{dA} \cdot \frac{\varepsilon_S dA}{dl} = \frac{\varepsilon_o \varepsilon_{r_{Si}}}{q(\mu_p p + \mu_n n)} \tag{5}$$

For low frequencies, substrate resistance, R_s, is more important and the associated capacitance, C_s, can be neglected. As pulsation ω increases, the capacitive effect rises to become equal to the resistive effect at the cut-off frequency, f_T, defined by:

$$\frac{1}{R_S} = \omega_T C_S = 2\pi f_T C_S \Rightarrow f_T = \frac{1}{2\pi T_S} = \frac{q(\mu_p p + \mu_n n)}{2\pi\varepsilon_o \varepsilon_{r_{Si}}} \tag{6}$$

The minimum f_T is achieved for a lightly-doped p-type substrate, because mobility is lower for holes than for electrons. A normal initial carrier concentration of 10^{15} cm^{-3} yields a minimum cut-off frequency of :

$$f_{T_{\min}} = \frac{q\mu_p p}{2\pi\varepsilon_o\varepsilon_{r_{Si}}} = \frac{1.602x10^{-19} \cdot 457x10^{15}}{2\pi \cdot 1.035x10^{-12}} = 11.3x10^9\ Hz \quad (7)$$

2.3 Depletion regions

More complex phenomena occur when a junction of two different material types is formed. These PN junctions are inherent to CMOS processes, since wells have to be created in the substrate.

The formation of a PN junction gives rise to a space-charge region due to impurity ionization and majority carrier diffusion. In the substrate, these junctions are reverse biased (i.e. there is no current flowing through the junction), and they behave as a variable capacitor. The capacitance, C_t, of such arrangements is given by:

$$C_t = \frac{A\varepsilon_{Si}}{X} \quad (8)$$

where A stands for the junction's area and X is the depletion region thickness. The latter depends on the doping profile of the junction and on the potential between the two difference. One can deduce from [4] a general relation to evaluate the thickness of the space-charge region :

$$X = [\alpha(\Delta\Psi_o + V_{NP})]^\gamma \quad (9)$$

where the parameters α and γ depend on the doping profile, as well as the junction's built-in voltage, $\Delta\psi_0$, see Table 1-1.

Table 1-1. Junction capacitance parameters α, γ and $|\Delta\psi_0|$ for two different doping profiles.

	asymmetric step junction	linearly graded junction (a = doping gradient)†
α	$\frac{2\varepsilon(N_A+N_D)}{qN_AN_D}$	$\frac{12\varepsilon}{qa}$
γ	$\frac{1}{2}$	$\frac{1}{3}$
$\lvert\Delta\psi_0\rvert$	$\frac{kT}{q}\ln\frac{N_AN_D}{n_i^2}$	$\frac{kT}{q}\ln\left[\frac{aX_0}{n_i}\right]^2 = \frac{qaX_0^3}{12\varepsilon}$

†. The computation of X_0 requires iterative solution.

In non-equilibrium situations, substrate junctions can momentarily become forward biased. For instance, this can occur while turning on an n-channel transistor. P-type majority carriers flowing through the substrate to build the channel will increase the potential near the drain or source junction. Since the junction becomes forward biased, electrons will diffuse into the substrate through the junction. When the minority carriers reach any substrate-well junction, they are pulled through the junction depletion region by the junction's electric field. The structure made by an n^+-drain/p^--substrate/n-well behaves as a parasitic npn bipolar transistor, as illustrated in *Figure 1-4*. Similarly, a p^+-drain/n-well/p^--substrate forms a parasitic pnp bipolar structure. Despite the poor characteristics of such devices, due to the large base dimension, they present a major risk if no precaution is taken to limit the voltage sweep near source or drain junctions.

2.4 Latch-up

All CMOS circuits have a potential problem called latch-up, which is related to parasitic bipolar transistors. *Figure 1-4* shows how, when fabricating CMOS devices, parasitic bipolar transistors are also created.

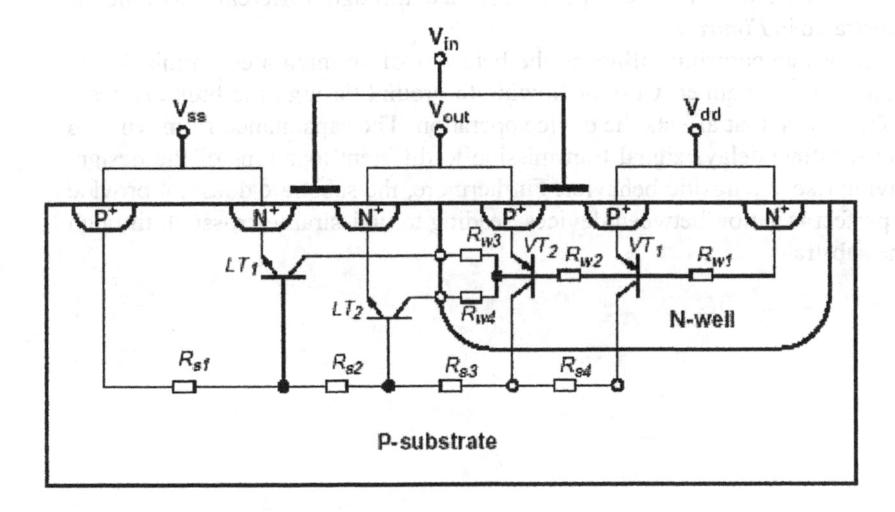

Figure 1-4. Parasitic bipolar transistors responsible for potential latch-up problems in an inverter fabricated with an N-well CMOS technology.

A latch-up structure is made by two pnp and npn parasitic bipolar transistors where the collector of one transistor is connected to the base of the other, forming a pnpn parasitic SCR (Semiconductor-Controlled Rectifier). Under certain conditions, such as terminal overvoltage stress, transient displacement currents or ionizing radiation, lateral currents can cause sufficient substrate or well voltage drop to forward bias emitter-base junctions and activate both bipolar devices. When the current-gain product is sufficient, the SCR will switch to a low impedance state. This condition is defined as latch-up. Latch-up can result in momentary or permanent loss of circuit function, depending on ability of the power supply to source the excess current.

Latch-up has been widely studied and the complex phenomena involved have been described, explained and modeled in depth [6]. Actually, latch-up can be avoided by observing appropriate technological and design rules. Hence, modeling of this parasitic substrate behavior is excluded from the present work.

3. PARASITIC SUBSTRATE EFFECTS

The increasing influence of parasitic effects occurring in mixed-signal ICs is seen in the significant degradation of system performances. IC parasitics have a very complex influence through different medium, as illustrated in *Figure 1-5*.

Substrate parasitic influence the behavior of an integrated circuit design in a negative manner. Current flowing to ground through the bulk creates a voltage drop that affects the device operation. The capacitance from wires to the substrate delays signal transmission to different locations of the design, giving rise to parasitic behavior. Furthermore, the substrate does not provide a perfect isolation between devices, leading to undesirable crosstalk through the substrate.

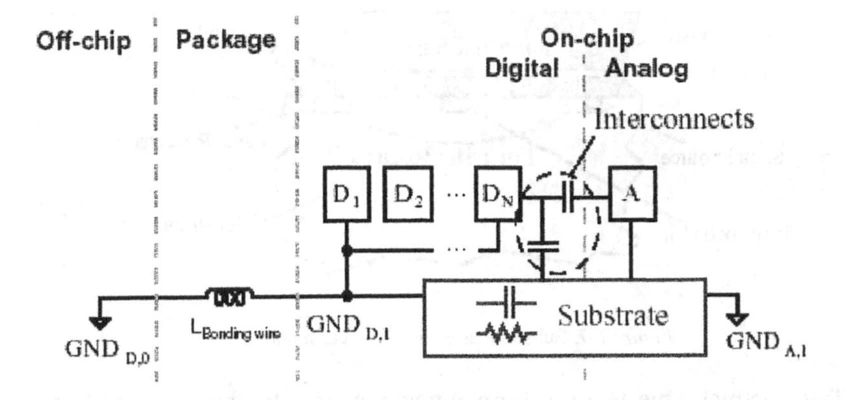

Figure 1-5. Simplified representation of IC parasitic effects.

Firstly, the bulk substrate affects integrated devices by adding parasitic resistances and capacitances. *Figure 1-6* provides an example where the substrate, together with bonding wires, adds a parasitic RLC structure to a capacitor. For example, a 10 pF capacitor together with a typical bond-wire inductance of 4 nH has a resonance frequency of 800 MHz, which is a source of instabilities and oscillations.

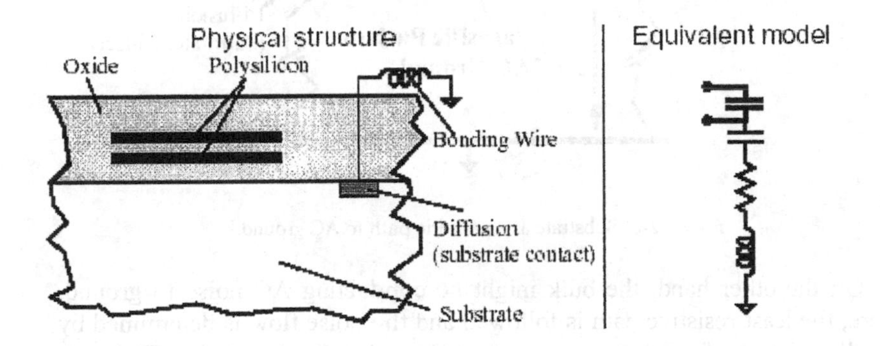

Figure 1-6. Parasitic effect on a capacitor made of two polysilicon layers.

Secondly, the substrate behaves as a vehicle carrying noise from one area to another. Noise is present on digital interconnects carrying switching signals as well as on digital power supply lines because of potential fluctuations due to the existence of bonding wires. This digital noise reaches sensitive analog cells through the interconnects and through the substrate. As the feature size of the technology decreases, substrate noise coupling is becoming a critical issue in high-end IC design.

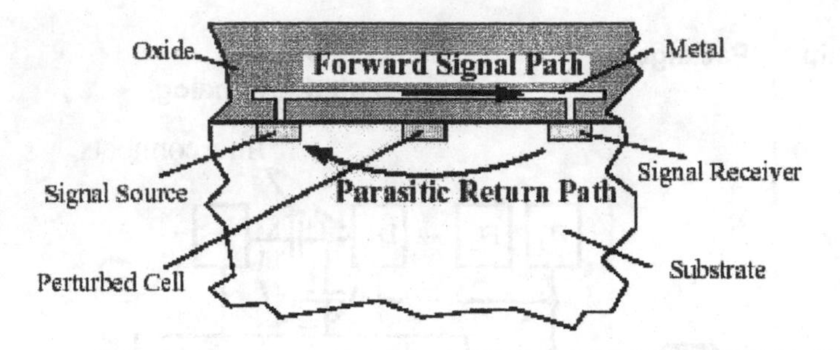

Figure 1-7. Substrate as a parasitic return path.

Two distinct aspects of substrate noise exist. On the one hand, as illustrated in *Figure 1-7*, the substrate is used as a parasitic return path for signals carrying relevant information. Crosstalk happens if a sensitive cell present along this parasitic path is perturbed by the signal. In this context, the parasitic current flow is parallel to the silicon surface.

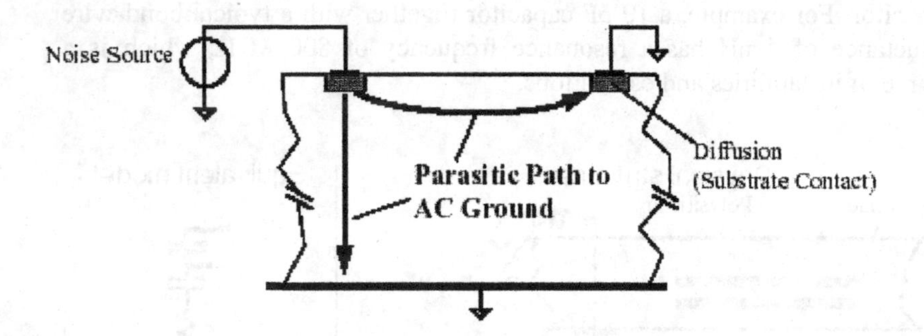

Figure 1-8. Substrate as a parasitic path to AC ground.

On the other hand, the bulk might be conducting AC noise to ground. Here, the least resistive path is followed and the noise flow is determined by the distribution of substrate contacts to AC ground. As depicted in *Figure 1-8*, the presence of a backside contact gives rise to a vertical current.

The occurrence of one or the other type of parasitic effect, together with the techniques available to reduce the parasitic impact, relates significantly to the kind of wafer and process used for the IC fabrication.

4. WAFER IMPACT

The substrate characteristics depend significantly on the type of wafer used together with the backside processing and packaging. Understanding the wafer impact as well as the backside influence is critical in choosing the correct noise reduction technique. See *Figure 1-9* for the basic substrate structure.

Two typical wafer types exist. A lightly doped wafer is made of a silicon material uniformly doped with a typical concentration of 10^{15} cm^{-3}. Epitaxial wafer consist of a similar lightly doped material grown by epitaxy on top of an heavily doped material — 10^{19} cm^{-3} is a normal doping concentration. The thickness of the epitaxial layer varies from 5 to 8 μm. The difference in resistivity of the two kinds of wafers results in a large variation of the electric field distribution. Subsections 4.1 and 4.2 present the distribution of the electric field for lightly doped and epitaxial wafers, and present simplified models to help understand the different behaviors of the two kinds of bulks.

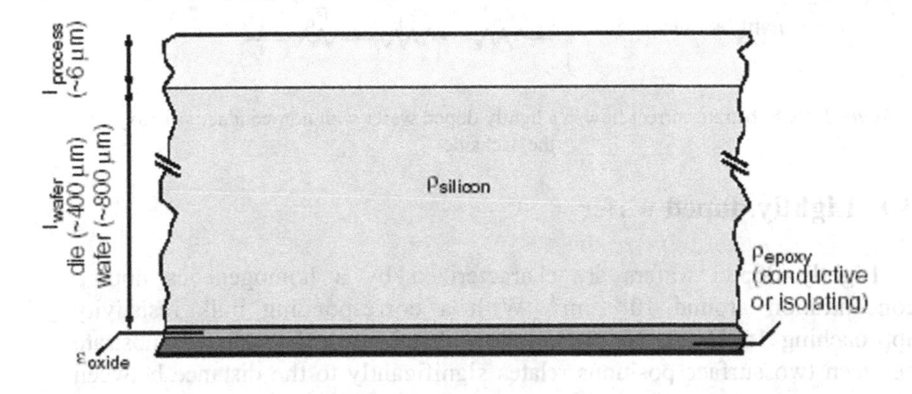

Figure 1-9. Basic substrate structure.

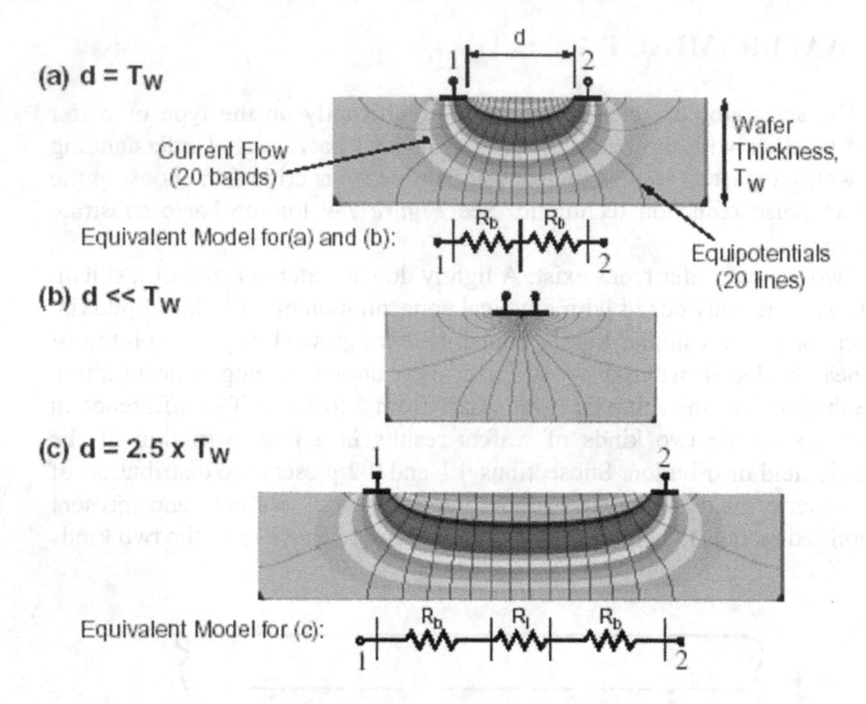

Figure 1-10. Substrate current flow in a lightly doped wafer with non-conductive epoxy on the backside.

4.1 Lightly doped wafer

Lightly doped wafers are characterized by a homogeneous doping concentration around 10^{15} cm^{-3}. With a corresponding bulk resistivity approaching 15 Ω·cm, the current flow distribution through the substrate between two surface positions relates significantly to the distance between the source and the receiver.

A MEDICI simulation of the current flow and equipotential lines between two surface contacts with respect to the distance is depicted in *Figure 1-10.* Each band holds approximately 5% of the total current. The backside is assumed non-conductive.

In *Figure 1-10 (a),* the contact distance has been set equal to the wafer thickness. The proposed modeling with two identical resistances R_b respects the structure symmetry. Therefore, the equivalent resistance between the surface contacts, $R_{12} = 2 \cdot R_b$, varies nonlinearly as the distance decreases, owing to the important bending of the electric field. As plotted in *Figure 1-10 (b),* the bending becomes greater as the contacts are getting closer. Adversely, when the distance increases beyond the wafer thickness as shown in *Figure 1-10 (c)*, the bending no longer changes with the distance. The equivalent resistance R_{12} is then made of two elements taking into account the bending, plus a medium component whose value increases linearly with the distance.

When the backside is made conductive and left floating, the behavior is very similar for small distances such as in *Figure 1-11 (a)*. However, as the space between the surface contacts reaches T_W, the low-resistivity backside becomes a preferred path for the current. For d close to the wafer thickness — see *Figure 1-11 (b)* — more than 50% of the total current is flowing through the backside contact. Therefore, the model is improved to separate the lateral component, R_l, from the vertical one, R_v. *Figure 1-11 (c)* shows how, when the distance further increases, the lateral effect becomes negligible causing the overall resistance value between the surface contacts to level off.

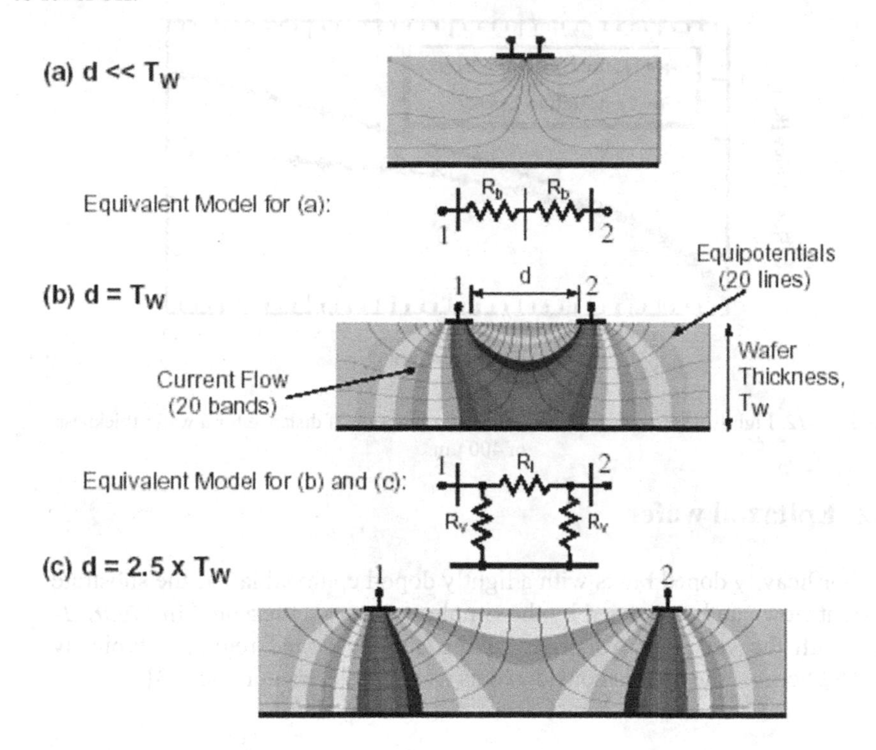

Figure 1-11. Substrate current flow in a lightly doped wafer with conductive epoxy on the backside.

The substrate resistance is plotted in *Figure 1-12* as a function of the distance between two surface contacts, for a wafer thickness of 400 μm. The bulk is made of P-doped silicon with a concentration of $7\cdot10^{14}$ cm^{-3}. The contacts have a length of 100 μm and the resistance values are provided for a structure width of 100 μm. For non-conductive epoxy, the linear behavior is clearly apparent beyond the wafer thickness. The use of a conductive die attachment causes a leveling off of the substrate resistance.

Independent of the type of epoxy used to attach the die to the package, the linear relationship applied to compute the resistance of interconnects or diffusion layers — i.e. $R = \rho_l \cdot A$, where ρ_l is the specific layer resistivity in Ω/square and A is the layer area in square — is not accurate for the substrate. If the layer width is taken into account, the resistance variation becomes more complex due to the 3-D nature of the current flow. On a real chip where many contacts are present, a complex resistive mesh is required to model the interaction through the substrate [9].

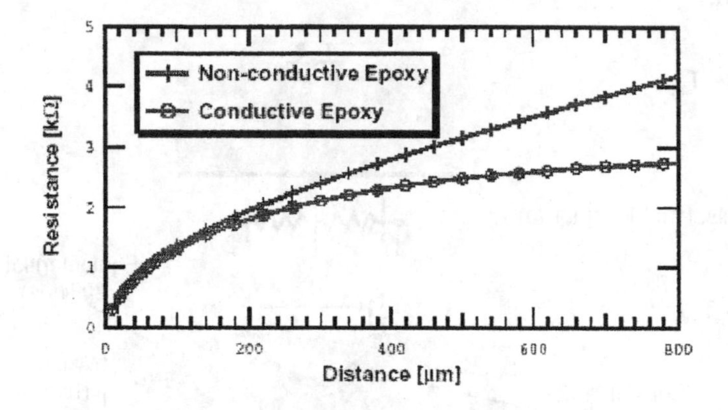

Figure 1-12. Lightly doped substrate resistance as a function of distance for a wafer thickness of 400 μm.

4.2 Epitaxial wafer

For heavily doped bulks with a lightly doped epitaxial layer, the substrate current flow can be modeled by the simple structure represented in *Figure 1-13*. With the very low resistivity of the heavily doped region— typically 0.015 Ω·cm — the bulk is considered as a single electrical node [13].

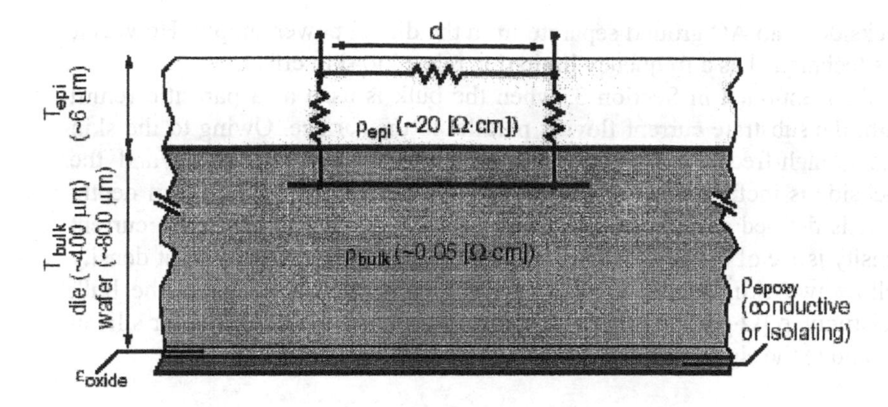

Figure 1-13. Structure of an epitaxial wafer.

The resistive model of an epitaxial wafer is as influenced by distance as one of a lightly doped wafer with conductive epoxy. For two contacts closer than $4 \cdot T_{epi}$, a significant portion of the total current flows in the epitaxial layer. The substrate model of *Figure 1-14 (a)* must be used to account for the lateral element, R_l, as well as the vertical component, R_v. Adversely, when the distance is larger than $4 \cdot T_{epi}$, the current in the epitaxial layer flows vertically and two resistances are sufficient to model the substrate, as illustrated in *Figure 14 (b)*. Typical values for R_v are around 2000 $\Omega/\mu m^2$.

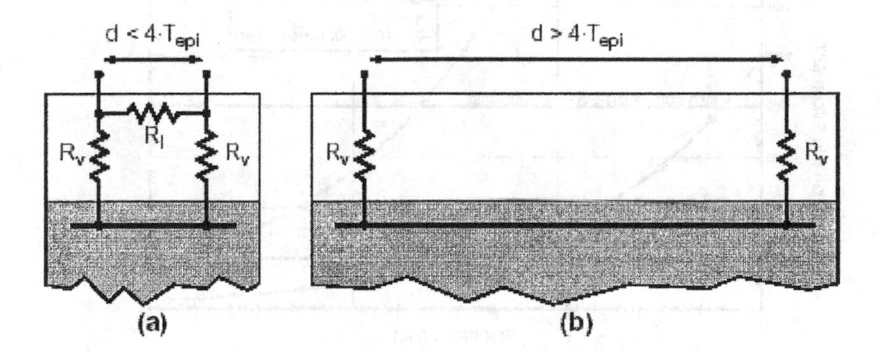

Figure 1-14. Resistive models for an epitaxial wafer.

As the resistance levels off beyond $4 \cdot T_{epi}$ — approximately 25 µm — distance cannot be used to isolate sensitive and perturbing cells. Therefore, the only efficient method to avoid substrate perturbations is to reduce as much as possible the amount of noise injected in the bulk (c.f. [7], Chapter 6). One possible solution is to use an additional pin to connect the substrate backside to an AC ground separate from the digital power supply. However, this technique has a frequency limitation related to skin effect.

As mentioned in Section 3, when the bulk is used as a parasitic return path, the substrate current flow is parallel to the surface. Owing to the skin effect, high-frequency signals have a limited penetration depth and the backside is inefficient in collecting such perturbations [1]. The skin depth, T_{skin}, is defined as the distance from the surface beyond which the current density is 1/e of the surface current density. Furthermore, the current density bellow two skin depths is negligible. The skin effect relates to the bulk resistivity, ρ [Ω·cm], to the permeability, μ [H/cm] — equal to μ_0 for silicon —, and to the signal frequency, f [Hz], and is calculated by :

$$T_{skin} = \sqrt{\frac{\rho}{\pi \mu f}} \; [cm] \tag{10}$$

The penetration depth for different resistivities is plotted in *Figure 1-15*. For a bulk resistivity around 0.01 Ω·cm, high-frequency perturbations above 600 MHz flowing at the silicon surface are not collected by the backside contact.

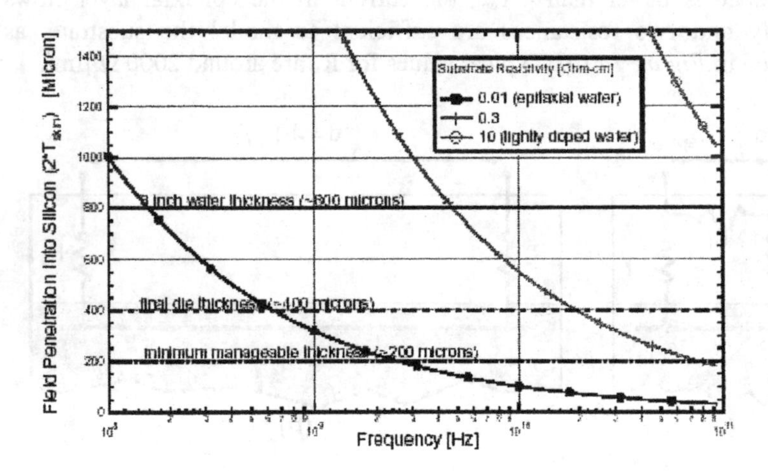

Figure 1-15. Penetration depth of a lateral current flow for different bulk resistivities.

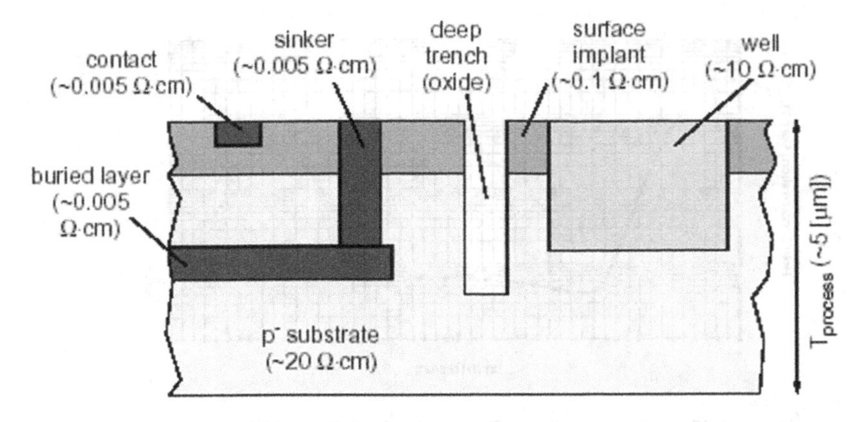

Figure 1-16. Typical process steps influencing substrate parasitics.

5. FABRICATION PROCESSES

Most of the current processes are based either on MOS or bipolar technologies. The first permits high levels of integration while the second is favored for the implementation of high speed functions. The process steps that have the most influence on substrate parasitics are summarized in *Figure 1-16.*

5.1 Surface implant

The surface implant, also known as p-tub or channel stop, has a thickness around 1 µm with a resistivity of 0.1 Ω·cm. The primary purpose of this layer is to avoid parasitic inversion of the silicon surface from the lowest metal layers. Additionally, the low resistivity decreases the risk of latch-up in MOS processes. This is the layer that affects the parasitic substrate behavior most significantly. The example surface implant provided in *Figure 1-17* exhibits a surface doping two orders of magnitude higher than the bulk concentration. Accordingly, the current density is significantly higher and much of the substrate noise is conducting in this layer.

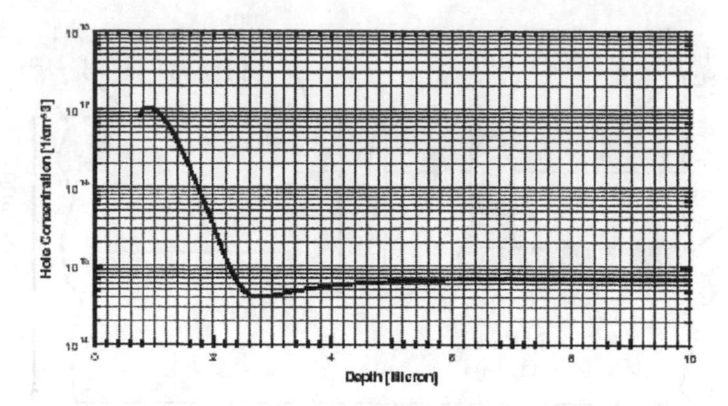

Figure 1-17. Surface implant profile in a standard 2μm MOS process.

The surface implant can be broken using the wells available in MOS processes (see *Figure 1-18*), or the deep trench existing in bipolar technologies. As a result, the noise is forced into the substrate depth where the resistivity is significantly higher. The use of a well is further improved by applying a reverse bias potential that increases the depletion width.

5.2 Buried layers

The fabrication process of BiCMOS technologies includes the implantation of heavily doped buried layers together with the epitaxial growth of a lightly doped silicon material. Lightly doped wafers are commonly used as bulk material to limit the parasitic capacitance between the collector and the substrate. The buried layers are essential for the vertical bipolar transistors and are also used to enhance the behavior of MOS transitors. Because of the low resistivity ot these layers, most of the substrate current flows through this region. If no particular precaution is taken to break the default p^+ buried layer, the substrate resistance is very small — typically a few Ω/square instead of 1000 Ω/square for the lightly doped bulk. Therefore, with unbroken buried layers, the substrate effect of a BiCMOS process compares to what happens with an epitaxial wafer.

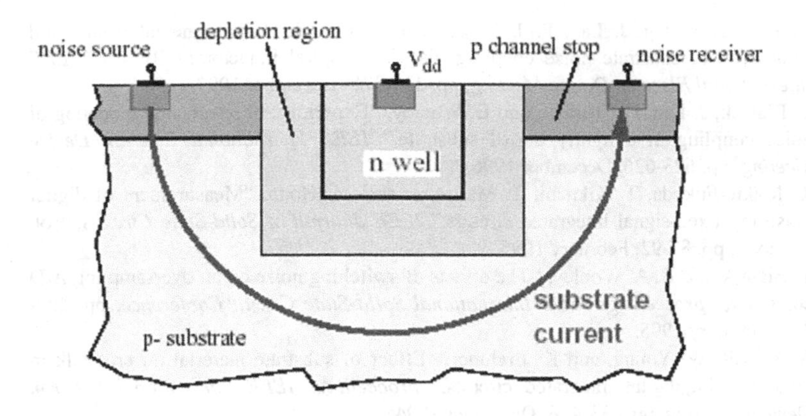

Figure 1-18. Using a well to break the surface implant.

6. CONCLUSIONS

The electromagnetic substrate behavior of integrated circuits (IC) has been reviewed and the significant parasitic phenomena have been presented. The technology impact has been examined from three complementary points of view. The respective influence of the lightly doped and epitaxial wafers was detailed. Fabrication process steps changing the substrate characteristics were addressed for CMOS and bipolar technologies. Die attachment was considered as a means to reduce substrate parasitics.

REFERENCES

[1] T. Schmerbeck, "Noise coupling in mixed-signal ASICs," *Low-power HF microelectronics: a unified approach*, pp. 373-430, G. Machado - Editor, IEE, 1996.

[2] N. Verghese, T. Schmerbeck, and D. Allstot, *Simulation techniques and solutions for mixed-signal coupling in integrated circuits*, Kluwer Academic Publishers, 1995.

[3] R. W. Dutton and Z. Yu, *CAD - Computer simulation of IC processes and devices*, Kluwer Academic Publishers, 1993.

[4] R. M. Warner and B. L. Grung, *Semiconductor device electronics*, Rinehart and Winston Inc., 1991.

[5] J. Y. Chen, *CMOS devices and technology for VLSI*, Prentice-Hall, 1990.

[6] R. R. Troutman, *Latchup on CMOS Technology - The problem and its cure*, Kluwer Academic Publishers, 1986.

[7] T. Blalack, "*Switching noise in mixed-signal integrated circuits*," Ph.D. Dissertation, Stanford University, Department of Electrical Engineering, December 1997.

[8] A. Pun, T. Yeung, J. Lau, F. J. R. Clément and D. Su, "Experimental results and simulation of substrate noise coupling via planar spiral inductor in RF Ics," *IEEE International Electron Device Meeting*, pp. 325-328, December 1997.
[9] T. Blalack, J. Lau, F. Clément, and B. Wooley, "Experimental results and modeling of noise coupling in a lightly doped substrate," *IEEE International Electron Device Meeting*, pp. 623-626, December 1996.
[10] K. Makie-Fukuda, T. Kikuchi, T. Matsuura, and M. Hotta, "Measurement of digital noise in mixed-signal integrated circuits," *IEEE Journal of Solid-State Circuits*, Vol. 30, no. 2, pp. 87-92, February 1995.
[11] T. Blalack and B. A. Wooley, "The effects of switching noise on an oversampling A/D converter," *proceedings IEEE International Solid-State Circuit Conference*, pp. 200-201, February 1995.
[12] R. Merrill, W. Young, and K. Brehmer, "Effect of substrate material on crosstalk in mixed analog/digital integrated circuits," *proceedings IEEE International Electron Devices Meeting*, pp. 433-436, December 1994.
[13] D. Su, M. Loinaz, S. Masui, and B. Wooley, "Experimental results and modeling techniques for substrate noise in mixed-signal integrated circuits," *IEEE Journal of Solid-State Circuits*, Vol. 28, no. 4, pp. 420-430, April 1993.
[14] R. Gharpurey, M. C. Chang, U. Erdogan, R. Aggarwal and J. P. Mattia, "RF MOSFET modeling accounting for distributed substrate and channel resistances with emphasis on the BSIM3v3 SPICE model," *proceedings IEEE International Electron Devices Meeting*, pp. 309-312, December 1997.
[15] R. Gharpurey and S. Hosur, "Transform domain techniques for efficient extraction of substrate parasitics," *proceedings IEEE International Conference on Computer-Aided Design*, pp. 461-467, December 1997
[16] J. Casalta, X. Aragones, and A. Rubio, "Substrate coupling evaluation in BiCMOS technology," *IEEE Journal of Solid-State Circuits*, Vol. 32, no. 4, pp. 598-603, April 1997.
[17] M. Pfost, H. Rein, and T. Holzwarth, "Modeling substrate effects in the design of high-speed Si-Bipolar ICs," *IEEE Journal of Solid-State Circuits*, Vol. 31, no. 10, pp. 1493-1501, October 1996.
[18] K. Kwan, I. Wemple, and A. Yang, "Simulation and analysis of substrate coupling in realistically-large mixed-A/D circuits," *proceedings IEEE Symposium on VLSI Circuits*, pp. 184-185, June 1996.
[19] R. Gharpurey and R. G. Meyer, "Modeling and analysis of substrate coupling in integrated circuits," *IEEE Journal of Solid-State Circuits*, Vol. 31, no. 3, pp. 344-353, March 1996.
[20] N. K. Verghese, D. J. Allstot and M. A.Wolfe, "Verification techniques for substrate coupling and their application to mixed-signal IC design," *IEEE Journal of Solid-State Circuits*, Vol. 31, no. 3, pp. 354-365, March 1996.
[21] A. Viviani, J. P. Raskin, D. Flandre, J. P. Colinge and D. Vanhoenacker, "Extended study of crosstalk in SOI-SIMOX substrates," *proceedings IEEE International Electron Devices Meeting*, pp. 713-716, December 1995.
[22] J. P. Raskin, D. Vanhoenacker, J. P. Colinge and D. Flandre, "Coupling effects in high-resistivity SIMOX substrates for VHF and microwaves applications," *proceedings IEEE International SOI Conference*, pp. 62-63, October 1995.
[23] I. L. Temple and A. T. Yang, "Mixed-signal switching noise analysis using Voronoi-tesselated substrate macromodels," *proceedings IEEE Design Automation Conference*, pp. 439-444, June 1995.

[24] T. Smedes, N.P. van der Meijs, A.J. van Genderen, P.J.H. Elias and R.R.J. Vanoppen, "Layout extraction of 3D models for interconnect and substrate parasitics," *proceedings European Solid-State Device Research Conference*, pp. 397-400, September 1995.
[25] S. Mitra, R. A. Rutenbar, L. R. Carley and D. J. Allstot, "A methodology for rapid estimation of substrate-coupled switching noise," *proceedings IEEE Custom Integrated Circuit Conference*, pp.129-132, May 1995.
[26] K. Joardar, "A simple approach to modeling crosstalk in integrated circuits," *IEEE Journal of Solid-State Circuits*, Vol. 29, no. 10, pp. 1212-1219, October 1994.

Chapter 2

SUBSTRATE NOISE GENERATION IN COMPLEX DIGITAL SYSTEMS

Analysis and experimental verification

Stéphane Donnay, Marc van Heijningen, Mustafa Badaroglu
IMEC - DESICS, Kapeldreef 75, B-3001 Leuven, Belgium

Abstract: More and more system-on-chip designs require the integration of analog circuits on large digital ASICs and will therefore suffer from substrate noise coupling. Accurate modeling and simulation are needed to investigate the generation, propagation, and impact of substrate noise. Recent studies were limited to the time-domain behavior of generated substrate noise and to noise injection from a single noise source. This chapter focuses on substrate noise generation by real digital circuits on low-ohmic substrates and on the spectral content of this noise. To simulate the noise generation, a SPICE substrate model for heavily doped epi-type substrates has been used. The accuracy of this model has been verified with measurements of substrate noise on a small experimental ASIC, using a wide-band substrate noise sensor amplifier, which allows accurate measurement of the spectral content of substrate noise. A second, more complex experimental ASIC has been designed, an 86-Kgate digital multirate filterbank with several noise sensor amplifiers, to analyze substrate noise generation in a real digital telecom ASIC and to investigate the influence of the different substrate noise coupling mechanisms.

1. INTRODUCTION

Substrate coupling in mixed-signal ICs can cause important performance degradation of the analog circuits integrated on the same die as large digital systems. Accurate simulation of the substrate voltage is necessary to analyze the proper functioning of these analog circuits [1]. Such simulations can give insight in the time and frequency domain behavior of substrate noise. This information is very useful when designing mixed-signal ASICs: time periods and frequency bands with less substrate noise can be identified and used for sensitive analog signal operations. In recent years a lot of research has been done on modeling the substrate and on substrate coupling reduction techniques [2],[3],[4]. Most substrate noise experiments, reported until now, have always been carried out on test chips with dedicated digital substrate noise generators [2],[5] or using only small digital CMOS circuits [6].

In this chapter, first a substrate modeling strategy is presented which allows accurate simulation of the time and frequency domain behavior of substrate noise generated by digital circuits. To verify these models and simulations a first, relatively small experimental ASIC was designed in a 0.5-μm CMOS technology on a low-ohmic epi substrate [7],[8],[9], containing substrate noise sensor amplifiers, which allows continuous-time wide-band measurement of substrate noise.

Next, we will present the design of a larger noise coupling experiment, an 86-Kgate digital multirate filterbank [10],[11], which allows us to measure substrate noise generation in a large, real-life digital telecom ASIC.

2. SOURCES OF SUBSTRATE NOISE

All current injected into the substrate will cause fluctuations of the substrate voltage. This is called substrate noise and is caused by coupling of switching or noisy signals to the substrate. In digital CMOS circuits this noise is caused by three mechanisms: (1) coupling from the digital power supply, (2) coupling from switching source–drain nodes and (3) impact ionization in the MOSFET channel. Noise on the digital power supply is caused by *di/dt* noise and resistive voltage drops due to the inductance and resistance in the power-supply connections to the chip. The combination of the inductance in the power-supply connection and the on-chip capacitance between power and ground will also cause ringing of the power-supply voltage. These effects are also called ground bounce or simultaneous switching noise [5],[13],[14] (see also chapter 12). Typically, the digital ground is connected to the substrate in every CMOS gate, which results in a very low resistance between digital ground and substrate, and all digital ground noise and ringing will also be present on the substrate. Therefore, this noise coupling mechanism is often the dominant cause of substrate noise.

The second source of substrate noise is capacitive coupling from switching source and drain nodes of the MOSFETs. The resulting substrate voltage waveform will show the same characteristics as the switching signals on the source–drain nodes. For noise coupling from the power supply this is not the case: a switching gate will cause an increase of the ground voltage, which causes a positive noise peak on the substrate. The third source of substrate noise is impact ionization [15]. Whether or not impact ionization is an important source of substrate noise depends on the technology, especially on the combination of the supply voltage and channel length.

3. SUBSTRATE MODELING

To simulate substrate noise, an accurate substrate model is necessary. These models vary from complicated electromagnetic models to simple lumped-element models. For epi-type substrates the heavily doped bulk can be considered as one electrical node and only the resistance of the epi layer has to be taken into account [2]. This results in a simple lumped-element model, as shown in *Figure 2-1*.

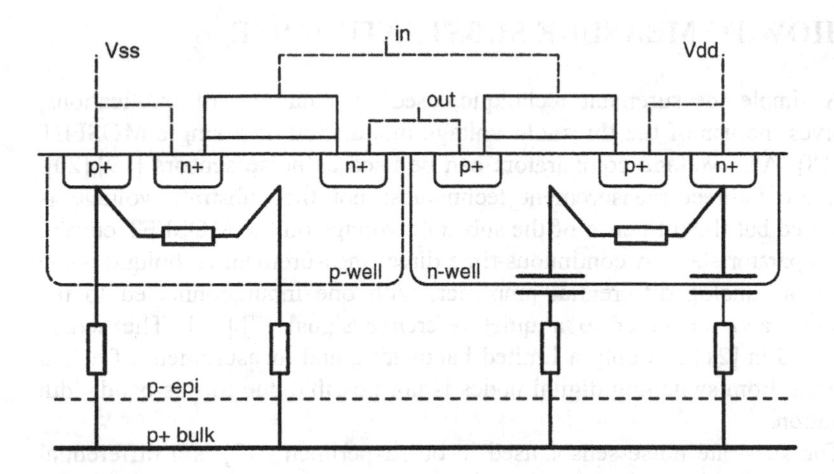

Figure 2-1. SPICE substrate model for a CMOS inverter on a low-ohmic substrate.

The coupling from the power supply has to be included in the substrate model by adding resistors from the substrate contacts connected to the digital ground to the substrate bulk node. Also the capacitive coupling via the n-well junction capacitance from the positive power supply to the substrate must be included. The coupling from switching source–drain nodes and impact ionization is handled by the used MOSFET model (the BSIM3v3 model). For the SPICE description of the digital circuit a layout parasitics extraction (LPE) file has been used that includes parasitics of the interconnect. Also the external parasitics in the power-supply connection, such as bondwire inductances and external decoupling capacitors are taken into account.

Our substrate model is based on the model presented in [2], but lateral resistances between MOSFET bulk nodes and nearby well contacts have been added. These lateral resistances are important, because they will reduce coupling from source–drain nodes and at the same time increase coupling from the power supply to the substrate. This is especially the case in twin-well technologies, which have an n-well and p-well that are more heavily doped than the epi layer. Noise coupling from other structures, like bondpads, can be easily added to this model by including a capacitor (for the field oxide) with series resistance (for the epi layer) connected to the substrate node. Such substrate models can be extracted with the assistance of tools like SPACE [16] (see also chapter 4) or SubstrateStorm [17].

4. HOW TO MEASURE SUBSTRATE NOISE

A simple measurement technique, used in a number of publications, involves the use of the threshold voltage modulation of a single MOSFET [2],[18]. Also voltage comparators can be used as noise sensors [19],[20]. Both are indirect measurement techniques: not the substrate voltage is measured but the influence of the substrate voltage on the MOSFET current or comparator state. A continuous-time direct measurement technique is the use of an analog differential amplifier, with one input connected to the substrate and the other to a quiet reference signal [7],[21]. The sensor presented in [21] has only a limited bandwidth, and measurement of actual coupling from switching digital nodes is not possible due to this bandwidth limitation.

The substrate noise sensor used in our experiments [7] is a differential amplifier with one input connected to a quiet ground and the other input connected to the substrate. Main objectives during the design have been a large bandwidth (over 500 MHz) and the ability to deliver a differential output signal in a 50 Ω external load. The schematic of the sensor is shown in *Figure 2-2*. The coupling capacitors C1 and C2 have been implemented as large finger-structured MOS capacitors (W/L = 2000/1). For the substrate voltage coupling capacitor C2, source and drain have been connected to a substrate contact, surrounding the transistor. The source and drain nodes of capacitor C1 have been connected via a dedicated connection, off-chip to the analog ground. Like the circuit, the layout has been made as symmetrical as possible.

The power dissipation is approximately 100 mW from a 3.0-V supply. To analyze the behavior of this amplifier it is necessary to include the substrate model in the simulations. Not only because this amplifier needs to sense the substrate voltage, but also because the substrate has a large influence on the common mode rejection.

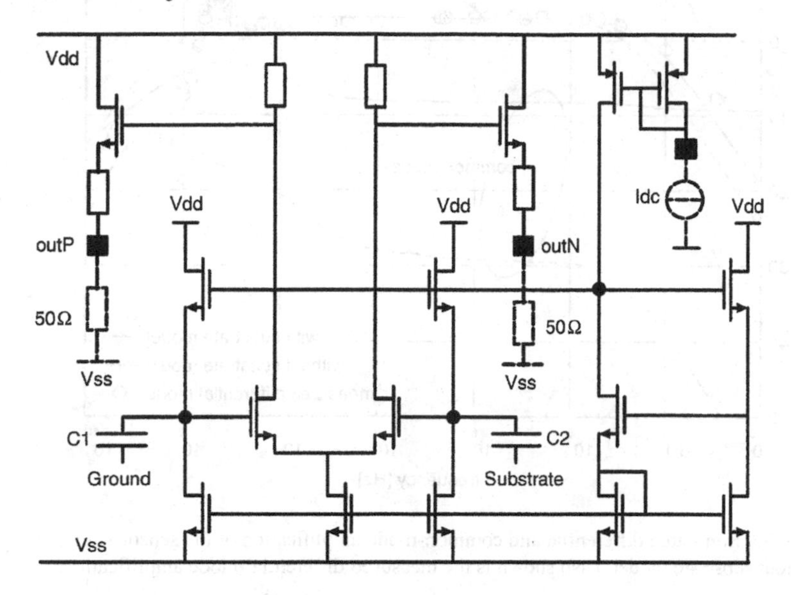

Figure 2-2. Circuit schematic of the noise sensor amplifier.

The simulated differential-mode and common-mode amplifications, for a circuit with and without substrate model, are shown in *Figure 2-3*. These simulations correspond very well with the measured differential-mode amplification of 3 dB and common-mode rejection of 8 dB at 100 kHz. It can be seen that the simulation without the substrate model severely underestimates the common-mode signal level. This common-mode signal caused by the substrate noise is rather high, but does not interfere with the functioning of the sensor when the differential output is measured. This analysis of the amplification of the noise sensor already shows the importance of taking the substrate into account, especially when simulating common-mode behavior. *Figure 2-3* also shows the measured differential-mode amplification. It can be seen that the bandwidth of the sensor is 20 kHz to 1 GHz, with an amplification around 3 dB. The peaking of the amplification around 500 MHz is caused by parasitics in the measurement setup, especially in the connection of the signal generator to the substrate. The substrate signal was injected via a digital ground connection and the return path for this signal went via the analog ground and measurement setup back to the generator, which caused a large inductance in the signal path. This behavior can be reproduced in SPICE by including this inductor and is not caused by instability of the amplifier.

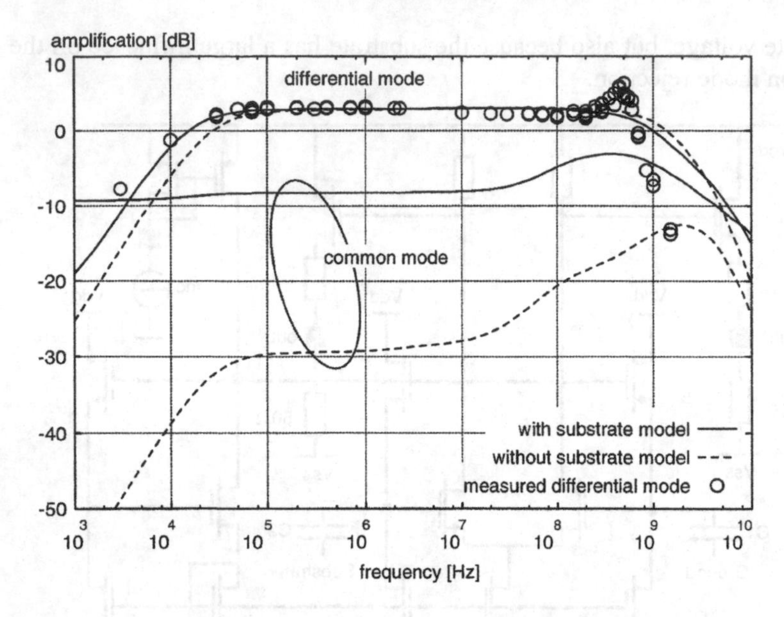

Figure 2-3. Simulated differential and common-mode amplification of the sensor with and without substrate model. Also shown is the measured differential-mode amplification.

5. FIRST MIXED-SIGNAL TEST CHIP WITH SIMPLE INVERTER CHAINS

A first experimental mixed-signal test chip was designed in a 0.5-μm CMOS technology on a low-ohmic epi substrate [7],[8],[9] to verify the SPICE-level substrate models. This ASIC contains several digital inverter chains for noise generation and substrate noise sensor amplifiers for the noise measurements. The microphotograph of this first noise coupling experiment is shown in *Figure 2-4*. Two noise sensors can be seen at the top and the inverter chains at the bottom.

Both the digital and analog signals and power supplies are directly connected to the chip, using multicontact wafer probes [22]. The probe needles for supplying the power to the chip contain a 22-nF decoupling capacitor, located near the point of the needles. This measurement setup makes it possible to measure the generated substrate noise of the digital circuits, without the influence of bondwires or other package parasitics.

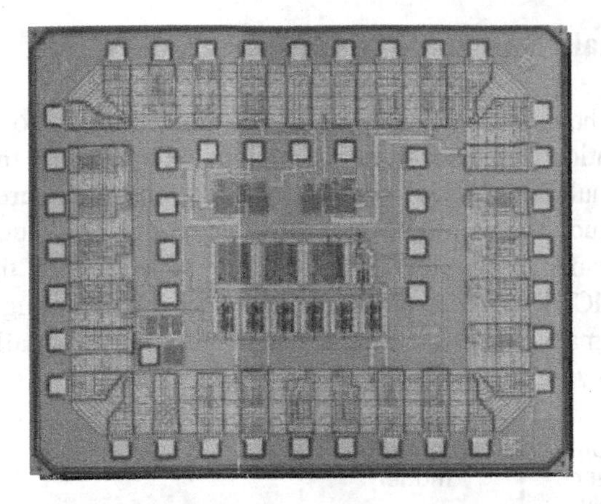

Figure 2-4. Microphotograph of the first noise coupling experiment.

As noise generation circuit, the 7-stage inverter chain was used, shown in *Figure 2-5*. The chain acts, for a short time period, as a ring oscillator, after a positive clock edge is given to the D-flipflop. Since this circuit requires two clock cycles to return to the original state, it acts like a divide-by-2 circuit and the periodicity of the generated noise corresponds to half the clock frequency. The inverter chain is loaded by extra, larger, inverters to decrease the switching frequency and to increase the noise coupling. Two versions of this inverter chain have been measured: a heavily loaded and a less heavily loaded version. In all experiments the differential output signal of the sensor is being measured and corrected for the 3-dB amplification of the sensor to derive the actual substrate voltage.

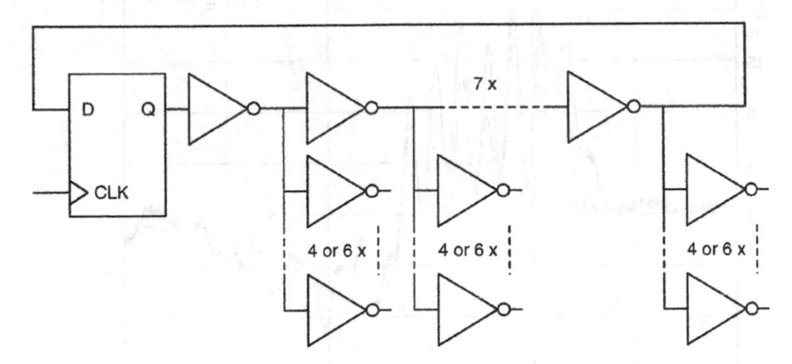

Figure 2-5. 7-stage inverter chain used as noise generator in first noise coupling experiment.

5.1 Time-Domain Substrate Noise

Measurements have been performed in the time domain to study the amplitude and duration of the substrate noise signal generated by the inverter chain. SPICE simulations have been performed and compared to the measurements to study the validity of the substrate model. For accurate noise simulations, the wafer probe elements have to be added to the SPICE description of the IC. The wafer probe elements are a decoupling capacitor of 22 nF with a small parasitic series resistance and some small parasitic inductances, as shown in *Figure 2-6*.

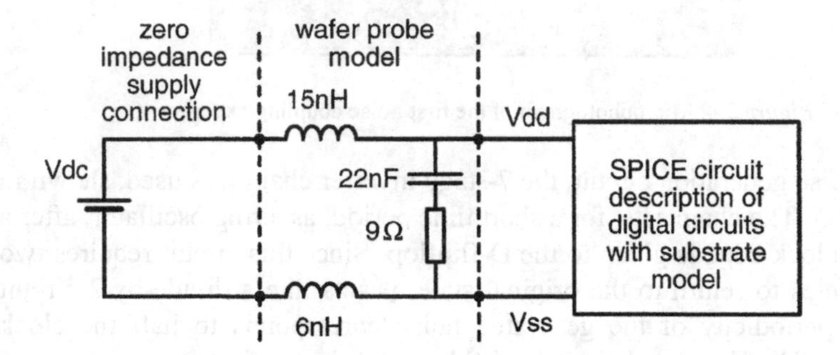

Figure 2-6. Parasitics of the measurement setup.

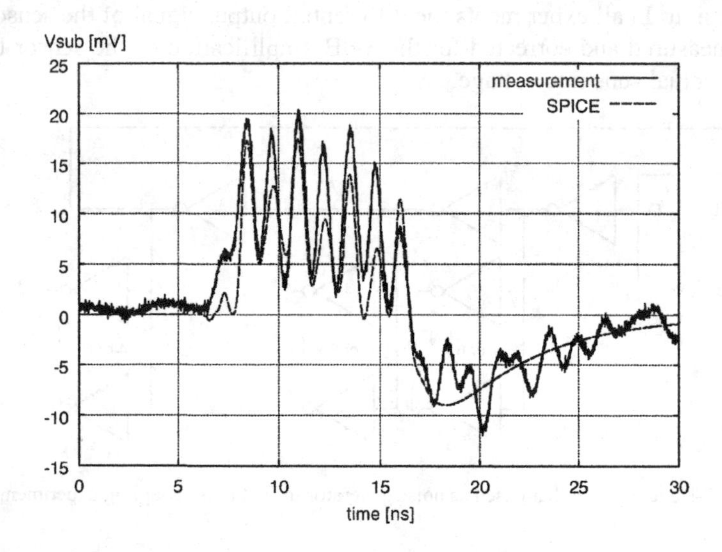

Figure 2-7. Comparison of on-chip measurements and simulations of substrate noise generated by the switching inverter chain.

The probe-to-supply connection is considered as a very low impedance circuit due to further decoupling. The parasitic component values are determined by fitting simulation and measurement results, but agree reasonably well with the expected parasitics of the wafer probe. *Figure 2-7* shows the measurement and the corresponding SPICE simulation.

From 8-ns to 16-ns substrate noise is generated by the switching inverters. Visible are the seven noise peaks corresponding to the switching of the seven stages in this inverter chain. The supply current for this switching activity is mainly delivered by the 22-nF decoupling capacitor. The dc-offset in the substrate noise signal is caused by the series resistance of this capacitor. Due to this resistor, power-supply noise coupling is dominant, as indicated by the seven noise peaks (instead of seven edges). The agreement between measurements and simulations is very good. Therefore further simulations can be done with reliable results.

5.2 Dominant Noise Coupling Source Analysis

To show the effect of only an external parasitic inductor (e.g., from a wirebond connection) on the shape and amplitude of the generated substrate noise signal, a SPICE simulation has been performed with a 0 nH, 1 nH, and 10 nH external inductor in the power-supply connection. The SPICE simulation model has been used, without the wafer probe model but with an equal inductance (of 0, 1, or 10 nH) in the power and ground lines. These simulations are shown in *Figure 2-8*. The 0-nH simulation shows the minimum amount of substrate noise that will be generated, only by capacitive coupling from the source and drain nodes. This substrate noise level can only be reduced by increasing the number of substrate contacts to the quiet digital ground. The simulation with the 1-nH inductor shows a much larger substrate noise signal, now dominated by noise coupling from the power supply. For even larger inductance values (10 nH), the maximum substrate noise will be caused by ringing of the damped *LC* tank, formed by the inductance and the on-chip capacitance with series resistance over the power supply.

To provide more insight in the dominant source of substrate noise, the peak-to-peak substrate voltage, generated by the inverter chain, has been analyzed as function of the inductance, for 3 different models: (1) a substrate model that only includes coupling from the MOSFET source and drain nodes, (2) for a model that only includes coupling from the power supply, and (3) for the total substrate model. The results are shown in *Figure 2-9*. For the simulations with only noise coupling from the MOSFETs, the substrate and wells have been connected to a quiet (dedicated) power supply.

For low inductances the generated noise is almost constant. When the noise on the power supply increases with increasing inductance, the substrate noise also increases due to capacitive noise coupling from source and drain nodes that are directly connected to the noisy Vdd or Vss. For the simulations with only power-supply noise coupling, the bulk nodes of the MOSFETs have been directly connected to a quiet power supply. For low inductances the substrate noise increases linearly with the inductance. At higher inductance values the maximum substrate noise is dominated by ringing of the power supply that couples to the substrate. The simulation with the complete substrate model clearly shows that for this circuit noise coupling from the MOSFETs is dominant up to 100-pH inductance and that the power-supply noise coupling is dominant for higher inductance values. This is important information when choosing between flip-chip connections, with a typical inductance around 10-30 pH, and wirebond connection, with an inductance between 2 nH and 10 nH.

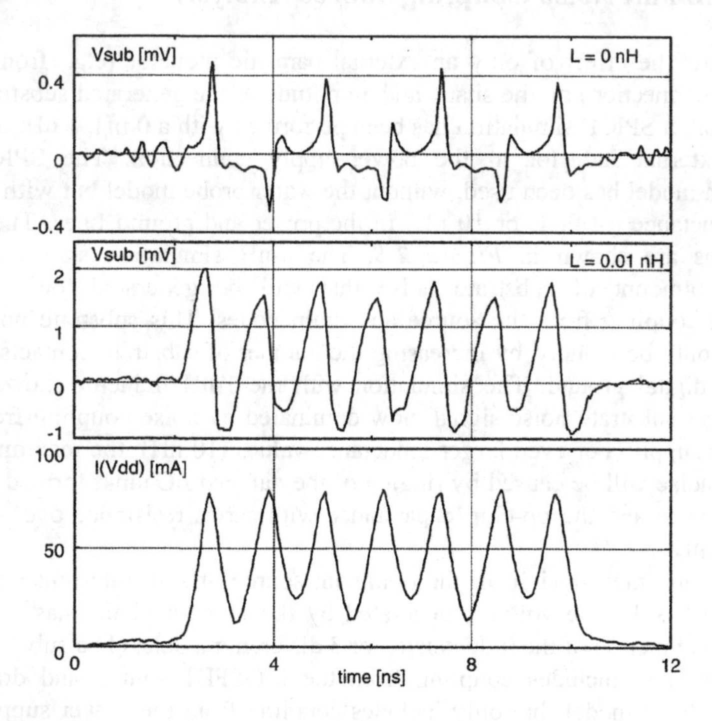

Figure 2-8. Simulated substrate noise generation by the inverter chain for 0 nH, 1 nH and 10nH power supply inductance.

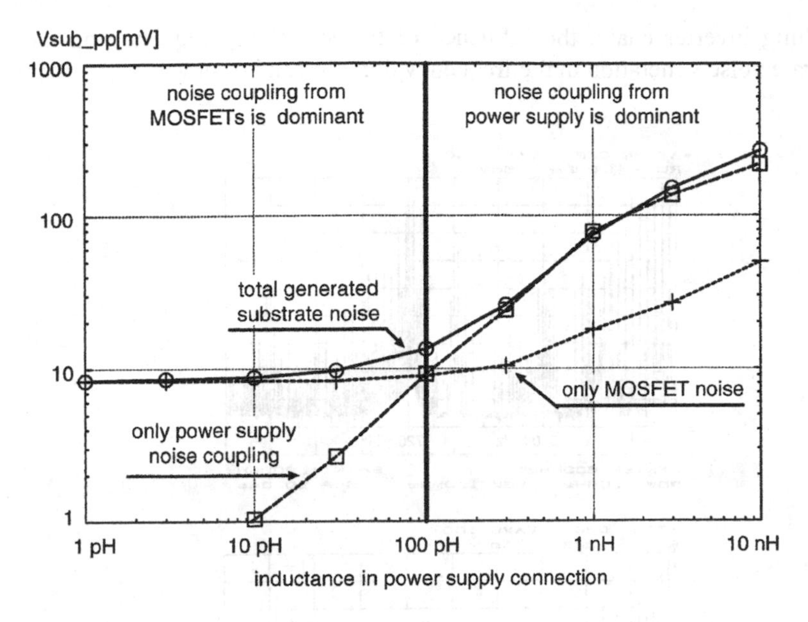

Figure 2-9. Simulation of substrate noise voltage (peak-to-peak) versus the power-supply inductance.

From this analysis it can be concluded that the substrate noise generation of a digital circuit can be reduced by reducing the value of the power-supply connection inductance. Only when the power-supply noise is not dominant anymore, the substrate noise can be further reduced by increasing the number of substrate contacts (i.e., reducing the resistance between the substrate and the digital ground). Increasing the number of substrate contacts, when power-supply noise coupling is dominant, will only increase the noise coupling from the noisy power supply to the substrate. Using a dedicated substrate and well bias, with low connection impedance, is also an effective way to reduce the noise generation.

5.3 Frequency Domain Substrate Noise

Studying substrate noise in the frequency domain can also reveal important information about the sources of the noise and can give useful information when designing mixed-signal circuits (e.g., during frequency planning for a receiver front-end). The spectral content measurements have been performed using the two versions of the inverter chain. By comparing the measurements of the slow (heavily loaded) and faster (less loaded) switching inverter chain, the influence of the switching frequency on the substrate noise generation in the frequency domain can be shown.

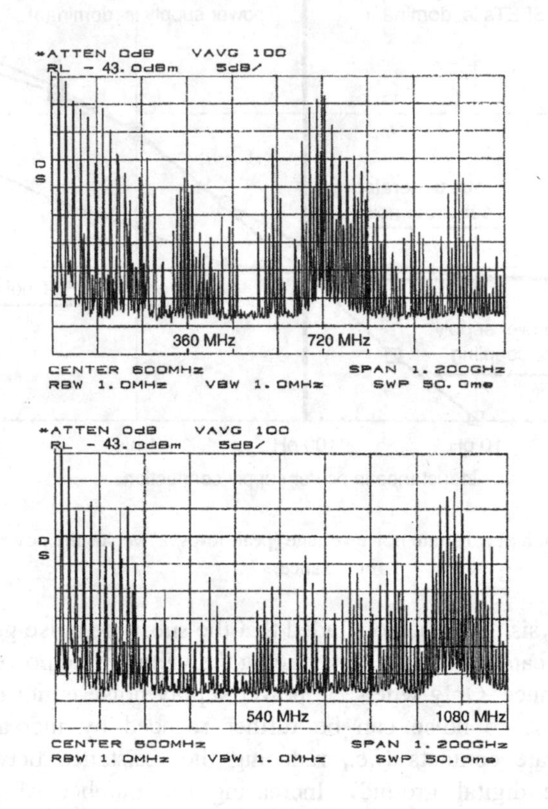

Figure 2-10. Measured spectral content of substrate noise generated by heavily loaded (top) and less loaded (bottom) inverter chains (reference level -43 dBm or 1.58 mV).

The two measured spectra are shown in *Figure 2-10*, with the heavily loaded version at the top and the less loaded at the bottom of the figure. In both cases the circuits are clocked at 20 MHz. At low frequencies, up to a few hundred MHz, the substrate noise is concentrated at multiples of the 20-MHz clock signal. Parasitic effects such as ringing of the power supply will cause an extra increase of substrate noise in this frequency range. Noise coupling from the inverter chains causes noise peaks at multiples of half the clock frequency, due to the divide-by-2 behavior of the circuit. SPICE simulations of the switching inverter chain show that, for the heavily loaded version, the noise coupling from the switching source–drain nodes is most dominant around 360 MHz, which corresponds to the switching frequency of the ring oscillator. The noise coupling from the power supply is most dominant at twice this frequency, 720 MHz. At both these regions a strong increase in substrate noise amplitude can be seen, but the largest contribution is from the power-supply coupling around 720 MHz. Also for the less loaded inverter chain the dominant source of substrate noise from the switching inverters comes from power-supply noise coupling, which occurs around 1080 MHz. A minor contribution from the switching source–drain nodes is visible at 540 MHz. Again, both switching frequencies are extracted from SPICE simulations.

From these measurements it can be concluded that substrate noise is concentrated at multiples of the digital clock frequency and multiples of the repetition frequency of the circuit (in our case half the clock frequency). These noise peaks occupy the entire spectrum, but the amplitude is influenced by the noise coupling mechanisms. When designing mixed-signal integrated circuits, such as an integrated analog IF or RF front-end stage together with a base-band digital modem, it is important to take the frequency and amplitude of the major substrate noise spectral components into account. These measurements show that substrate noise signals as high as 1.5 mV are generated at multiples of the clock frequency and that the substrate noise peaks are 40 dB above the measurement noise floor, which can seriously degrade analog amplifier behavior.

6. SECOND TEST CHIP: A 86-KGATE DIGITAL FILTER BANK

To analyze substrate noise generation in real-life telecom ASICs, we designed an 86 Kgate digital signal processing circuit combined with analog substrate noise sensors to measure the substrate noise voltage, realized in a low-ohmic epi-type 0.5 μm CMOS technology. The digital circuit is a multi-rate up/down converter and channel select filter for cable modem applications [23]. This chip can upconvert or downconvert 12 bit I/Q data by a factor of 16 and perform channel selection. *Figure 2-11* shows the microphotograph of this chip and also the location of the analog substrate noise sensors. For the substrate noise measurements a 12 bit I/Q random data stream is provided to the ASIC by a digital pattern generator. The output data is observed with a logic analyzer. The load of this logic analyzer, seen by each output buffer of the ASIC, is around 12 pF in parallel with 100 kΩ.

The experimental ASIC has been mounted in a 120 pin Ceramic Pin Grid Array (CPGA) package. The package parasitics of the power supply connections have been obtained by measuring the impedance of a power supply pin pair with a network analyzer. An average inductance value of

12nH for one connection from chip to package pin has been measured. Since for each supply 8 parallel connections have been used, an inductance value of 1.5 nH has been used in the package model.

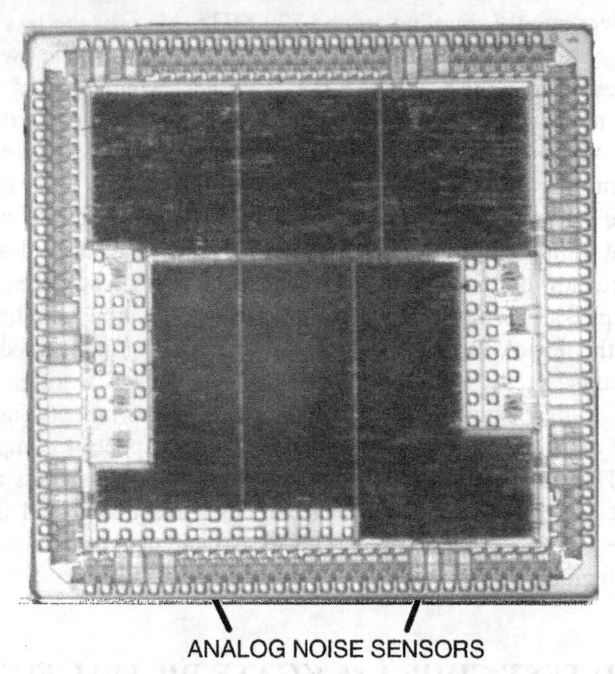

Figure 2-11. Microphotograph of the 86 Kgate digital filter bank ASIC, showing the location of the substrate noise sensors.

Since it is not possible to simulate noise generation of large digital circuits using the SPICE-level models presented in section 3, a new simulation methodology is needed. Simulation methodologies have been presented that make it possible to simulate the total noise current that is injected into the substrate by a digital design [24] or that can estimate the total noise power that is generated [25]. These techniques do not simulate the actual waveform of the voltage noise on the substrate, which is needed to simulate performance degradation of integrated analog circuits. Also no verification of these methodologies with substrate noise measurements on realistic large ASICs have been shown. In chapter 6, we will describe a simulation methodology that makes it possible to accurately and efficiently simulate the substrate noise voltage, and apply it to this 86Kgate digital filterbank ASIC and verify against measurements. In this chapter we will also use some simulation results generated with the methodology described in detail in chapter 6.

6.1 Measurement results

Figure 2-12 shows a comparison of the measured and simulated substrate noise voltage when the digital circuit is operating in 16 times up-conversion mode. An external clock source of 50 MHz is used that is internally divided by 16, which results in data being up-converted from a sample rate of 3.125 MHz to a sample rate of 50 MHz. The RMS value of the measured substrate noise voltage in a 5 μs time period is 13.3 mV. The maximum measured peak-to-peak substrate noise voltage is 80.6 mV.

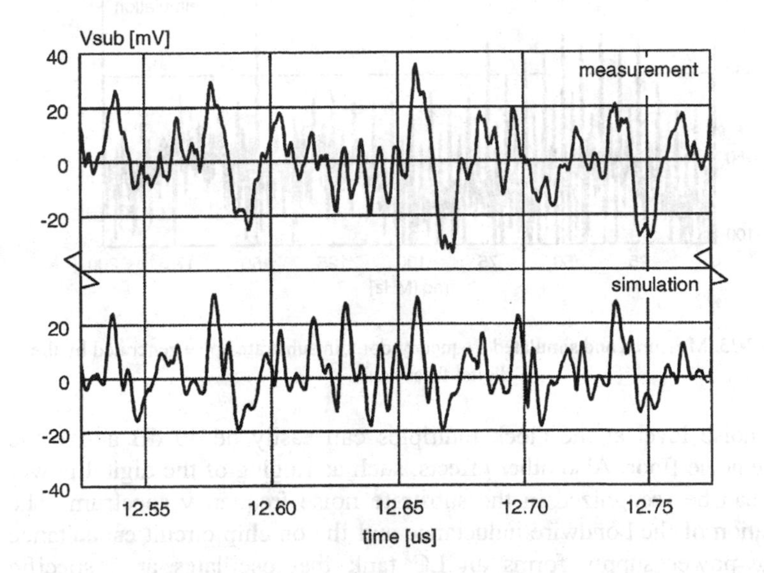

Figure 2-12. Measured and simulated substrate noise generation when the digital filter bank is performing 16-times upconversion from 3.125 to 50 MHz.

Figure 2-13 shows an FFT of the measured and simulated substrate noise voltage. Good agreement between measurements and simulations can be seen both in the time domain and in the frequency domain. As shown in *Figure 2-13*, most noise is generated at multiples of the lowest clock frequency of 3.125 MHz.

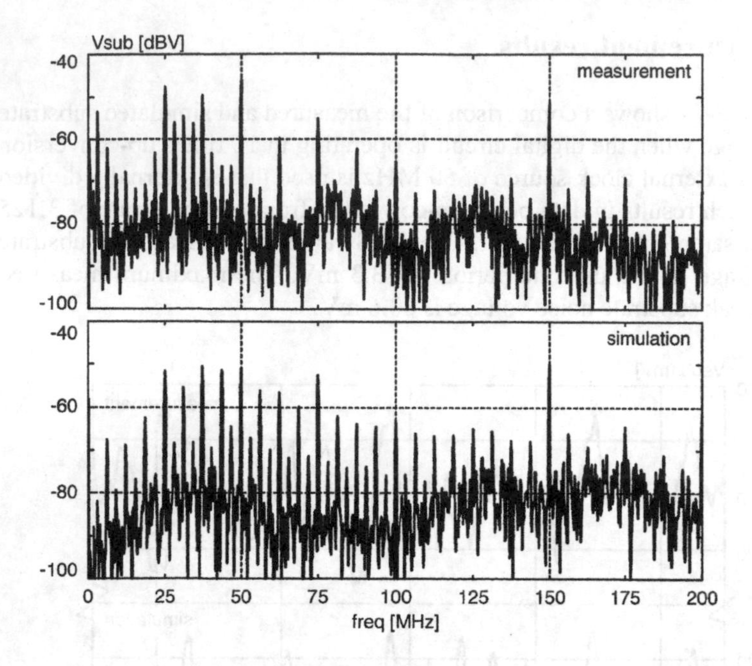

Figure 2-13. Measured and simulated frequency-domain substrate noise generated by the digital filter bank.

The noise level at the clock multiples can easily be 40 dB above the substrate noise floor. Also other effects, such as ringing of the digital power supply, can be recognized in the substrate noise frequency spectrum. The combination of the bondwire inductance and the on-chip circuit capacitance over the power supply forms an LC tank that oscillates at a specific frequency. This oscillation or ringing frequency can be recognized in the frequency spectrum of the substrate noise signal. From the extracted bondwire inductance and on-chip capacitance it follows that this frequency is around 40 MHz, and in the measured and simulated frequency spectra there is indeed an increase of substrate noise with 10 to 20 dB around this frequency.

6.2 Substrate noise analysis

The high-level substrate noise simulation model (see chapter 6) can be used to analyze the major sources of substrate noise generation. This can be done by disabling certain noise current sources or changing the package model. This section describes a number of experiments that analyze the substrate noise generation by changing the simulation model.

6.2.1. Core cell versus I/O cell switching

Figure 2-14 shows the measured and simulated substrate noise voltage for 16 times up-conversion of data with a 3.125 MHz sample rate to a 50 MHz sample rate.

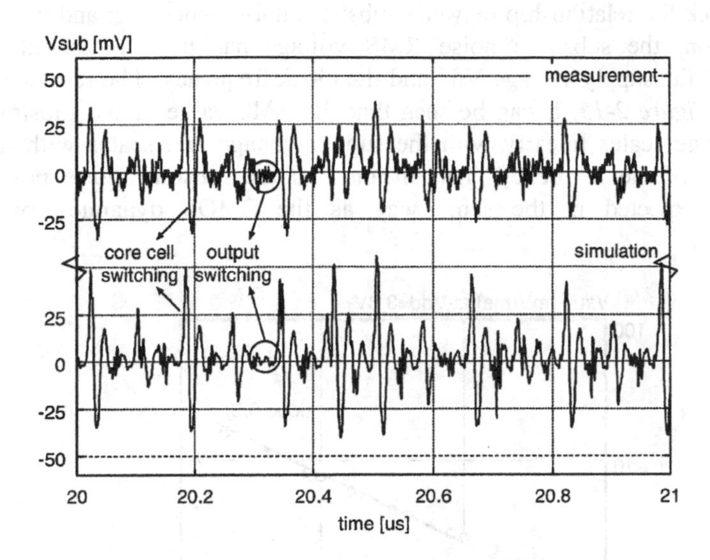

Figure 2-14. Measured and simulated substrate noise voltage with indication of the noise contributions.

By comparing this measured signal with the gate-level simulation, the major noise contributors can be identified. From this comparison it can be concluded that the major noise spikes are generated by simultaneous switching of a large number of core cells (mainly flip-flops), at each clock edge of the 3.125 MHz clock. For this operation mode of the circuit the switching of the output buffers has a much smaller noise contribution. When the substrate noise simulation is performed with only the noise current sources for the core part active, or only the noise current sources for the I/O cells active, it is possible to quantify the noise contribution of the I/O cells. For this simulation the I/O cells generate 18% of the total RMS substrate noise voltage. When the circuit is operating in down-conversion mode, with data output at the slow clock frequency, the I/O cells only contribute 7% of the total substrate noise power. These results indicate that simultaneous switching activity of a large number of core cells can be a dominant source of substrate noise generation. The noise generation of the I/O cells is very much dependent on technology, circuit operation and even PCB design (e.g. output impedance) and cannot be neglected.

6.2.2. Substrate noise versus clock frequency and supply voltage

To check the relationship between substrate noise generation and power consumption, the substrate noise RMS voltage has been measured as function of the supply voltage Vdd and the clock frequency. The results are shown in *Figure 2-15*. It can be seen that the RMS value of the substrate noise voltage scales linearly with the supply voltage and scales with the square-root of the frequency. This means that the substrate noise power scales as expected in the same way as the CMOS dynamic power consumption.

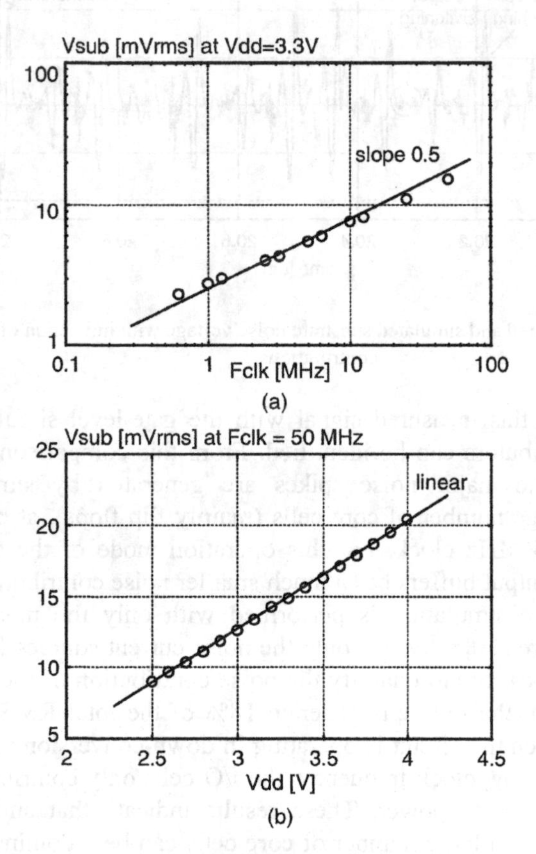

Figure 2-15. Measured rms substrate noise voltage versus (top) clock frequency and (bottom) supply voltage.

6.2.3. Substrate noise versus package parasitics

The parameters of the package model (the value of the inductance and resistance) can be easily changed in the simulations, to analyze the effect of these parasitics on the total substrate noise generation. In this way, it is also possible to explore different packaging options for a chip. *Figure 2-16* shows the simulated RMS substrate voltage for the experimental ASIC versus the inductor and resistor value of these package parasitics. Indicated in this figure with a "W" is the approximate location of the parasitics from the CPGA package in which the chip was wirebonded (12 nH in series with 1Ω for every connection). Indicated with a "C" is the approximate location of the package parasitics (1 nH in series with 0.1Ω) for a Ball Grid Array (BGA) Chip Scale Package (CSP) [26]. Also indicated with an "F" is the approximate location of the parasitics for an ideal flip-chip mounting in which only the parasitics of the flip-chip bumps (10 pH in series with 30 mΩ) have been taken into account [27].

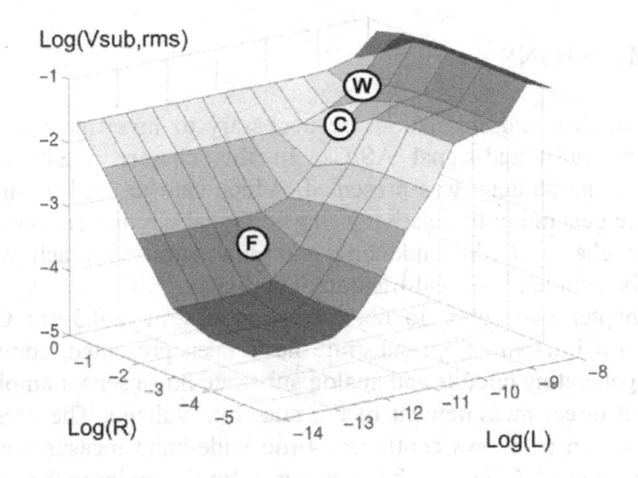

Figure 2-16. Simulated rms voltage versus parasitic package inductance and resistance.

It can be seen that packaging with lower parasitic inductance can offer a large decrease of substrate noise. Despite the flip-chip mounting that is used in the CSP, this package still has rather large parasitics from the redistribution traces and the substrate noise generation is only about 3 times less than for the CPGA package. In the most ideal case, with only package parasitics from the flip-chip bumps, the substrate noise is reduced by 2 orders of magnitude. But even for this ideal case, the substrate noise is still dominated by noise coupling from the power supply, since the absolute minimum level of substrate noise is still around 1 order of magnitude lower. This minimum substrate noise level corresponds to a situation without any power supply noise. All substrate noise is then generated by the switching transistors.

It can also be seen that at high inductance values the RMS substrate noise can be reduced somewhat by optimizing the parasitic series resistance. This can be explained by the fact that with these high inductance values the substrate noise is dominated by coupling of ringing from the power supply. This ringing can be reduced (damped) when the series resistance is set to a certain optimum value [28]. For higher series resistance values the resistive voltage drop starts dominating the noise generation. These simulations show the importance of package parasitics in substrate noise generation and also indicate that simulating substrate noise without power supply noise coupling will result in noise levels that are much too low. Therefore the problem of substrate noise coupling should always be examined for packaged ASICs and even external parasitics from a PCB need to be taken into account.

7. CONCLUSIONS

Accurate models and simulation are necessary to investigate substrate noise coupling in mixed-signal ASICs. In this chapter a SPICE-level substrate modeling strategy was presented, which can be used to simulate substrate noise generation by relatively simple digital circuits on low-ohmic substrates. In chapter 6 this modeling and simulation approach will be extended to the simulation of real-life complex digital ASICs.

In this chapter two substrate noise experiments in a 0.5-μm CMOS technology on a low-ohmic epi substrate have been presented, containing digital noise generating circuits and analog substrate noise sensor amplifiers, which allow a direct measurement of the substrate voltage. The presented substrate noise sensor allows continuous-time wide-band measurements of substrate noise up to 1 GHz, which is necessary for determining the spectral content of substrate noise and checking the validity of the SPICE substrate model.

The first substrate noise experiment is a relatively simple mixed-signal ASIC, containing two inverter chains and analog noise sensors. The simulated substrate noise waveforms have shown good correspondence with the measurements, although the results are still dominated by external parasitics. It has been shown that, for small power-supply inductances, the substrate noise has a certain minimum value, and that for larger inductance values, power-supply noise will be the dominant source of substrate noise.

The measured spectra of the substrate noise have shown that most substrate noise is concentrated at multiples of the digital clock frequency and repetition frequencies of the circuit. At these frequencies substrate noise signals are generated as high as 1.5 mV, and the substrate noise peaks are 40 dB above the measurement noise floor. This indicates the importance of a good selection of digital clock frequency and analog IF frequencies (frequency planning) in integrated transceiver front-ends.

The second substrate noise experiment is a 86-Kgate digital multirate filterbank. From the measurements and simulations on this ASIC it can be concluded that most substrate noise is generated from direct coupling of on-chip power supply noise to the substrate. Because the power supply noise determines the substrate noise it is very important to take the complete packaged chip into account when analyzing the substrate noise levels.

By varying the package parasitics it has been observed that low-parasitics packaging techniques can reduce the RMS substrate voltage level by 2 orders of magnitude. But even for the most ideal flip-chip mounting technique, the power supply noise still dominates the substrate noise generation.

REFERENCES

[1] T. J. Schmerbeck, *Low-power HF microelectronics: A unified approach*, G. A. S. Machado, Ed. London, U.K.: Institution of Electrical Engineers (IEE), 1996, chapter 10.

[2] D. K. Su, M. J. Loinaz, S. Masui, and B. A. Wooley, "Experimental results and modeling techniques for substrate noise in mixed-signal integrated circuits," *IEEE J. Solid-State Circuits*, vol. 28, pp. 420–430, Apr. 1993.

[3] R. Gharpurey and R. G. Meyer, "Modeling and analysis of substrate coupling in integrated circuits," *IEEE J. Solid-State Circuits*, vol. 31, pp. 344–353, Mar. 1996.

[4] N. K. Verghese and D. J. Allstot, "Verification of rf and mixed-signal integrated circuits for substrate coupling effects," in *Proc. 1997 IEEE Custom Integrated Circuits Conf.*, 1997, pp. 363–370.

[5] A. Samavedam, A. Sadate, K. Mayaram, and T. S. Fiez, "A scalable substrate noise coupling model for design of mixed-signal ic's," *IEEE J. Solid-State Circuits*, vol. 35, no. 6, pp. 895-904, June 2000.

[6] M. Nagata, J. Nagai, K. Hijikata, T. Morie, and A. Iwata, "Physical design guides for substrate noise reduction in cmos digital circuits," *IEEE J. Solid-State Circuits*, vol. 36, no. 3, pp. 539-549, Mar. 2001.

[7] M. van Heijningen, J. Compiet, P. Wambacq, S. Donnay, and I. Bolsens, "A design experiment for measurement of the spectral content of substrate noise in mixed-signal integrated circuits," in *Proc. 1999 Southwest Symp. Mixed-Signal Design*, Tucson, AZ, Apr. 11–13, 1999, pp. 27–32.

[8] M. van Heijningen, J. Compiet, P. Wambacq, S. Donnay, M. Engels, I. Bolsens, "Modeling of Digital Substrate Noise Generation and Experimental Verification Using

a Novel Substrate Noise Sensor," in *Proc. of European Solid-State Circuits Conference*, pp. 186-189, September 1999.

[9] M. van Heijningen, J. Compiet, P. Wambacq, S. Donnay, M. Engels, and I. Bolsens, "Analysis and experimental verification of digital substrate noise generation for epi-type substrates," *IEEE J. Solid-State Circuits*, vol. 35, no. 7, pp. 1002-1008, July 2000.

[10] M. van Heijningen, M. Badaroglu, S. Donnay, H. De Man, G. Gielen, M. Engels, and I. Bolsens, "Substrate Noise Generation in Complex Digital Systems: Efficient Modeling and Simulation Methodology and Experimental Verification," in *ISSCC Digest of Technical Papers*, pp.342-343, 463, February 2001.

[11] M. van Heijningen, M. Badaroglu, S. Donnay, G. Gielen, and H. De Man, "Substrate noise generation in complex digital systems: efficient modeling and simulation methodology and experimental verification," *IEEE J. of Solid-State Circuits*, vol. 37, pp. 1065-1072, August 2002.

[12] P. Larsson, "di/dt noise in cmos integrated circuits," *Analog Integrated Circuits and Signal Processing*, vol. 14, pp. 113–129, 1997.

[13] T. Gabara, "Reduced ground bounce and improved latch-up suppression through substrate conduction," *IEEE J. Solid-State Circuits*, vol. 23, pp. 1224–1232, Oct. 1988.

[14] R. Senthinathan and J. L. Prince, "Simultaneous switching ground noise calculation for packaged cmos devices," *IEEE J. Solid-State Circuits*, vol. 26, pp. 1724–1728, Nov. 1991.

[15] J. Briaire and K. S. Krisch, "Substrate injection and crosstalk in cmos circuits," in *Proc. 1999 IEEE Custom Integrated Circuits Conf.*, 1999, pp. 483–486.

[16] A. J. van Genderen and N. P. van der Meijs, "Modeling substrate coupling effects using a layout-to-circuit extraction program," in *Proc. ProRISC/IEEE Benelux Workshop Circuits, Systems and Signal Processing*, Nov. 1997, pp. 193–200.

[17] SubstrateStorm from Cadence: *http://www.cadence.com/products/substrate_noise_analysis.html*

[18] T. Blalack, J. Lau, F. J. R. Clément, and B. A. Wooley, "Experimental results and modeling of noise coupling in a lightly doped substrate," in *IEDM '96 Tech. Dig.*, Dec. 1996, pp. 623–626.

[19] K. Makie-Fukuda, T. Ando, T. Tsukada, T. Matsuura, and M. Hotta, "Voltage-comparator-based measurements of equivalently sampled sub-strate noise waveforms in mixed-signal integrated circuits," in *IEEE J. Solid-State Circuits*, vol. 31, May 1996, pp. 726–731.

[20] M. Nagata, Y. Kashima, D. Tamura, T. Morie, and A. Iwata, "Measurements and analyzes of substrate noise waveform in mixed-signal ic environment," in *Proc. 1999 IEEE Custom Integrated Circuits Conf.*, pp. 575–578, 1999.

[21] M. Nagata and A. Iwata, "Substrate noise simulation techniques for analog-digital mixed lsi design," *IEICE Trans. Fundamentals*, vol. E82-A, no. 2, pp. 271–277, Feb. 1999.

[22] Y. Rolain, W. van Moer, G. Vandersteen, and M. van Heijningen, "Measuring mixed signal substrate coupling," in *Proc. IMTC 2000*, Balti-more, MD, May 2000.

[23] R. Pasko, L. Rijnders, P. Schaumont, S. Vernalde, and D. Durackova, "High-performance flexible all-digital quadrature up and down converter chip," in *Proc. IEEE Custom Integrated Circuits Conf.*, 2000, pp. 43-46.

[24] E. Charbon, P. Miliozzi, L.P. Carloni, A. Ferrari, and A. Sangiovanni-Vincentelli, "Modeling digital substrate noise injection in mixed-signal ic's," *IEEE Trans. Computer-Aided Design of Integrated Circuits*, vol. 18, no. 3, pp. 301-310, Mar. 1999.

[25] S. Mitra, R. A. Rutenbar, L. R. Carley, and D. J. Allstot, "A methodology for rapid estimation of substrate-coupled switching noise," in *Proc. 1995 IEEE Custom Integrated Circuits Conf.*, 1995, pp. 129-132.

[26] D. Light, A. Faraci, and J. Fjelstad, "Chip-size package technology for semiconductors," *Microwave Journal*, vol. 41, no.5, pp. 280-294, May 1998.

[27] Z. Feng, W. Zhang, B. Su, K. C. Gupta, and Y. C. Lee, "Rf and mechanical characterization of flip-chip interconnects in cpw circuits with underfill," *IEEE Trans. Microwave Theory Tech.*, vol. 46, no. 12, pp. 2269-2275, Dec. 1998.

[28] P. Larsson, "Resonance and damping in cmos circuits with on-chip decoupling capacitance," *IEEE Trans. Circuits and Systems - I*, vol. 45, no. 8, pp. 849-858, Aug. 1998.

Chapter 3

MODELING AND ANALYSIS OF SUBSTRATE NOISE COUPLING IN MIXED-SIGNAL ICS

Nishath Verghese, Wen Kung Chu and Jim McCanny
CadMOS Group, Cadence Design Systems, 555 River Oaks Parkway, San Jose, CA 95134.

Abstract: Methods for the modeling and analysis of substrate noise coupling in mixed-signal integrated circuits are presented. The physical phenomena responsible for the creation of the undesired signals as well as the transmission mechanisms and media are described. Modeling and analysis techniques to quantify the noise coupling phenomena are presented. A computer-aided design methodology based on the modeling approaches and developed for the analysis and design of noise coupling in mixed-signal integrated circuits is described.

1. INTRODUCTION

The advent of ultra deep submicron CMOS technology (0.25µm or below) has made it possible to integrate many existing system components onto a single chip. This is desirable as it dramatically reduces the overall system cost, area and power while enhancing performance. However, creating a "system-on-chip" (SOC) design is very challenging, especially when it involves the integration of both analog and digital components on the same chip. This is because as digital circuits switch, they inject noise into the common substrate that can easily corrupt sensitive analog signals. As feature sizes decrease and clock frequencies increase, the amount of substrate noise created by digital switching increases dramatically. Indeed, substrate noise is a key reason for inexplicable design failures and poor yields of mixed-signal SOC designs.

In order to manage the substrate noise problem, designers resort to a number of expensive techniques to ensure the noise immunity of their design. These include the strict partitioning of analog and digital functions, a special semiconductor process, and a full custom design effort. However, without the ability to analyze the true effects of substrate noise, many of these techniques are often over-deployed, resulting in longer design cycles and increased manufacturing costs. The main source of on-chip noise in mixed-signal ICs is the digital switching noise. In purely digital applications, the CMOS static logic family offers several attractive features including very low static power dissipation, high packing densities, wide noise margins and high operating frequencies. For high-frequency mixed-signal applications, however, its major drawback is the generation of a large amount of digital switching noise [1-3]. When many digital gates change state, a large cumulative current spike flows through the parasitic networks representing the power lines. Due to the inductance of the package and bond wires these currents get reflected back into the circuit creating power supply spikes known as *Vdd bounce* or *Gnd bounce*. Some fraction of this supply noise is injected into the substrate via substrate contacts or well taps. In addition, when transistors switch, currents get injected into the substrate via their bulk terminal connections. Once noise is present in the substrate, it can cause fluctuations in the bulk terminal voltage of any transistor it reaches. This voltage variation can be enough to cause sensitive analog portions of the design to malfunction. As shown in *Figure 3-1*, substrate noise results from currents injected into the substrate by neighboring switching devices or through well taps or substrate contacts.

A variety of design decisions can be made to reduce the risk of a substrate noise induced chip failure. These include choices that focus on reducing substrate noise and protecting known sensitive areas. For example, by reducing the package inductance visible to the power supplies, the amount of noise injected into the substrate will be reduced. This can be achieved by using a more expensive package e.g. BGA or C4 versus plastic, or by using parallel bonding of the package pins that connect to the power supplies. Another technique to reduce substrate noise is to use a process technology where the substrate is more resistive i.e. using a non-EPI process versus a EPI process. In an EPI process, the substrate behaves as a single node and isolation cannot be achieved using layout techniques. To protect against substrate noise in a non-EPI process, however, choices include use of separate supplies to isolate sensitive blocks from noisy blocks, guard rings, a Kelvin reference (in which the substrate or well contacts are separate from the circuit power or ground connections), or changes to the floorplan where sensitive analog components are isolated from their noisy digital counterparts.

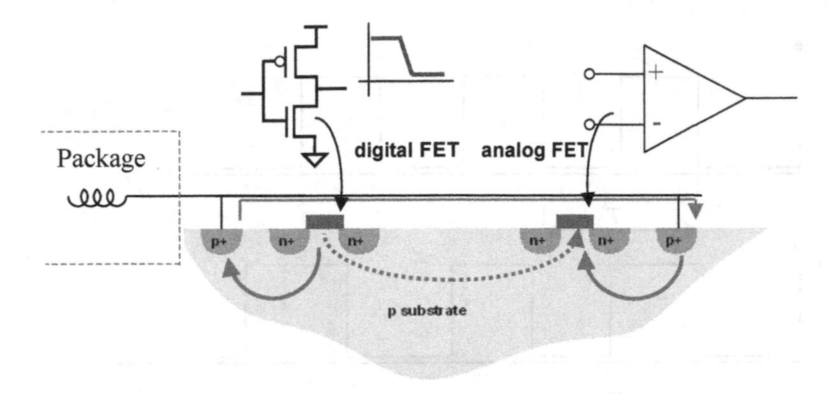

Figure 3-1. Mechanisms for substrate noise coupling.

An example application where substrate noise is a severe problem is in the design of a phase-locked loop (PLL). The PLL is a fundamental building block in many different application areas – data recovery in disk drives, wired and wireless communications, and even in predominantly digital circuits such as high-speed microprocessors and memories. Wireless communication circuits employ PLLs to generate the local oscillator (LO) input to the mixer that performs up-conversion or down-conversion of the input signal. Microprocessor and memory circuits employ phase locking to suppress timing skews between the on-chip clock and the system clock. The PLL consists of a high-speed divider/counter circuit that can produce significant switching noise, as well as noise sensitive circuits such as a voltage-controlled oscillator (VCO) and a charge pump. Moreover, since the PLL is embedded with other digital circuitry on the same chip substrate, it is exposed to substrate noise. This noise manifests itself as phase noise or jitter at the output of the PLL, primarily through mechanisms in the VCO. Phase noise and jitter are inter-related in that they are frequency-domain and time-domain representations respectively of the same phenomenon. For an ideal oscillator the period of oscillation is independent of time. However, due to substrate noise in the circuit, the oscillator period varies as a function of time resulting in a deviation from the mean period that is indicative of jitter. Proper operation of the PLL is highly contingent on its ability to reject noise. To overcome this tough design challenge, sophisticated design automation tools are required to model the effects of substrate noise coupling and its impact on sensitive analog circuitry.

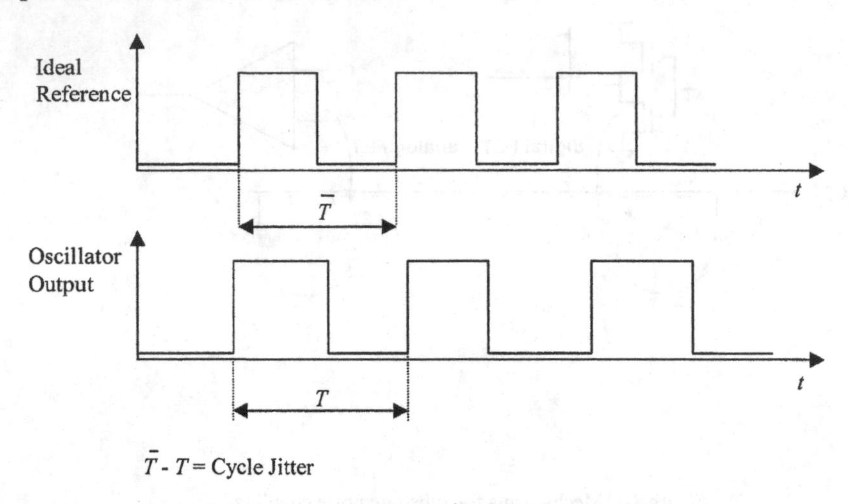

Figure 3-2. PLL jitter.

2. SUBSTRATE NOISE ANALYSIS METHODOLOGY

A complete analysis of substrate noise requires both extraction of the substrate as well as simulation. Any substrate noise analysis technique has to include some form of circuit simulation to assess the impact of substrate noise on a particular parameter of interest for an analog function of interest. Extraction is the process by which an electrical equivalent model of the substrate, possibly including resistance, capacitance or inductance, is determined. To accurately extract a substrate, the complex geometries of wells, contacts, well taps, diffusions, trenches etc. need to be extracted. Once extraction has been completed, simulation can be performed on a circuit including the three-dimensional extracted RC network for the substrate. Simulation to predict substrate noise requires some knowledge of the equivalent extracted network, as well as the nature and location of noise injectors which are causing the noise. If a SPICE simulation is performed with devices and substrate parasitics present, the time required explodes very quickly; hence this approach is tractable only for analyzing small components of the order of a few hundred devices. Instead, one can perform a noise simulation without the presence of devices (using equivalent noise sources) in order to compute the time- or frequency-domain substrate noise waveforms at the bulk nodes of interest. Such a methodology can be utilized to analyze chips with 1 million or more devices. After the substrate noise waveforms have been computed, a simulation with devices can be performed to assess the impact of the noise on the subcircuits of interest. A typical substrate noise analysis methodology for verification of mixed-signal designs is shown in *Figure 3-3*.

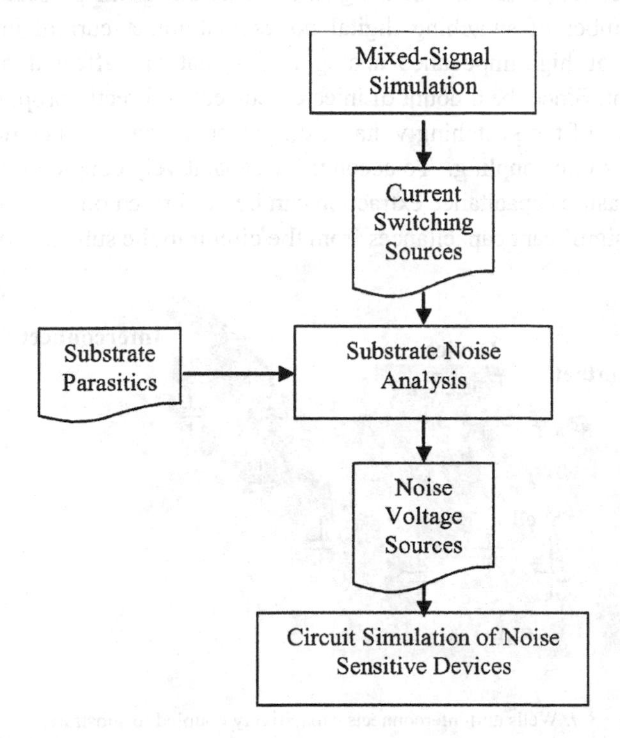

Figure 3-3. Substrate noise analysis methodology for mixed-signal designs.

3. MODELING PARASITICS

Several types of parasitics must be modeled in order to analyze mixed-signal noise coupling problems. These include device/interconnect capacitance, package/bondwire inductance and substrate resistance and capacitance.

3.1 Device/Well/Interconnect Parasitics

As shown in *Figure 3-4*, every transistor, well and interconnect on an IC die can couple capacitively to the substrate. When digital circuits switch, current is injected into the substrate via these capacitances. This current is of significant consequence in mixed-signal circuits, due to the presence both of a large number of switching digital nodes that inject current into the substrate and of high-impedance analog nodes that are affected by this injected current. Since the amount of injected current is directly proportional to the slew rate of the switching voltage, the faster the circuit operation the greater the substrate coupling. To account for capacitively coupled substrate currents, a parasitic capacitance extraction can be performed on the design to determine all significant capacitances from the circuit to the substrate [9].

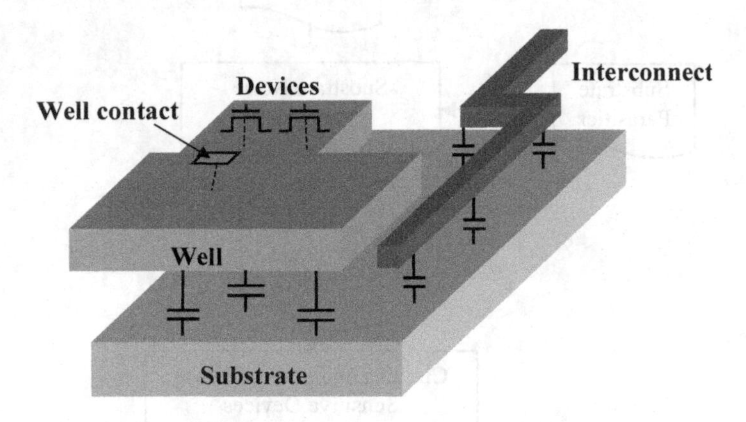

Figure 3-4. Wells and interconnects capacitively coupled to substrate.

3.2 Package Parasitics

The effect of non-ideal (inductive) power supplies has a significant impact on the amount of substrate noise in an IC design. Since the bondwires and package pins associated with the substrate supplies have finite and often large inductances, any substrate current picked up by these supplies can cause large glitches in the value of the substrate supply bias.

This phenomenon is referred to as inductive or Ldi/dt noise. The presence of parasitic inductances in the substrate supplies can severely aggravate the noise coupling problem. Hence, it is necessary to use package inductance models in the supply leads when analyzing substrate noise.

A simple chip-package model [4] is illustrated in *Figure 3-5* where RAPV, LAPV and RACV represent the package resistance, package inductance and on-chip resistance respectively in the analog VDD line while RAPG, LAPG and RACG represent similar parasitics in the analog ground line. CAC represents the chip VDD to GND capacitance while CAV and CAG represent the capacitances from the analog VDD and GND lines to substrate. Similar parasitics are also illustrated for the digital GND and VDD lines.

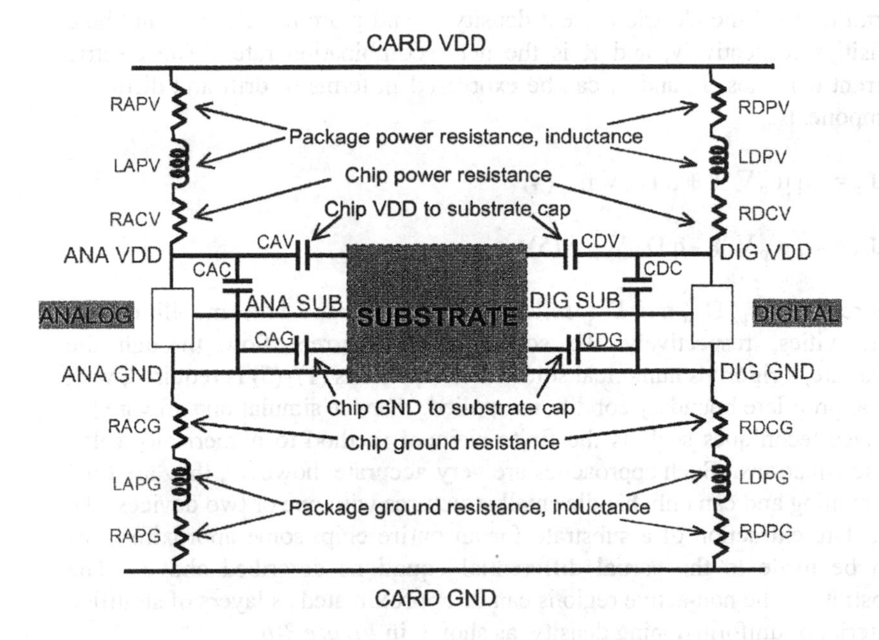

Figure 3-5. Parasitic model of a chip in its package[4].

4. SUBSTRATE PARASITICS

Modeling current flow in the substrate requires modeling active areas such as devices and substrate contacts, as well as inactive areas in the semi-conductor material. The physics of current propagation through the semi-conductor material are described by Poisson's equation and the continuity equation for electrons and holes:

$$\nabla^2\Psi = -\frac{\rho}{\varepsilon} \qquad (1)$$

$$\frac{\partial n}{\partial t} - \frac{1}{q}\nabla \cdot J_n = -R \qquad (2)$$

$$\frac{\partial p}{\partial t} + \frac{1}{q}\nabla \cdot J_p = -R \qquad (3)$$

where Ψ is the electrostatic potential, ρ the charge density, ε the dielectric permittivity, J the electric current density, n and p are the electron and hole densities respectively, and R is the net recombination rate. The electric current densities, J_n and J_p can be expressed in terms of drift and diffusion components:

$$J_n = -q\mu_n \nabla\Psi + qD_n\nabla n \quad (4)$$

$$J_p = -q\mu_p \nabla\Psi - qD_p\nabla p \quad (5)$$

where μ_n, μ_p, D_n and D_p reflect the electron and hole mobilities and diffusivities, respectively. To solve for the current flow through the substrate, a rigorous numerical solution of equations (1)–(5) is required, with the appropriate boundary conditions applied. Device simulation software [8] utilizes techniques such as the finite element method to numerically solve these equations. Such approaches are very accurate, however, they are time consuming and can only handle small structures with one or two devices. To facilitate extraction of a substrate for an entire chip, some approximations can be made to the partial differential equations described above. The substrate in the non-active regions can be approximated as layers of stratified material of uniform doping density, as shown in *Figure 3-6*.

Figure 3-6. Stratified substrate approximation for rapid analysis.

Under this approximation, for an n-type substrate and p-type substrate, respectively, the electric current density can be given by [6]:

$$J_n + J_p \cong q\mu_n n E \quad (6)$$

$$J_n + J_p \cong q\mu_p p E \quad (7)$$

where E (=∇ Ψ) is the electric field intensity. Using equations (6) and (7) in conjunction with equations (1)-(3) results in the following simplified equation that describes the substrate behavior outside the active device areas [6]:

$$\varepsilon \frac{\delta}{\delta t}(\nabla.E) + \frac{1}{\rho}\nabla.E = 0 \qquad (8)$$

Equation (8) can be solved in differential form or integral form. To solve the differential form of equation (8), nodes are defined across the entire substrate volume and the electric field vector between adjacent nodes is approximated using a finite difference formula and this results in a 3-D RC network. Alternatively, an integral form of equation (8) can be formulated as follows:

$$\Psi(r) = \int_V J(r')\, G(r,r') d^3r \qquad (9)$$

where Ψ is the electrostatic potential, J (= $J_n + J_p$) is the current density and G(r, r') is the Green's function satisfying the boundary conditions of the substrate. Solving equation (9) results in a matrix relating current to electrostatic potential. The resulting matrix can be physically realized using a RC network. For extraction of large multi-layered substrates, it has been shown that the integral form of equation (9) can be solved with far greater speed for a given level of accuracy [6]. However, to model wells and trenches that impinge the top layer of the substrate, a finite difference model produces more accurate results. A combination of the two techniques can be used for accurate substrate extraction with low computational overhead.

5. ANALYSIS OF SUBSTRATE NOISE

Once an accurate substrate extraction has been performed, the location and magnitude of noise injectors needs to be determined to facilitate simulation of the substrate noise waveforms. The location of noise injectors can be determined from the layout and the schematic netlist information. To determine the magnitude and phase of injected currents, some form of simulation input is required, under assumed switching activity. Once this has been ascertained, the problem is reduced to solving a very large RC network with active current sources, as shown in *Figure 3-7*. The number of current sources can be extremely large, for example, a million-transistor mixed-signal design may have a million current sources. To see how a large RC network driven by active current sources is analyzed, assume that the voltage response at a bulk node of interest, vb, is desired. The voltage response can be written as follows:

$$vb(s) = z1(s).i1(s) + z2(s).i2(s) + z3(s).i3(s) + \ldots \tag{10}$$

where i1, i2, i3 etc. are the current sources at various locations on the substrate and z1, z2, z3 etc. are their corresponding impedances to the bulk node of interest. The current source values, i1, i2, i3 etc. can be determined from a simulation of the original circuit (without parasitics) by observing the currents flowing in the power /ground nodes and the device bulk terminals. This can be accomplished either with a transistor-level circuit simulator or a gate-level event-driven simulator in conjunction with pre-characterized cell libraries [6]. The currents can be either time-domain waveforms or a composition of spectral values at every frequency ($s = j\omega$) of interest. The impedances, z1, z2 etc. can be obtained by inverting the admittance matrix formed by the RC substrate network and package inductances (*Figure 3-7*) at every frequency ($s = j\omega$) of interest. The frequency-domain response of vb can be obtained by solving (10) at every frequency of interest. Applying the inverse Laplace transform to this response results in the corresponding time-domain waveform.

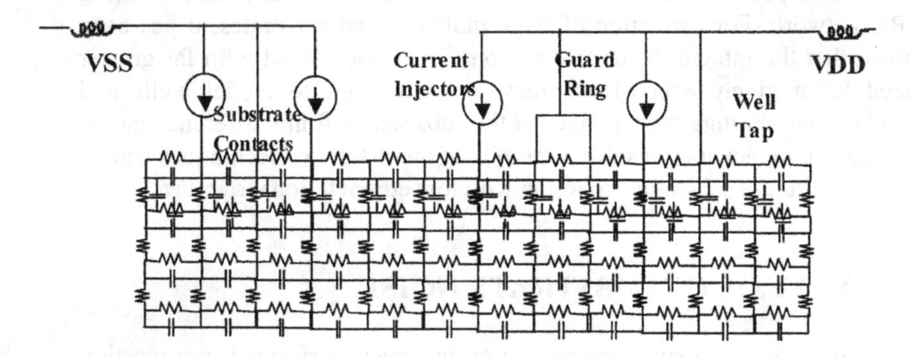

Figure 3-7. Simulation model for full-chip substrate with a large number of noise injectors.

One advantage of using (10) to calculate the noise response of a bulk node of interest is that each individual noise contributor can be calculated independently. Hence, from (10), the noise contribution at the bulk node of interest from injector 1 is z1(s).i1(s). Similarly, z2(s).i2(s) is the contribution from injector 2, z3(s).i3(s) is the injector 3, and so on. Thus, the most significant noise contributors can be identified and appropriate measures can be taken to minimize their impact.

6. ANALYSIS OF IMPACT OF SUBSTRATE NOISE

Once substrate noise at a sensitive analog device location is calculated, its impact on the behavior of that circuit can be determined in one of several ways. The easiest and most general-purpose technique is to model the substrate noise on the analog circuit substrate as an equivalent noise voltage source. This simple noise source model can be included in a transistor-level simulation of the analog circuit to determine its impact on circuit behavior.

For some circuits, particularly PLLs, special-purpose techniques can be employed to analyze the impact of substrate noise. In digital PLLs used in clock and frequency synthesis applications, timing jitter is usually the critical design parameter. Since the VCO (typically implemented as a ring oscillator) generates the output clock, its jitter tends to dominate the overall jitter performance of the PLL. Coupled noise in the substrate of the oscillator causes device capacitances in the oscillator to vary, which in turn causes the output clock frequency to vary, introducing jitter. A simple technique to calculate jitter involves the use of a transient circuit simulator to determine a DC transfer function that models the variation of oscillator clock frequency with small perturbations in the substrate voltage. Multiplying this transfer function with the previously calculated substrate noise, and applying the auto-correlation function on the result gives the cycle jitter and cycle-to-cycle jitter of the VCO.

In analog PLLs, used in wireless receivers, the critical design parameter is phase noise in the VCO since it often limits the adjacent channel-to-channel spacing. Phase noise (measured in dBc/Hz) quantifies the spectral purity of the VCO output. To calculate phase noise in a VCO due to substrate noise, one can utilize a periodic steady-state circuit simulator to determine periodic transfer functions from the substrate of the oscillator to its output. A periodic transfer function relates an output to an input in a circuit that is biased under a periodically time-varying operating point. Multiplying the periodic transfer functions with the previously calculated coupled noise gives the phase noise of the VCO at each frequency of interest.

7. SUBSTRATE NOISE ANALYSIS DATA FLOW

A data flow for mixed-signal substrate noise coupling analysis is outlined in *Figure 3-8.* The layout geometry information and the extracted circuit netlist are used to determine device, well and substrate contact geometry locations. A process technology file encapsulating the substrate doping information is input to a field solver which generates analytical equations for substrate parasitics for the given circuit structure. Switching noise macro-models of logic circuitry are created with the aid of a simulator. An analysis module simulates the effect of the substrate parasitics, noise macro-models and package parasitics working in tandem. The output of the analysis module is the substrate noise at all the sensitive analog circuit locations on chip. In order to efficiently model and analyze the substrate of large designs, an adaptive modeling approach must be used. This can be achieved through the use of sensitivity analysis to determine which areas of the chip need high model accuracy and where the model accuracy can be relaxed without impacting the accuracy of the overall analysis. Noise sensitivity analysis can also be used to measure the impact on substrate noise with respect to a change in any given parameter. By calculating the sensitivity to various layout, process and package parameters, appropriate measures to minimize substrate noise can be determined. For instance, *guard rings* or resistive connections that shield sensitive analog devices are commonly used to mitigate against the effects of substrate noise. Guard rings work by absorbing noise from the substrate and deflecting it away from sensitive areas. However, if guard rings are over-used, they can aggravate the problem by inadvertently bringing noisy supplies closer to sensitive devices. By using noise sensitivity analysis the true impact of guard rings can be predicted and hence unnecessary damaging layout changes can be avoided.

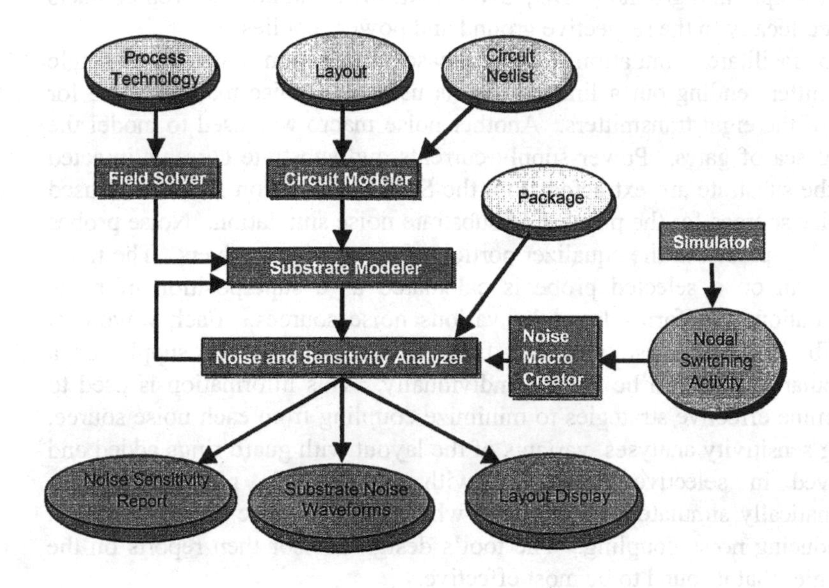

Figure 3-8. Substrate noise analysis data flow.

8. A DESIGN EXAMPLE

An example of using the substrate noise analysis flow described above is presented here in the analysis of a dual-speed eight-port transceiver. In creating this dual-speed repeater chip family, a key goal was to maximize cable length performance. To achieve this, substrate noise as well as other noise needs to be carefully minimized. A commercial tool for substrate noise analysis [7] using the methodology described above was utilized to develop strategies for minimizing substrate noise and creating a robust design. In the following section a description is given of how such an analysis aided in optimizing the design for noise immunity.

The transceiver consists of 8 channels including the receive and transmit circuitry placed adjacent to one another, with neighboring channels mirroring one another on the top half of the design. The digital portion of the chip responsible for the repeater is implemented as a placed and routed sea of gates located in the bottom half of the chip. The design consists of over 1 million transistors. The power distribution network includes thirty-two power supplies with unique power and ground connections to each transmit and receive section. The rest of the design feeds of a common power supply and ground (VDD, GND). All the substrate and well contacts are tied locally to the respective ground and power supplies.

To facilitate simulation for this case, simulation data of a single transmitter sending out a link pulse was used as a noise macro source for each of the eight transmitters. Another noise macro was used to model the digital sea of gates. Power supply currents and substrate currents injected into the substrate are extracted from the SPICE simulation results and used as noise sources for the purpose of substrate noise simulation. Noise probes were then placed in the equalizer portion of one of the receivers. The noise waveform on a selected probe is calculated as a superposition of noise contribution waveforms from the various noise sources. Each waveform contribution (i.e. noise coupling from a particular power supply or a particular macro) can be viewed individually. This information is used to determine effective strategies to minimize coupling from each noise source. Using sensitivity analyses, variants of the layout with guard rings added and removed in selective areas and with power pads redistributed are automatically simulated to determine which changes have the most impact on reducing noise coupling. The tool's design advisor then reports on the strategies that it found to be most effective.

For such a large design, roughly 4 hours of computation time was required to prepare the data and extract a substrate model (Sun Sparc 60). Transient simulation of the noise took about 2 ½ hours. Memory usage was 450 Mbytes.

The simulation results shown in *Figure 3-9* indicate that the noise waveform at the equalizer consists mostly of contributions from the adjoining transmitters. For the equalizers, the major noise contribution comes from the adjoining transmitter's ground currents (> 80%) while the contribution from the transmitter's junction currents account for almost 10%.

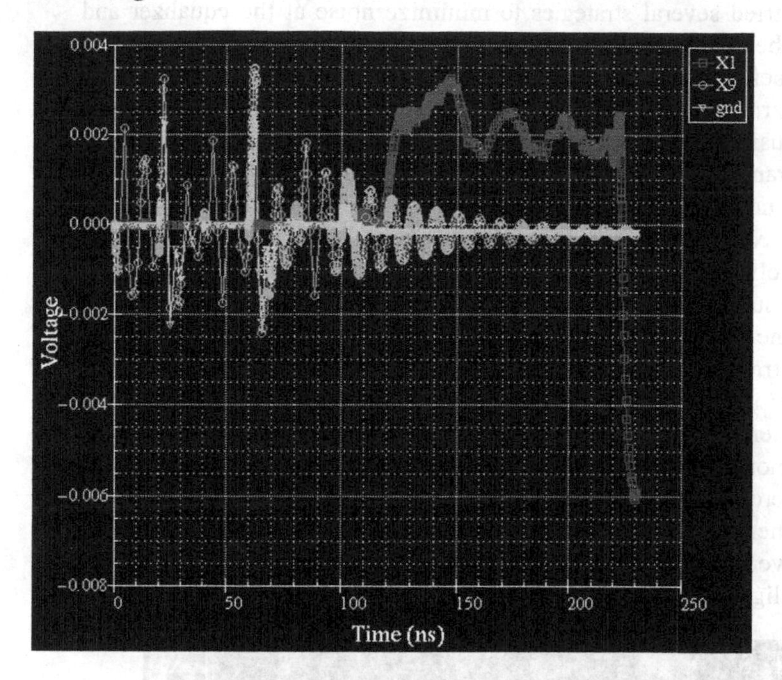

Figure 3-9. Noise contributions at the equalizer with contributions from the digital sea of gates (X9), the digital ground supply (GND) and the adjoining transmitter (X1).

Surprisingly, the digital switching from the sea of gates, despite a worst-case simulation, accounts for less than 10% of the total peak-to-peak noise. This was because of the very low inductance in the digital VDD and GND lines due to the use of thirty-five pads.

The analysis identified that most of the coupling from the transmitter to the equalizer is between the supplies, i.e. currents leaking from the substrate contacts in the transmitter through the substrate contacts in the receiver to the equalizer devices (see *Figure 3-10*).

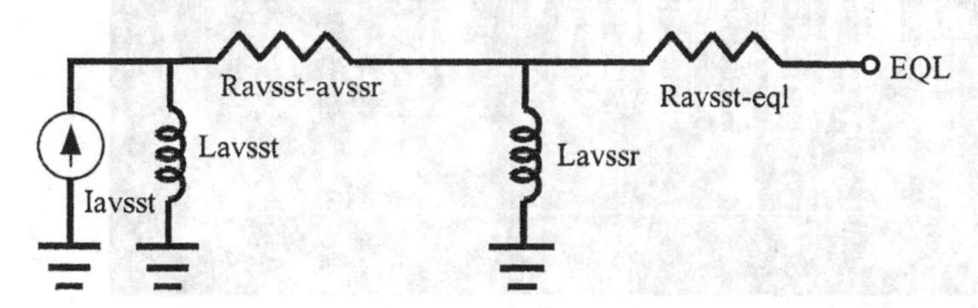

Figure 3-10. Current flows from the transmitter ground through the substrate contacts to the equalizer.

The tool tried several strategies to minimize noise at the equalizer and reported on their effectiveness. It was found that adding guard rings in this case is not useful since this makes the supplies noisier. On the other hand, removing guard rings in the driver is helpful in increasing the Ravsst-avssr resistance causing more attenuation on the receive side. Adding more bond pads to the transmit ground, avsst, is moderately helpful. Adding two extra pads reduces noise at the equalizer side by 30%. However, the best strategy that the tool recommends is to use a Kelvin ground for the transmit ground (avsst). A Kelvin ground is different from the circuit ground and is used to only bias the substrate. By splitting the transmit ground into two, one for the circuit and one for the substrate, the power supply currents flowing directly into the substrate from the transmit side could be eliminated. To verify the effectiveness of this approach, a second simulation was run with Kelvin grounds implemented. The new simulation results shown in *Figure 3-11* show a total noise reduction on the receive side of about eighty percent when a Kelvin reference was used. Now most of the noise is from the junction currents of the adjoining transmitters. (Note that there is little contribution from the power supply currents of the transmitters). Noise from the digital side is only slightly higher than before (about 10%).

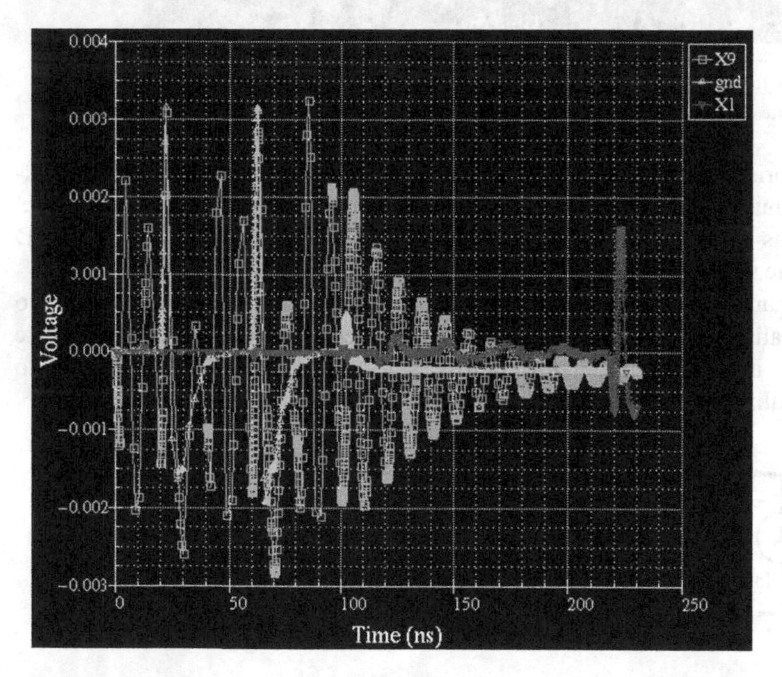

Figure 3-11. Noise contributions at the Equalizer after inserting Kelvin Grounds.

9. SUMMARY

An overview has been provided of the substrate noise problem, as well as a description of substrate noise analysis. Methods to perform substrate extraction were briefly described. A methodology and a data flow for simulating the full-chip substrate of mixed-signal designs combining the substrate model and switching noise models were described. The simulation flow was demonstrated for an ethernet transceiver chip containing one million devices. This example shows the utility of using substrate noise analysis in the debug phase of a design, the value of identifying the worst noise contributors for a given design, and using analysis in the design phase to optimize the noise immunity of a design.

ACKNOWLEDGMENT

The authors would like to thank Dr. Saila Ponnapalli for her contribution to this work.

REFERENCES

[1] G.H. Warren and C. Jungo, "Noise, crosstalk and distortion in mixed analog/digital integrated circuits," *Proc. of the IEEE Custom Integrated Circuits Conference*, pp. 12.1.1-12.1.4., May 1988.

[2] B.J. Hosticka and W. Brockherde, "The art of analog circuit design in a digital VLSI world," *Proc. of the IEEE International Symposium on Circuits and Systems*, pp. 1347-1350, April 1990.

[3] D.J. Allstot, S. Kiaei, and R.H. Zele, "Analog logic techniques steer around the noise," *IEEE Circuits and Devices Magazine*, vol. 9, pp. 18- 21, Sept. 1993.

[4] N.K. Verghese, T.J. Schmerbeck and D.J. Allstot, *Simulation Techniques and Solutions for Mixed-Signal Coupling in Integrated Circuits*, Kluwer Academic Publishers, 1995.

[5] N.K. Verghese, "*Extraction and Simulation Techniques for Substrate-Coupled Noise in Mixed-Signal Integrated Circtuits*", Ph.D. Dissertation, Carnegie Mellon University, 1995.

[6] E. Charbon, P. Miliozzi, L. P. Carloni, A. Ferrari and A. Sangiovanni-Vincentelli, "Modeling Digital Substrate Noise Injection in Mixed-Signal IC's", *IEEE Transactions on Computer-Aided Design of Integrated Circuits and Systems*, vol. 18, no. 3, March 1999.

[7] SeismIC User's Guide, Cadence Design Systems, 2002.

[8] Medici User's Guide, Avanti Corporation.

[9] Assura RCX User's Guide, Cadence Design Systems, 2002.

Chapter 4

SPACE FOR SUBSTRATE RESISTANCE EXTRACTION

Modeling and Verification of Substrate Coupling Problems

N.P. van der Meijs
TU-Delft

Abstract: In this chapter, we first present a general introduction to the problem of substrate-resistance extraction and give an overview of possible extraction techniques. Subsequently we will describe three methods for substrate-resistance extraction. In particular, we will describe a boundary-element method (BEM), an empirical parametric method and a combination of a BEM with a finite-element method (FEM). All three methods exhibit a different but useful accuracy/performance trade-off and suit different situations in the design flow. We will also show how to produce reduced-order equivalent circuits (rather than full detailed models that mandate a-posteriori model-order reduction techniques to be useful) and how this can actually reduce extraction time and memory. We finally introduce the SPACE layout-to-circuit extractor as a comprehensive tool for transforming a layout into a netlist with all relevant parasitics, including the substrate resistances, which can be computed using any of the three methods that are presented.

1. INTRODUCTION

The substrate of integrated circuits is a significant factor in determining the electrical and functional characteristics of integrated circuits. The "ideal ground-plane" model of the substrate of VLSI chips is invalid for a rapidly growing fraction of all the chips that are designed. This is especially true for mixed-signal and RF circuits. Generally, this situation calls for improved tool support in all phases of the design flow, from specification and architecture to physical verification.

Figure 4-1 presents an illustration of the substrate-coupling problem. It shows a mixed-signal circuit, with the digital part introducing substrate noise via the supply line that is connected to the substrate. If the digital circuit switches, the substrate contact exhibits a noise signal because of the resistive and inductive impedance of the supply line combined with the (rate of change of the) supply current. The resulting noise is, in this example, transferred through the substrate and picked up by the analog circuit, which consequently exhibits degraded performance.

However, many other substrate noise sources can potentially be identified, also including device junctions and channels that inject noisy currents into the substrate. Furthermore, noise consequences can vary, possible effects might even include disturbances induced in digital circuits (e.g. clock jitter) by (high-power) analog circuits.

Focusing on the steps later in the design flow, substrate-noise verification is often subdivided in modeling and extraction on the one hand and analysis on the other hand. Analysis can then be further subdivided into static and dynamic analysis. With static analysis electrical consistency is checked without explicitly applying stimuli or signal waveforms, aiming at results of general validity. Dynamic analysis is usually called simulation and is aimed at calculating signal behavior as a function of time under certain input stimuli. Different levels of accuracy can be achieved using different simulation principles, as a trade-off against the CPU time required.

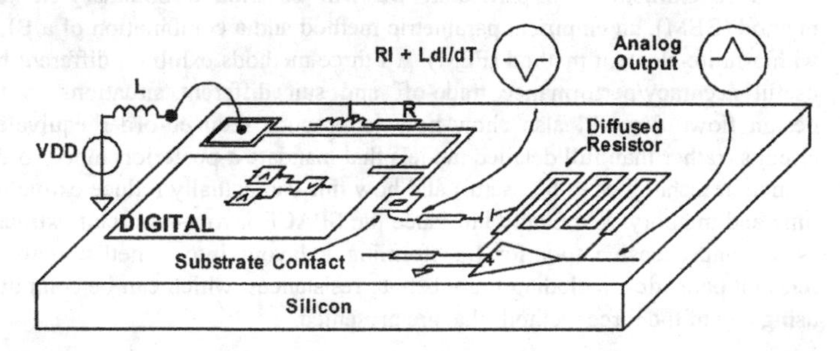

Figure 4-1. Sketch of a substrate coupling problem in a mixed-signal circuit.

In this chapter, we will focus on the step that precedes the analysis, which will produce a model of the circuit in a form suitable for further static or dynamic analysis. This step is typically called extraction because an electrical model is derived, or extracted, from the relevant physical design data of the chip. The model should represent all relevant physical effects that would co-determine the behavior and performance of the chip after fabrication. A typical form of the model is a netlist (e.g. a spice netlist) that contains all intended components, properly interconnected, but also the unintended parasitic elements that can only approximately be taken into account during design but that are introduced by the physical implementation of the design and therefore depend on the layout and the packaging of the chip.

The relevant physical design data from which the model is extracted, is in our cased formed by the layout data and the technology data. The latter would describe the details of the fabrication process to which the layout is submitted, and which together with the layout determines the final electrical properties of the chip.

As almost all IC physical verification tasks, the substrate resistance extraction problem is computationally difficult and complex. This complexity is associated with the large amount of data to be handled as well as with the 'intrinsic complexity' related to the basic structural details of the chip that is being designed and the target fabrication technology. Consequently, software tools for substrate resistance extraction present formidable challenges with respect to their efficiency. They can only succeed by making certain assumptions about the nature of the problem to be solved, and such assumptions will obviously (co-)determine their applicability for each specific design situation. Apart from describing some different extraction techniques, this chapter aims at showing how such assumptions can result in different solutions for the extraction problem with correspondingly different accuracy-efficiency trade-offs.

Although the integration of the basic interconnect and device extraction procedures and techniques with those for substrate resistance extraction in a seamless way, that properly accounts for all intended and unintended electrical effects, can be very challenging, we will not go into such details too much. Instead, we will study the substrate problem almost in isolation.

This chapter is structured as follows. We will start in Section 2 with a general introduction to the problem and an overview of methods for substrate resistance extraction and associated issues. Subsequently, in Section 3, we will focus on the Boundary Element Method (BEM) for substrate resistance extraction, and how such a method actually produces a reduced-order model and can achieve a linear time complexity by applying a special method for matrix inversion, called the Schur algorithm. Despite such and other methods for improving the extraction speed, this speed remains insufficient for many potential applications for today's (and tomorrow's) chips. Therefore, we will also describe a faster, but unfortunately less accurate, parametric modeling method in Section 4. Also this method actually produces a reduced-order circuit, while saving computation time by avoiding the calculation of irrelevant detail. While in certain situations the greater speed of this method is preferable over the reduced accuracy, in other situations the BEM itself might not even be accurate enough. As will be explained, this is related to the inability of the BEM to model layout-dependent doping variations. Therefore, we will in

Section 5 describe a way to increase the flexibility and applicability of the BEM by combining it with a FEM-based technique. Then, in Section 6 we will briefly describe the SPACE layout-to-circuit extractor, in which the described techniques for substrate resistance extraction have been implemented. Finally, we will conclude in Section 7.

2. SUBSTRATE ANALYSIS OVERVIEW

In this section, we will consider the substrate analysis problem from a general perspective. We will distinguish two subproblems, namely modeling and extraction. Modeling entails the mathematical analysis of the physical/electrical problem, and extraction entails the computational procedure to compute the parameters of a model for each specific instance of the problem.

2.1 Modeling

At not too high frequencies, depending on the substrate doping, the substrate behaves resistively. Consider for example a homogeneous block of silicon as shown in *Figure 4-2*. It has two contacts at opposite sides, which are resistively and capacitively coupled. The admittance Y between both contacts is in fact given by

$$Y = G + sC \quad with \quad G = \sigma \frac{LW}{t} \quad and \quad C = \varepsilon_0 \varepsilon_r \frac{LW}{t}$$

Then, when $\sigma << |s\sigma_0\varepsilon_r|$, the substrate predominantly behaves resistively. This is independent of the geometry because the formulas for G and C have the same shape factor. Using $s = j\omega = |2\pi f|$, this is equivalent to $f << \sigma / \varepsilon_0\varepsilon_r$. This latter quantity is actually equal to the inverse of the so-called relaxation time constant $\Delta\tau$ of the medium, and $f_c = 1/2\pi\tau$ is a kind of crossover frequency. Typical substrate resistivities in real IC technologies can range from 10 S/m (lowly-doped) to 10000 S/m (highly doped). Lowly doped, high-ohmic substrates thus have an f_c around 15 GHz. This means that only when $f <<$ 15GHz, the substrate behaves resistively. It is clear that, on such substrates, the reactance may not be ignored blindly for state-of-the-art circuits with RF frequencies or harmonics. Nevertheless, in this work we focus on the resistance only. Reactance can be ignored until much higher frequencies on highly doped substrates. Moreover, just considering complex material properties can at least in principle extend the verification methodology to simultaneously include resistive and capacitive effects.

For highly doped substrates, f_c can be much higher. However, for frequencies >> 15 GHz, other effects such as the inductive skin effect and the slow-wave effect [10] can arise. These frequencies and effects will require completely different modeling and verification techniques, possibly abandoning quasi-static approximations.

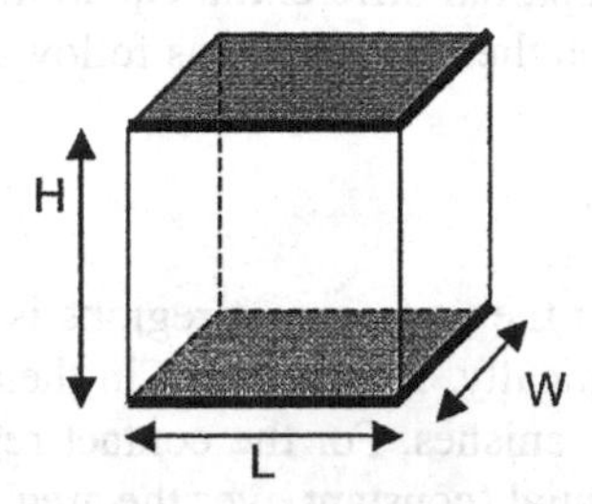

Figure 4-2. Homogenous block of silicon with two contacts.

Thus, in this chapter, we will consider the low-frequency resistive behavior of the substrate of a silicon chip. The substrate will be modeled as a 3D box of semiconductor material, as shown in *Figure 4-3*. For such a model, we will discuss how to compute a resistive network connected to a set of contact areas on top of it. We will ignore any frequency-dependent effects. Moreover, we will ignore other effects such as minority carrier conduction.

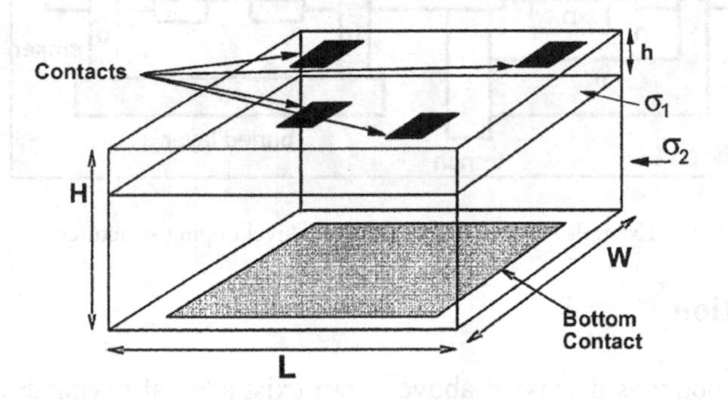

Figure 4-3. 3D view of the epi-layer and substrate with contacts.

Because of the doping profile, the resistivity varies in the vertical direction, perpendicular to the $Si\text{-}SiO_2$ interface. The contact areas are usually assumed equipotential. They represent the areas where the circuit may interact with the substrate, and can model e.g. MOS back gates, source and drain junctions, and substrate taps. The equipotential assumption is valid when the contact areas are not too large, potentially they can be subdivided. Sometimes there can also be a single large contact at the bottom.

Again, consider the structure of *Figure 4-3*. In order to model the behavior of the substrate, we aim at computing the multi-terminal admittance matrix with the contacts (assumed equipotential) as ports. Mathematically, the problem is described by a partial differential equation in a domain *O* with boundary conditions (BC's) on the boundary *G* as follows:

$$\nabla\sigma(x)\nabla\varphi(x)=0 \tag{1}$$

The boundary condition at the non-contact regions is that no currents are flowing through it. Mathematically, it means that in these places the normal derivative of the potential *f* vanishes. For the contact regions, the boundary condition prescribes the potential (constant over the area for ideal contacts).

The simplified model as depicted in *Figure 4-3* may not be suited for all occasions. Typically, there are some layout-dependent doping variations in the top layer of the substrate too. As illustrated in *Figure 4-4*, examples include channel stoppers, trenches and deep diffusions, but also the wells and source/drain junctions. Depending on the characteristics of the circuit and on the desired accuracy, it might be necessary to properly include these top layer structures in the modeling. This subject is discussed further in Section 5.

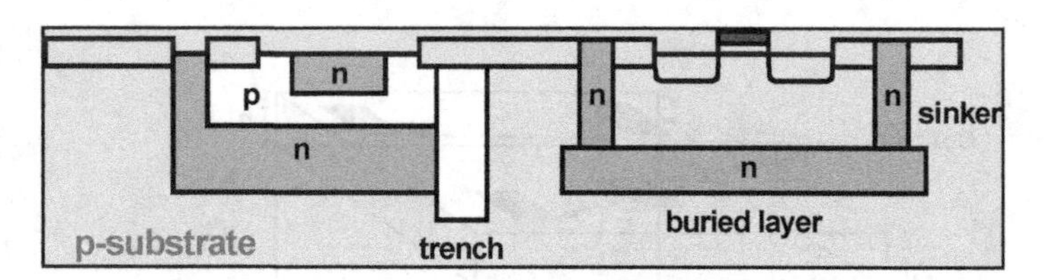

Figure 4-4. Example of layout-dependent top-level doping structures.

2.2 Extraction

Given the model as discussed above, there exist several reasonable ways to compute the substrate resistance network, each with a different trade-off between accuracy, flexibility and computation speed. Perhaps the most basic approach is using a device simulator [5][22]. This approach can be the most accurate, although it usually does not produce a resistive model directly. As the name implies, it already produces voltage and current waveforms, based on applied stimuli. These results can include any nonlinear effects that would arise from the changing width of depletion layers, but also other effects such as minority carriers and latch-up.

If desired, one could derive a model by combining results from multiple simulations, although in practice the approach is limited to the linear case. For a linear network with N ports, N linearly independent stimuli can be applied to arrive at N port currents each. The resulting N^2 different currents enable the calculation of the NxN admittance matrix. For the case of symmetric networks, however, only $N(N+1)/2$ different currents are actually necessary.

Perhaps the least basic approach is parametric modeling, where fitting formulas are being used to estimate the resistances directly from the geometric dimensions of the contact structures [11][24]. The fitting formulas are optimized once for each technology, i.e. doping profile. This approach is further discussed in Section 4.

In between the device simulator method and the parametric modeling method, approximately ordered from slower to faster and from more accurate to less accurate as will be explained below, are the finite difference (FDM) and finite element (FEM) techniques (e.g. [4]) and boundary element (BEM) techniques (e.g. [21]). Similar to the device simulator techniques, they are also physically based. The difference is that a device simulator also includes the equations for the semiconductor physics behavior, while these latter methods only model equation (1) or its integral equation equivalent.

The differences in modeling accuracy of these methods are mainly related to the ability to incorporate the structures present in the top layer of the substrate. In principle, a 3D device simulator should be able to model these structures fairly accurately, including any nonlinear effects that would arise from changing widths of depletion layers. Such effects can not accurately be captured in the linear resistance models that we normally aim for, but FDM and FEM can still model local variations in doping patterns. This is because these methods perform a 3D discretization of the medium, where each discretization element can have specific material properties assigned.

The BEM, on the other hand, only discretizes the boundary of the region (or sometimes only the contacts) and must assume regular material properties. Typically, only a vertical doping profile with constant properties in planes parallel to the Si-SiO_2 boundary is sufficiently regular to be included in the BEM formulation. Therefore, the BEM cannot model lateral layout-dependent top-level doping patterns.

The differences in discretization methodology also give rise to differences in computation speed. The 3D discretization of the FEM results in a dramatically larger system of equations to be solved than the (essentially

2D) discretization of the BEM. However, the FEM system is sparse, while the BEM system is full. For each method, highly optimized specific solution methods are typically used. For the FEM, popular solution methods include GMRES-like methods [18] and for the BEM they include the so-called fast multipole method [15] and the Schur method [16]. However, using state of the art solution techniques for both FEM and BEM, BEM approaches are usually significantly faster and can handle larger problems. As already noted, this is typically at the expense of modeling flexibility.

Alternatively, hybrid BEM/FEM methods could combine the advantages of both methods [19]. These methods can provide a way to incorporate the top-level layout-dependent doping patterns in a predominantly BEM-based substrate model by applying a FEM technique for the top level, and properly combining both methods. This method is further discussed in Section 5.

3. THE BOUNDARY ELEMENT METHOD

3.1 Introduction

In this section, we will introduce the BEM for substrate resistance extraction. This method could start with a bounded 3D model as indicated in *Figure 4-3*. However, the development of the method, as well as the computational complexity of the subsequent implementation, is significantly simplified if we model the finite domain O as an infinite semi-space extending towards infinity in the horizontal and negative z-directions. The only boundary that remains is the Si-SiO_2 interface, and the corresponding boundary conditions (BC's) should impose that no current can flow through it, except via the contact areas.

As already noted, in practice the BEM requires that the substrate doping exhibits a vertical variation only. We will in fact assume a layered doping profile. That is, the chip is subdivided along the vertical dimension into a number of uniform layers with interfaces parallel to the x-y plane. In each layer, the doping level is considered constant. Then, in each layer the potential $f(x)$ is governed by the Laplace equation:

$$\nabla^2 \varphi(x) = 0 \qquad (2)$$

The different resistive layers are coupled through interface conditions (IC's), which specify that both the potential and the normal component of the current density are continuous.

We can apply Green's theorem to transform the set of PDE's (2) to a Boundary Integral Equation [3] that with the above simplifications can take a particularly simple form:

$$\varphi(x) = \int_{\Gamma_c} G(x;x_0) j(x_0) dx_0 \tag{3}$$

Here, $j(x_0)$ is the current density at x_0, G_c is the part of the boundary where the current density $j(x_0)$ is non-zero, i.e. the contacts, and $G(x;x_0)$ is the so-called Green's function. Basically, the Green's function is the solution of the fundamental PDE, corresponding to equation (2):

$$\sigma(x)\nabla^2 G(x;x_0) = -\delta(x - x_0) \tag{4}$$

For the analogous electrostatic problem, the Green's function may be interpreted as the potential at point x (observation point) in a domain, induced by a unit point charge at position x_0 (source point). The simplest form of a fundamental solution is the Green's function of the free space electrostatic problem, where ε_0 replaces σ. Thus,

$$\varphi(x) = G(x;x_0) = \frac{1}{4\pi\varepsilon_0 \|x - x_0\|} \tag{5}$$

For stratified media, the Green's function is more involved. In fact, for the stratified substrate resistance problem, the Green's function can capture all the interface conditions as well as the Neumann boundary conditions. It is this latter property that actually enables restriction of the integration in equation (3) to the contact boundaries G_c instead of the complete top-level Si-SiO_2 interface. For our situation, where we model the chip as an infinite layered semi-space (infinite lateral dimensions and thickness), the Green's function shows cylindrical symmetry. It can then be found by separation of variables in cylinder coordinates. This leads to series expansions of the following form:

$$G = \sum_{n=0}^{\infty} \frac{a_n \left(\dfrac{\sigma_1 - \sigma_2}{\sigma_1 + \sigma_2} \right)^n}{\sqrt{\rho^2 + (b_n + c_n z)^2}} \tag{6}$$

Each term can be interpreted as originating from the method of images [25]. Here, a_n, b_n and c_n are constants, s_i is the conductivity of layer *i*, ρ the lateral Euclidian distance between *x* and x_0, and *z* the vertical distance. For more layers, similar formulas can be developed, although other formulations can often be more appropriate for domains with more layers.

3.2 Discretization

For actual computation, equation (3) must be discretized. That is, the contact regions forming the integration boundary are subdivided into *N* elemental areas G_j, $j = 1...N$. Then, there are several ways to construct a linear system of equations, but the so-called collocation method is the simplest. With this method, a unit source current density is assumed over element G_j, and the potential that results from this current in the collocation point (typically at the center) of element *i* is computed via

$$G_{ij} = \frac{1}{A_j} \int_{\Gamma_j} G(x_i; x_j) d\Gamma(x_j), \tag{7}$$

where A_j denotes the area of element G_j. Upon assembling these influences in a matrix *G*, we can write

$$\Phi_e = G \cdot J_e \tag{8}$$

Here, F_e and J_e are two N-dimensional vectors collecting the boundary element potentials and currents, respectively. Subsequently, we can relate elemental quantities to contact quantities by defining an incidence matrix *F* such that F_{ij} is 1 if element *i* lies on contact *j* and is 0 otherwise. Furthermore, let F and *J* be two M-dimensional vectors collecting contact potentials and currents, respectively. It follows that

$$J = F^T J_e = F^T G^{-1} \Phi_e = F^T G^{-1} F \Phi = Y \Phi \tag{9}$$

Here

$$Y = F^T G^{-1} F \tag{10}$$

is the admittance matrix to be determined. For *N* contacts, an *NxN* matrix *Y* corresponds to a network with *N+1* nodes: one for each contact plus a (virtual) reference node. A circuit can be defined as

$$y_{ij} = -Y_{ij} \qquad i \neq j$$
$$y_{i\infty} = \sum_j Y_{ij} \qquad i = 1 \ldots N \tag{11}$$

where y_{ij} denotes a conductance between two nodes and $y_{i\infty}$ denotes the conductance of a contact to the reference node. It is tempting to refer to this virtual reference node as the substrate node. This is not strictly correct, but is sometimes justified by noting that in practice the contact potentials are measured to 'the' potential of the substrate. Thus, the virtual reference node in the extracted admittance network superficially performs the same role as the substrate node in the chip. Especially in the case of a highly doped substrate with a backside metallization contact, it might be allowed to identify the virtual reference node with this contact.

3.3 Matrix inversion

The matrix inversion in equation (10) is the main computational bottleneck of the method. Exact inversion requires $O(N^3)$ time, and actually is unacceptable. Therefore, a significant amount of research has been performed to alleviate this problem. Although this research was mostly performed with BEM-based interconnect capacitance extraction in mind, the resulting methods also work for BEM-based substrate resistance extraction. A particularly effective method is the Schur method [16], which we will describe first. Subsequently, we will point out some other techniques.

The Schur method actually delivers a 'windowed' approximation of the admittance network corresponding to equation (10). That is, given a window of size w, it only includes admittances between contacts that are separated by a distance of $O(w)$ or less. Thus, widely separated contacts are not directly coupled, but only indirectly via intermediate contacts and the reference node. This is a reasonable model, since for widely separated contacts the direct resistance (i.e. inverse of the admittance) between both contacts is much larger than the sum of the resistances from both contacts to the reference node. This windowing is a very effective form of a-priori model-order reduction, which, as we will see, can greatly reduce the computational complexity of extracting the model.

Mathematically, the windowed admittance network corresponds to a sparse admittance matrix Y in (10). However, one cannot start with sparsifying G, since the inverse of a sparse matrix is in general not sparse. The Schur algorithm instead produces a sparse approximation to Y based on a partially specified version of G. The result is termed Y_{me}, because it is actually the inverse of the so-called *maximum entropy* extension of the partially specified matrix G. Y_{me} is the unique matrix that contains zeros on the places corresponding to the unspecified entries in G, and which upon exact inversion (Y_{me} is fully specified, but sparse) would coincide with G on its specified part. This is illustrated in *Figure 4-5*.

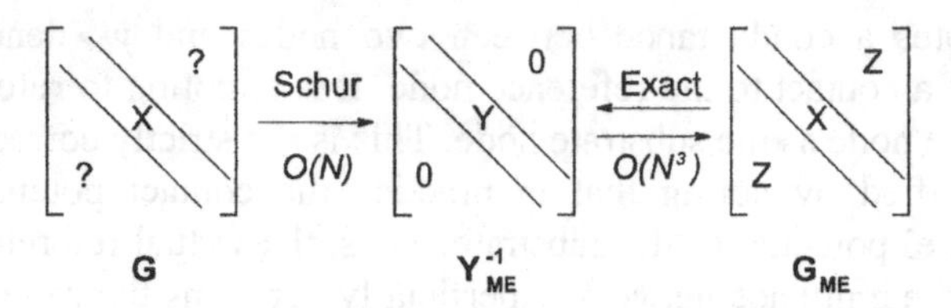

Figure 4-5. Illustration of the Schur matrix extension algorithm properties.

If we assume a one-dimensional contact configuration (i.e. a linear row of substrate contacts), the matrix will be a band matrix with bandwidth $b=O(w^2)$ where w is the width of the window. The computational complexity of the Schur algorithm is actually $O(Nb^2)$. Asymptotically for large w, G becomes fully specified, and the Schur algorithm produces the exact inverse of G. This is consistent with the computational complexity since with $b=O(N)$, we have $O(Nb^2)=O(N^3)$. These computational complexities are also annotated in *Figure 4-5*. The memory complexity of the Schur algorithm is actually $O(b^2)$, independent of N.

The Schur algorithm requires a stair-case-banded specification support, but this cannot generally be achieved. That is, the positions of the specified entries in the matrix should form a stair-case structure centered around the main diagonal of the matrix. Such a structure arises naturally when the contacts are located along a one-dimensional row, and when only interactions over distances smaller than the window threshold are included. For 2-dimensional contact configurations, a suitable ordering of the BE's that produces a stair-case specification support does not exist.

However, [16] has presented a hierarchical extension of the Schur algorithm suitable for matrices with multiple-band support. This hierarchical Schur algorithm is compatible with a 2D contact layout, but requires a relaxed window constraint. That is, all couplings at distance $d = w$ are included and all couplings with $d = 2w$ are excluded. However, some couplings with $w < d < 2w$ are included and some are excluded [13][14].

Table *4-1* shows the model reduction behavior of the Schur algorithm. The layout is formed by a regular array of 30 x 30 substrate contacts of size 2μ x 2μ with a pitch of 10μ, on a substrate with a 4μ, 10 *S/m* epi-layer on a 10000 *S/m* substrate. The column labeled w displays the window size parameter. This layout was extracted with SPACE [2][12], using the BEM method.

The results refer to an interior contact, on array position (5,5). The d column gives the number of resistors connected to this contact (the degree), the columns R_i, $i \in \{1, 2, 5, 10, 20\}$, the value of the resistance to other contacts on position (5+i,5+i), R_0 the resistance to the virtual substrate node and R_s the so-called short-circuit resistance. This is the effective resistance seen from a contact when all the others are short-circuited to ground.

The d and R_i columns clearly show the sparsity of the extracted networks as a function of the window size. With increasing w, the extracted network will become less and less sparse. Moreover, it can be observed that the Schur algorithm provides some 'compensation behavior'. That is, R_s is nearly constant, independent of the window size. This is desirable, because this value largely determines the sensitivity (or aggressiveness) of the contact for crosstalk. This compensation behavior is also shown in the R_i and R_0 columns.

Table 4-1. Schur modeling properties for an epi-type substrate.

w	d	R_1 [MΩ]	R_2 [MΩ]	R_5 [MΩ]	R_{10} [MΩ]	R_{20} [MΩ]	R_0 [kΩ]	R_s [kΩ]
5	1	-	-	-	-	-	20.97	20.97
15	15	25.69	-	-	-	-	21.47	20.97
25	35	25.72	759.3	-	-	-	21.48	20.97
55	121	25.73	767.6	2009	-	-	21.51	20.97
105	336	25.73	768.1	2050	4055	-	21.55	20.97
205	780	25.73	768.3	2052	4124	8139	21.58	20.97
∞	900	25.73	768.3	2052	4126	8270	21.59	20.97

Table 4-2 shows the behavior of the Schur algorithm for the same layout on a uniform substrate (10 S/m, 500 μm thick). Clearly, the approximation is more difficult but still reasonable. The data would suggest that a larger window size would be necessary for the same accuracy. The first 'inclusion' of a specific R_i when w has been enlarged has not yet converged.

Table 4-2. Schur modeling properties for an uni-type substrate.

w	d	R_1 [MΩ]	R_2 [MΩ]	R_5 [MΩ]	R_{10} [MΩ]	R_{20} [MΩ]	R_0 [kΩ]	R_s [kΩ]
5	1	-	-	-	-	-	23.77	23.77
15	15	0.5691	-	-	-	-	49.29	23.04
25	35	0.8101	1.786	-	-	-	70.83	22.99
55	121	0.899	3.195	5.381	-	-	149.2	22.96
105	336	0.9085	3.459	25.19	19.82	-	233.0	22.96
205	780	0.9101	3.485	26.17	110.4	45.85	338.7	22.96
∞	900	0.91	3.484	26.16	110.3	166.7	345.3	22.96

Note that contacts without a direct coupling resistance are still indirectly coupled, not only via other contacts but also via the reference node. If this node would be assumed floating, it could be removed from the model by a

Gaussian elimination procedure without changing the I/U relations for the remaining nodes. However, after elimination the resistance network would be full, irrespective of the Schur window size. Thus, the reference node as produced by the solution of equation (10) is also a very effective model-order reduction effect.

Table 4-3 shows the CPU time and memory that was necessary, as a function of the window size. For reference, also the total number of resistances extracted (*N*) and the degree (*d*) of an interior contact are shown. Clearly, small window sizes greatly improve the efficiency of the method and produce much-reduced models.

Table 4-3. Memory and CPU time data for the Schur algorithm on an HP 9000/800 computer.

w	N	d	Mem [Mbyte]	CPU [min:sec]
5	900	1	0.511	0.3
15	5994	15	1.12	1.3
25	15570	35	1.92	4
55	64350	121	5.68	23
105	182970	336	17.8	01:21.2
205	364950	780	49.3	03:08.3
∞	405450	900	70.8	03:39.7

Note again that with a fixed window size the CPU time becomes linear in the size of the layout. Also, the memory requirements will be $O(N^{0.5})$, assuming a scanline technique that only keeps layout and circuit data around the scanline in core – see Section 6. Consequently, this method allows comparatively huge designs with millions of contacts to be handled.

Furthermore, note that the resistance network would become full upon elimination of the substrate node. That is, if the substrate node were floating, it could be removed from the model by a Gaussian elimination procedure without changing the *I/V* relations for the remaining nodes. However, after elimination the resistance network would be full. Thus, from a model-order reduction perspective, it would be very inefficient to eliminate this node.

Instead of using the Schur method for solving equation (10), this equation could alternatively be written as

$$FX = G \qquad and \qquad Y = F^T X \tag{12}$$

and we can use an iterative algorithm to solve for *X*, and then obtain *Y* by adding columns of *X*. The main computational cost of such algorithms is caused by the associated matrix-vector multiplications, and these can be accelerated using some techniques that compress the dense matrices involved. Some general techniques for this include the fast multipole method [8] [15], the clustering method [9] and the wavelet transform method [1].

The advantage of the Schur algorithm over the alternative techniques, based on equation (10) respectively (12), is that the former produces a sparse, reduced-order Y, while the latter techniques produce a full Y.

3.4 Results

Figure 4-6 shows the layout of a metallization dummy structure for HF admittance parameter measurements for MOSFET characterization. The bottom-left bondpad is connected by vias with the epi-layer. The metal interconnect is coupled capacitively to the substrate resistance network by its ground capacitances. The substrate via is essentially an ohmic connection between the interconnect network and the substrate network.

Measurement and simulation results are shown in *Figures 4-7* and *4-8.* Note that the measurements are not accurate at low frequency, due to the measurement set-up. Extraction of parasitics was done with and without accounting for substrate crosstalk. Neglecting the substrate coupling clearly gives a completely wrong prediction of the behavior. Here the transadmittance is mainly capacitive, i.e. just the coupling capacitance from one metal line to the other. However, with substrate resistance included via our BEM method, we find an excellent agreement with the measurements. While there is no DC connection, we still observe that the real part of the admittance is nearly as important as the imaginary part.

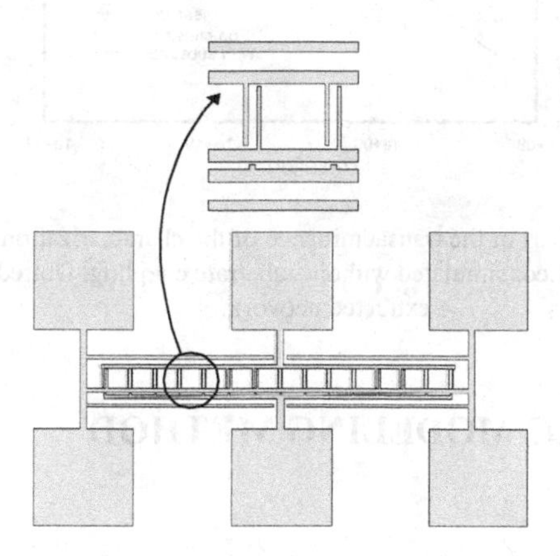

Figure 4-6. Layout of metallization dummy structure.

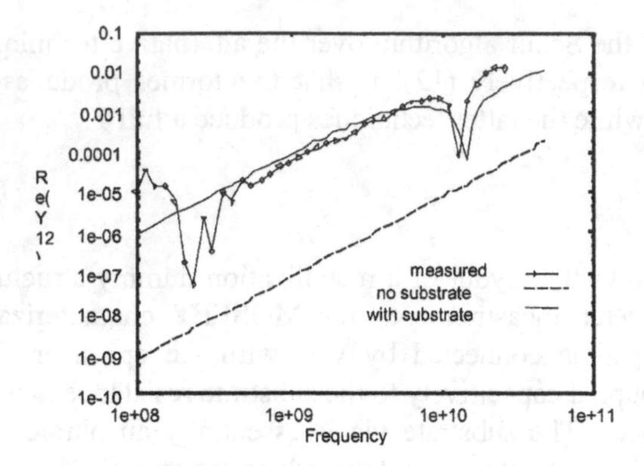

Figure 4-7. Real part of the transadmittance of the characterization structure. With markers: measured. Dashed: simulated without substrate coupling. Dotted: simulated with extracted network.

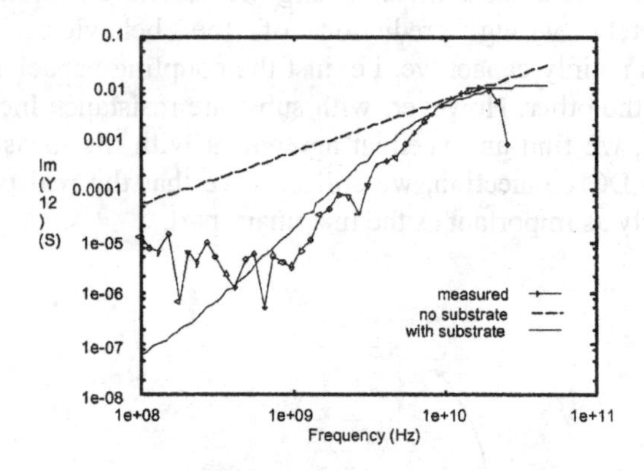

Figure 4-8. Imaginary part of the transadmittance of the characterization structure. With markers: measured. Dashed: simulated without substrate coupling. Dotted: simulated with extracted network.

4. PARAMETRIC MODELING METHOD

4.1 Methodology

A BEM works from first principles. That is, it starts from a physically based mathematical description of the problem to be solved, and solves the resulting set of equations. The advantages of the BEM and other physically based approaches include robustness, flexibility, extrapolation capabilities and relatively simple calibration requirements. The main disadvantage of such methods, despite all efforts to balance this, is however a relatively high computational complexity. Consequently, there will always be a need for faster methods. Non-physically-based methods can potentially fulfill this need, at the expense however of some or all of the advantages mentioned above for the more fundamental techniques. Some of such techniques have been presented before [7][6][11][24]. We collectively refer to such methods as parametric modeling techniques.

In [24] the speed-up is obtained by pre-computing point-to-point impedances, which are then used to find the admittance matrix for the actual contact configuration. Also hierarchy and delimitation are used in [24] to reduce computational complexity. However, this method still requires matrix inversion.

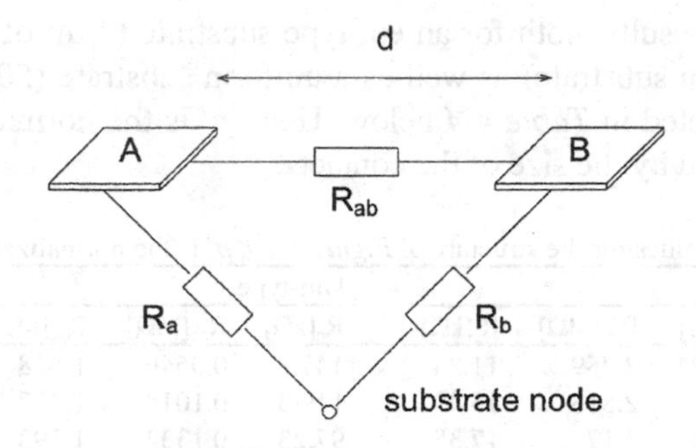

Figure 4-9. Substrate model for a configuration with two contacts.

In this section, we will present the method of [7]. The starting point of this method is the value of the resistance between two contacts as a function of their distance. We will use a resistance model as shown in *Figure* 4-9. This is motivated in part by noting that, for a 2-contact configuration, the BEM from Section 3 would produce exactly the same model. It is also motivated by the value of the resistance as a function of the contact separation distance *d*.

The model is obviously very appropriate for epi-type substrates where the resistances R_a and R_b would model the current that would flow from each contact to the highly doped, almost equipotential, substrate layer. On such substrates, the value of R_a (R_b) would be relatively independent from the presence of other contacts, but would mainly depend on the individual shapes of the contacts.

Only when other contacts would be very close, there would be a relevant lateral current flow. This current is modeled with R_{ab}, the value of which increases monotonically with increasing *d*.

We will investigate this behavior by extracting the resistances of a 3x5 contact configuration with varying pitch *p* as illustrated in *Figure 4-10*. In particular, we will study the pair-wise resistances associated with contacts *a*, *b*, and *c*. The contacts are of dimension 2μm x 2μm.

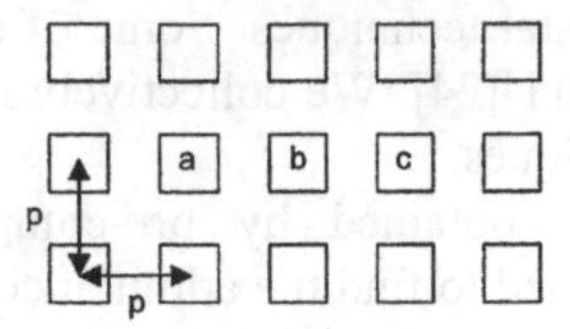

Figure 4-10. 3x5 contact configuration for studying pair-wise resistances.

The extraction results, both for an epi-type substrate (4μm of 10 S/m on top of a 10,000 S/m substrate) as well as a uniform substrate (500μm thick, 10 S/m), are presented in *Table 4-4* below. Here, *p'* is the normalized pitch, i.e. the pitch divided by the size of the contacts.

Table 4-4. Resistance values for the structure of *Figure 4-10*, *p'* is the normalized pitch.

	Epi-type				Uni-type			
p'	R_a[kΩ]	R_{ab} [kΩ]	R_{ac}[MΩ]	R_s[kΩ]	R_a[kΩ]	R_{ab}[MΩ]	R_{ac}[MΩ]	R_s[kΩ]
1.05	77.74	0.06604	2.759	11.84	147.8	0.0646	1.638	11.88
1.25	58.36	0.1074	2.815	15.49	119.3	0.1014	1.377	15.63
1.5	44.55	0.1506	3.17	17.38	97.23	0.1334	1.193	17.69
2.5	25.9	0.398	8.916	19.28	61.3	0.2249	0.984	20.39
5	20.05	3.661	219.1	19.54	41	0.3917	1.056	21.67
10	19.56	166.6	950	19.55	32.53	0.6502	1.331	22.1
25	19.55	1192	2397	19.55	27.89	1.176	1.855	22.26
50	19.55	2396	4799	19.55	26.41	1.664	2.247	22.3
100	19.55	4798	9600	19.55	25.66	2.122	2.563	22.31

The results for an epi-type substrate clearly indicate that R_a is only increasing for small pitch, and that R_{ab} rapidly increases with larger pitch. On a uniform substrate, similar behavior can be observed. However, on such a substrate the value of R_a more strongly depends on the pitch and R_{ab} increases less rapidly.

For parametric modeling purposes, it might be especially relevant to consider the so-called short-circuit resistance R_s, also shown for both cases. R_s is defined as the resistance from a contact (in this case *a*) to a short-circuit connection of all the other contacts. This R_s is almost constant, except for very small pitches. This behavior can be intuitively understood by considering a field line 'picture'. The total 'amount' of field lines emanating from a contact is more or less constant, but of some of them their end point changes from the substrate to other contacts when they are brought in proximity. This constant-R_s property can help in defining good empirical substrate resistance formulas, in a manner analogous to empirical interconnect capacitance formulas [12].

The simple empirical model of [7] actually ignores this behavior, as well as the dependence of R_a (the resistance of any contact a to the substrate node) on the proximity of other contacts. Also, the lateral resistance R_{ab} between two contacts is taken independently from the proximity of other contacts, except that the model to be extracted is made sparse by only including nearest-neighbor couplings. That is, R_{ab} is only included in the model when contacts a and b are direct neighbors of each other. Two contacts are considered each other's neighbor if they are adjacent in the Delaunay triangulation of the contact geometry as will be explained below. When two contacts are not Delaunay neighbors, we will call them 'screened' contacts.

Note that we only exclude the direct coupling resistance between any two screened contacts, but we always include the coupling to the substrate node. Thus, depending on the impedance of this node, there remains some coupling between screened contacts, via the substrate node as well as via other contacts.

The exclusion of direct coupling resistances actually is an effective form of model reduction. Since R_{ab} goes to infinity if p increases, R_{ab} might be ignored for p sufficiently large. The result would actually be a sparse, reduced-order model since it would only contain much less than $O(N^2)$ resistances for N contacts.

The Delaunay triangulation and related concepts are defined in the field of computational geometry, see e.g. [17]. Consider a finite set of points embedded in a 2D plane (the concepts extend to more dimensions, but this is not relevant for our purposes). For each given point from this set, its Voronoi polygon is defined as the loci of all points closer to this point than to any other point from the point set. The Voronoi diagram is defined, in its turn, as the union of all Voronoi polygons. Then, the Delaunay triangulation is the dual of the Voronoi diagram. It is a planar graph embedded in the plane with an edge between two vertices if and only if the corresponding Voronoi polygons have a common edge. Stated otherwise, there is an edge in the Delaunay triangulation if and only if the two points are neighbors of each other in the Voronoi diagram.

If the contacts in substrate-resistance extraction would be point contacts, they would define a Delaunay triangulation in a straightforward way. This triangulation would present an obvious heuristic for deciding whether two contacts are screened or not: they are not screened if, and only if, they are connected by a Delaunay edge. In that case, we will compute a direct resistance between them. Otherwise, we will omit this resistance from the model and only include both resistances from each contact to the common substrate node.

For finite sized contacts, this heuristic is modified as follows: We will create a point in the Delaunay triangulation for each corner of the contact, but we exclude from the triangulation any edges in the interior of the contact polygons. An example of such a modified Delaunay triangulation for a set of finite-sized contacts is given in *Figure 4-11*. Then, a direct resistance is computed between two contacts if, and only if, there is *at least one* line of the Delaunay triangulation that directly connects the contacts.

A suitable Delaunay triangulation algorithm was given in [23]. It iteratively constructs the triangulation by adding points one at a time. This algorithm could easily be modified so that it omits edges in the interior of the contact polygons and implemented in our scanline-based layout-to-circuit extraction program SPACE, that will be described in Section 6. Apart from a sorting step, the algorithm runs in almost linear time.

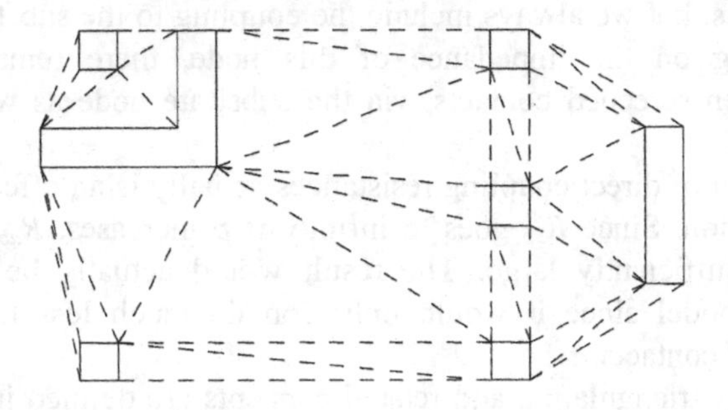

Figure 4-11. Example of a Delaunay triangulation (drawn in dashed lines) for a set of contacts (solid lines). A direct coupling resistance is computed between two contacts if there is at least one line of the triangulation that directly connects them.

Now, with the topology of the resistive model for an arbitrary set of contacts defined, by the triangular contact pair model of *Figure 4-9* and the Delaunay-based pair selection of *Figure 4-11*, we will present the formulas for the resistor values. They are empirically defined fitting formulas as follows, where R_a is the resistance of a single contact to the substrate node and R_{ab} is the direct resistance between contacts:

$$R_a = \frac{1}{k_1 + k_2 P_a + k_3 A_a} \qquad R_{ab} = \frac{k_4 D_{ab}^{k_5}}{\sqrt{A_a} + \sqrt{A_b}} \tag{13}$$

Here, A_a and A_b denote the contact areas, P_a the perimeter of contact a, D_{ab} the distance between both contacts, and $k_1...k_5$ are fitting parameters which are, in an off-line preprocessing step, calibrated for each specific doping profile / technology.

4.2 Implementation and results

The parametric modeling method has been implemented in the SPACE layout to circuit extractor, see Section 6. We will study the high-frequency behavior of a bipolar amplifier on a substrate consisting of a 1.4 μm 0.15 Ωcm top layer and a 300 μm 4 Ωcm bottom layer. The circuit in *Figure 4-12* was extracted without substrate resistances, using the BEM method for substrate-resistance extraction and using the parametric modeling method from this section. In all cases, the resulting circuit was simulated using SPICE. The simulation results are presented in *Figure 4-13*. They show that the substrate-coupling effects that are estimated using the parametric modeling method, are almost identical to the BEM results. On an HP 9000/735 computer, extraction of the amplifier using the BEM method took 3 minutes and 4 seconds (248 elements were used). Extraction on the same computer, using the parametric modeling method, took less than 1 second. For more results, see [7].

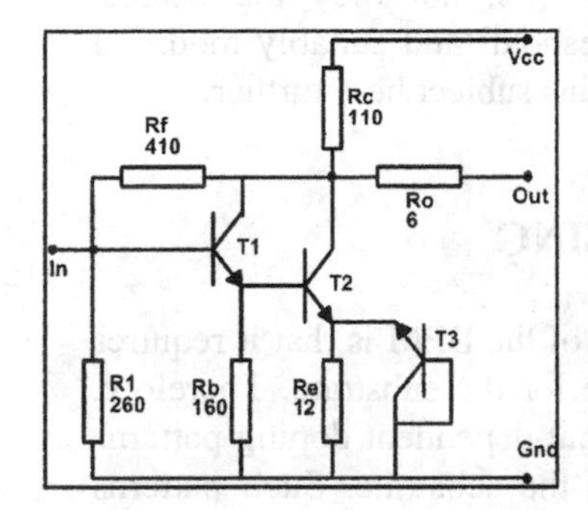

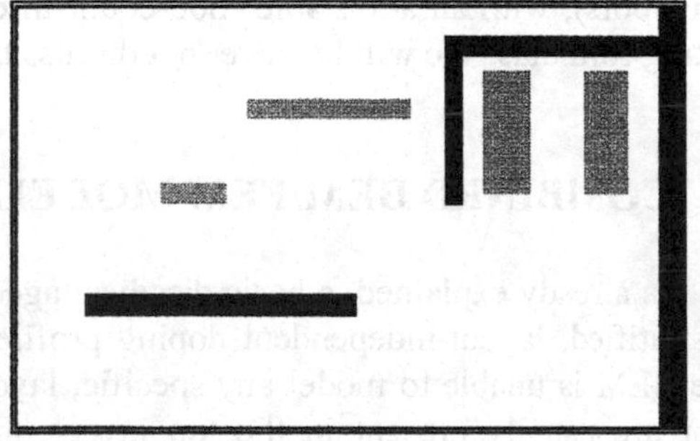

Figure 4-12. Schematic and simplified layout of a bipolar amplifier. Grey areas model transistors, black areas model substrate contacts.

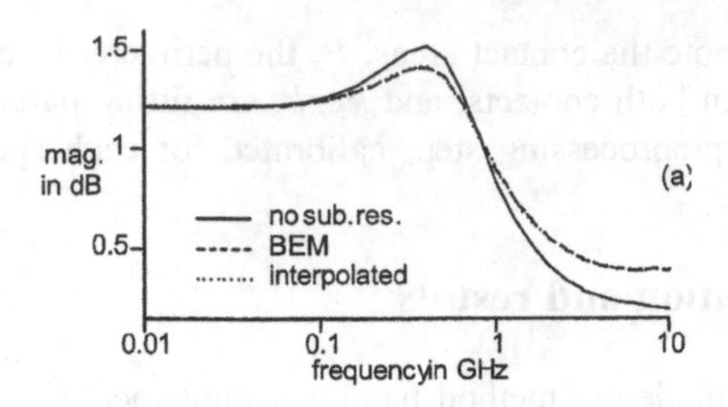

Figure 4-13. Simulated magnitude of the transfer function of the amplifier vs. frequency.

4.3 Conclusion

In practice, the outlined approach works reasonably well, see the results in Section 4.2. However, some more obvious and some less obvious issues are present. For example, only including the corners of the contacts in the triangulation point set can actually find too few or too many couplings. Also, the modeling of screening with 'on-off behavior' of the direct resistances introduces further errors. In practice, the effect of screening is smoother. Solutions to such problems can be envisaged, e.g. by adding more points of a contact to the triangulation and by including resistances corresponding to multiple hops in the Delaunay triangulation (i.e. not only the nearest neighbors), with an adjustable 'hop-count threshold' and suitably modified fitting formulas. We will however not discuss this subject here further.

5. COMBINED BEM/FEM MODELING

As already explained, a basic disadvantage of the BEM is that it requires a stratified, layout-independent doping profile for the substrate. Therefore, the BEM is unable to model any specific, layout-dependent doping patterns that are usually present in the top layers of the substrate. Such patterns include wells, channel stop layers, buried layers, trenches and sinkers. These features can make BEM-based methods inaccurate for state of the art technologies and applications (like RF CMOS). At the same time, the speed of the FEM-based methods is becoming inadequate.

Therefore, we proposed in [19] a method to incorporate top-layer doping patterns in a BEM-based extractor by combining it with a 2D FEM-based approach. This method basically relies on the fact that the current patterns around such artifacts can be modeled as predominantly lateral, i.e. parallel to the Si-SiO2 interface. Such a model can especially be expected to be valid for channel stoppers, which are thin and have a relatively low resistance.

Otherwise, the approach being described here could be extended to a BEM/3D-FEM combination, although many issues that are related to the efficient implementation and the accuracy of the method still need to be investigated. We will see that even for the BEM/2D-FEM combination, some specific issues will appear that tend to combine not only the favorable but also the unfavorable properties of both methods.

With the Finite Element Method (FEM) the interior of the domain is discretized (instead of only the boundary, as in the case of the BEM). This is, in general, a 3D prismatic mesh or a 2D triangular mesh. In [6] we have shown that for a 2D resistive problem (i.e. for the Laplace equation), the mesh is equivalent to a planar resistance network. Solving the FEM model can than be interpreted in a circuit-theoretic sense as generalized star-delta transformation, which is known to be equivalent to Gaussian elimination of the admittance matrix.

Furthermore, remember that with the BEM, the initial result before application of the incidence matrix, is also a detailed resistance network on the BE's. Then a FEM and BEM network can, based on their common representation as an elemental resistance network, be joined together if both meshes are compatible. This approach is schematically illustrated in *Figure 4-14*.

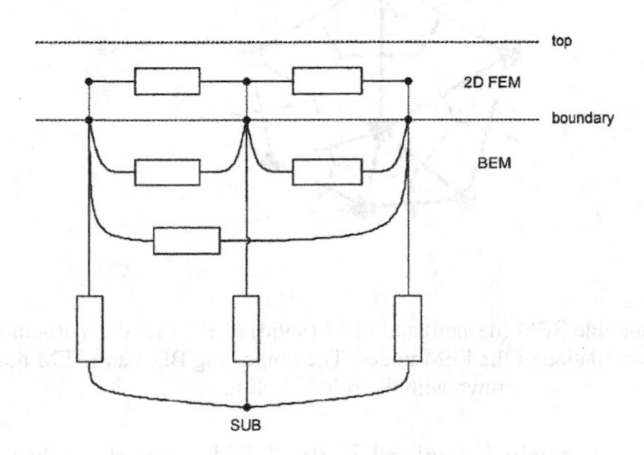

Figure 4-14. Schematic representation of the combined BEM/2D-FEM modeling approach.

Thus, the method calls for compatible meshes along the BEM/FEM boundary. Given that the nodes of the FEM network correspond to the vertices in the mesh, and that the nodes of the BEM network correspond to the collocation points of the BEM elements, we actually require a correspondence of FEM mesh vertices and BEM collocation points. This can be accomplished if the BEM elements are the dual of the FEM triangles, much the same as the duality between the Voronoi diagram and the

Delaunay triangulation. That is, the Voronoi diagram of the FE vertices would define the BEM discretization. This is illustrated in *Figure 4-15*.

Obviously, the BEM elements are in this case non-rectangular. However, this is not a conceptual issue but it only affects the integrations of the Green's function over the element. Breaking down each Voronoi polygon into triangles can easily solve this issue.

Furthermore, please note that other formulations of the BEM (e.g. piecewise-linear current distributions or Galerkin-type discretizations [3]) would give rise to other results for mesh compatibility, but we will not consider this issue here any further.

After combination of both networks as suggested in *Figure 4-14*, the resulting network can be simplified by star-delta transformation (which is identical to Gaussian elimination). Although straightforward in theory, it can be cumbersome in practice because the combined network inherits the size of the FE meshes and the density of the BE meshes. Therefore, the degree of all nodes in the combined network is high, and because elimination of a node with degree x takes $O(x^2)$ time, the method can become unattractive from a computational point of view.

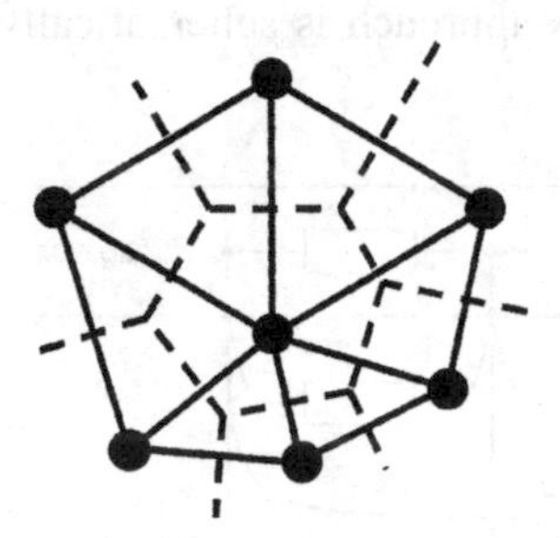

Figure 4-15. Compatible BEM (dashed) and FEM (solid) meshes via the Voronoi diagram and Delaunay triangulation of the FEM nodes. The coinciding BEM and FEM nodes are drawn with the filled circles.

The problem can partly be solved if the BEM network underneath the FEM network is made as sparse as possible. In any case, an algorithm such as the Schur algorithm should be used to avoid full matrices, but furthermore the bandwidth of the Schur algorithm could locally be reduced as much as possible. Alternatively, heuristic sparsification methods could be applied by noting that, because the substrate resistivity is usually much higher than the channel stop resistivity, the resistances originating form the BEM region are much larger than those from the FEM region.

In [20] we have theoretically shown the consistency and convergence behavior of the hybrid modeling method. Experimental results have been presented in [19] and [20].

6. THE SPACE LAYOUT TO CIRCUIT EXTRACTOR

The results as described in this chapter have for a large part been implemented in the SPACE [2][12] layout-to-circuit extractor. This is a comprehensive tool, incorporating not only the above-described methods for substrate-resistance extraction, but also some advanced methods for interconnect-resistance extraction, interconnect-capacitance extraction, device recognition and model-order reduction. The tool is a full-fledged environment for transforming a layout plus technology description into a netlist. All extraction methods are fully integrated and mutually compatible. As a result, the extracted netlist can be an accurate electrical model for the physical structure to be fabricated. SPACE is very efficient, achieving linear extraction speed and sublinear memory requirements. Depending on the extraction options it can extract over 750 transistors per second on a PIII/500, or a fully flattened 2,800,000 transistor chip in less than 1 hour.

The interconnect-resistance extraction employs a finite-element technique that efficiently employs the properties of typical interconnect structures to run in practically linear time (see [12] for additional references). Interconnect-capacitance extraction can either employ a fast parametric modeling technique, or a 3D-BEM method with Schur acceleration. Device recognition is not only well developed for MOS technologies (e.g. with accurate modeling of irregularly shaped transistors and support for an arbitrary number of transistor types) but also for bipolar technologies, including matching and interpolation of bipolar devices to a template library of predefined topologies.

Model-order reduction is an important strength of the methods that are implemented. This might already be apparent from the discussion so far, e.g. referring to the Schur algorithm and the Delaunay topology for the method of Section 4, but SPACE also incorporates other model-order reduction methods. One of these is a technique called Selective Node Elimination, in which a detailed fine-granularity RC network obtained from a discretized layout representation is simplified using elimination techniques. Non-terminal nodes are iteratively and selectively eliminated in order of least significance for the transfer of the network, until further elimination would violate a stop criterion related to the user-specified maximum operating frequency f_s. The result is a reduced network, close to the coarsest network that still accurately models the behavior of the original netlist until f_s.

One important advantage of this circuit-reduction technique is that all internal nodes are eliminated on the fly, as soon as possible after they have been created during extraction. The final reduced circuit is also written to the disk as soon as possible, and core memory is reclaimed. In fact, this is a primary design feature of SPACE, for all extraction algorithms. In particular, SPACE operates with a vertical scanline moving over the layout from left to right, and only the layout and circuit data around the scanline position are kept in core memory. As a result, the total memory requirements are much reduced, allowing handling of the largest designs.

Moreover, SPACE incorporates many 'little' features aimed at integration of the tool in a design flow, including back-annotation capabilities, support for IP-based blocks with only a black-box layout and a library simulation model (SPICE or other), various input and output formats, hierarchical, flat or mixed extraction, incremental extraction and 45-degree capabilities. Finally, SPACE incorporates an intuitive, state of the art GUI.

7. CONCLUSION

In this chapter we have discussed three different methods for substrate-resistance extraction. Each of these methods presents a different accuracy/ performance tradeoff, fulfilling different needs in typical design flows. We have argued that on-the-fly a-priori reduced-order modeling is important to optimize the efficiency, not only of the tools and design steps that take the extracted netlists as input, but also of the extraction process itself.

Future IC's will be denser and faster, and will require more effects to be modeled. For example, it will become invalid to model the interconnect independently from the substrate. Moreover, future IC's will still be larger, requiring intrinsically larger models. Both trends will require much improved model-order reduction techniques to be able to simulate the resulting models in reasonable time, even on future computers.

We believe that such detailed verification will continue to play an important role, because the economics of future sub-tenth-micron technologies won't allow for non-aggressive design styles with sufficient guard-banding to effectively guarantee correct-by-construction designs.

ACKNOWLEDGEMENT

This work was supported in part by the Dutch Technology Foundation and in part by Agilent. Their support is gratefully acknowledged.

REFERENCES

[1] Alpert, B., G. Beylkin, R. Coifman and V. Rohklin, 'Wavelets for the Fast Solution of Second-Kind Integral Equations, *SIAM J. Sci. Comp.*, vol. 14, pp. 159-184.

[2] Beeftink F., A.J. van Genderen, N.P. van der Meijs, and J. Poltz, 'Deep-Submicron ULSI Parasitics Extraction Using SPACE,' in *Design, Automation and Test in Europe Conference 1998*, Designer Track, pp. 81--86, Feb. 1998.

[3] Brebbia C.A., *The Boundary Element Method for Engineers*. Plymouth: Pentech Press, 1978.

[4] Clement F.J.R., E. Zysman, M. Kayal, and M. Declercq, 'LAYIN: Toward a global solution for parasitic coupling modeling and visualization,' in *Proc. IEEE Custom Integrated Circuits Conference*, pp. 537 -- 540, May 1994.

[5] Dutton R.W., 'The role of TCAD in Parasitic Analysis of ICs', *Proceedings ESSDERC* 1993, pp. 7581, 1993.

[6] Genderen A.J. van and N.P. van der Meijs. 'Extracting Simple but Accurate RC models for VLSI interconnect'. In *Proc. Int. Symp. on Circuits and Systems*, pages 2351--2354, Helsinki, Finland, June 79 1988.

[7] Genderen, A.J. van, N. P. van der Meijs and T. Smedes, Fast Computation of Substrate Resistances in Large Circuits, In *Proc. European Design and Test Conf.*, pp. 560-565, Paris, France, March 1996.

[8] Greengard, L., *The Rapid Evaluation of Potential Fields in Particle Systems*, MIT Press, Cambridge, Massachusetts, 1988.

[9] Hackbush, W., and Z. Nowak, 'On the Fast Matrix Multiplication in the Boundary Element Method by Panel Clustering', *Numer. Math*, vol 54, pp. 463-491.

[10] Hasegawa, H., M. Furukawa, and H. Yanai, 'Properties of microstrip-lines on Si-SiO2 system', *IEEE trans. on Microwave Theory and Techniques* MTT-19, pp. 869-881, Nov. 1971

[11] Joarder K., 'A Simple Approach to Modeling Cross-Talk in Integrated Circuits', *IEEE Journal of Solid-State Circuits*, Vol. SCC-29, No. 10, pp. 1212-1219, Oct. 1994.

[12] Meijs N.P. van der, A.J. van Genderen et. al. The SPACE layout to circuit extractor, see http://cas.et.tudelft.nl/space. A commercial version can be obtained from OptEM, http://www.optem.com.

[13] Meijs N.P. van der, and A.J. van Genderen, 'An Efficient Finite Element Method for Submicron IC Capacitance Extraction', IEEE *Proc. 26th Design Automation Conference*, pp. 678-681, June 1989

[14] Meijs N.P. van der. '*Accurate and Efficient Layout Extraction*'. PhD thesis, Delft University of Technology, Delft, The Netherlands, January 1992.

[15] Nabors K. and J. White. 'Fastcap: A multipole accelerated 3d capacitance extraction program', *IEEE Trans. on Computer-Aided Design*, 10(11):1447--1459, November 1991.

[16] Nelis H., E. Deprettere, and P. Dewilde. 'Approximate inversion of positive definite matrices, specified on a multiple band', In *Proc. SPIE 88*, San Diego, California, August 1988.

[17] Preparata F.P. and M.I. Shamos, *Computational Geometry: An Introduction*, Springer Verlag, 1985.

[18] Saad Y. and H. A. van der Vorst, 'Iterative Solution of Linear Systems in the 20-th Century', *J. Comp Appl. Math*, 123, pp. 1-33, 2000.

[19] Schrik, E., N.P. van der Meijs, 'Combined BEM/FEM substrate resistance modeling', *Design Automation Conference*, Proceedings. 39th, pp. 771 -776.

[20] Schrik, E., N.P. van der Meijs, 'Theoretical and Practical Validation of Combined BEM/FEM Substrate Resistance Modeling', Accepted for publication, *Proc. ICCAD* 2002.

[21] Smedes T., N. P. van der Meijs, and A. J. van Genderen, 'Extraction of Circuit Models for Substrate Crosstalk,' in *Proc. Int. Conf. on Computer-Aided Design*, (San Jose, California), pp. 199--206, Nov. 1995.

[22] Su D.K., M.J. Loinaz, S. Masui, and B.A. Wooley, 'Experimental Results and Modeling Techniques for Substrate Noise in Mixed-Signal Integrated Circuits,' *IEEE Journal of Solid-State Electronics*, vol. 28, pp. 420--430, Apr. 1993.

[23] Tipper J.C., 'A Straightforward Iterative Algorithm for the Planar Voronoi Diagram', *Information Processing Letters* 34, Elsevier, pp. 155-160, April 9, 1990.

[24] Verghese N.K., D.J. Allstot, M.A. Wolfe, 'Fast Parasitic Extraction for Substrate Coupling in Mixed-Signal ICs', *Proc. CICC*, pp. 121-124, May 1-4, 1995.

[25] Weber, E. *Electromagnetic Fields Theory and Applications*, John Wiley and sons, 1957.

Chapter 5

MODELS AND PARAMETERS FOR CROSSTALK SIMULATION

Valentino Liberali
Department of Information Technologies, University of Milano, Italy

Abstract: This chapter illustrates a simplified model for analysis of crosstalk effects in deep submicron CMOS technologies. Most parameters for parasitic element values can be easily obtained from technology information contained in the physical design rules. However, the substrate bias resistance, which is one of the most important parasitic elements in CMOS technologies with highly-doped substrate with epitaxial layer, is usually neglected in the silicon foundry documentation. The substrate bias resistance value can be obtained either from technology parameters or by experimental measurements on a test structure, and crosstalk effects can then be easily estimated through a SPICE-level simulation. The proposed approach has been validated by comparing results with simulations after extracting parasitics with a commercial tool and with experimental measurements on a test chip.

1. INTRODUCTION

For over 30 years, Moore's law has ruled the development of CMOS technology [1]. Main targets are: low cost, high performance, and high integration density. To achieve these goals, deep changes have been incorporated in new fabrication technologies, while maintaining the design rules as similar as possible to previous technologies [2].

To overcome latch-up problems, modern CMOS technologies use a heavily doped substrate (p+) covered by a lightly doped epitaxial (epi) layer having the same polarity (p) [3]. The epi layer is very thin (< 10 μm) and has a high resistivity, in the range of 0.1 Ωm, while the resistivity of the substrate can be less than 0.1 mΩm.

Such a technology is optimized for digital design. However, when analog/digital interfaces have to be implemented together with a digital processing core, accuracy limitations arise, due to technology aspects.

Indeed, disturbances generated by the digital part can severely affect the performance of the analog circuits [4], [5].

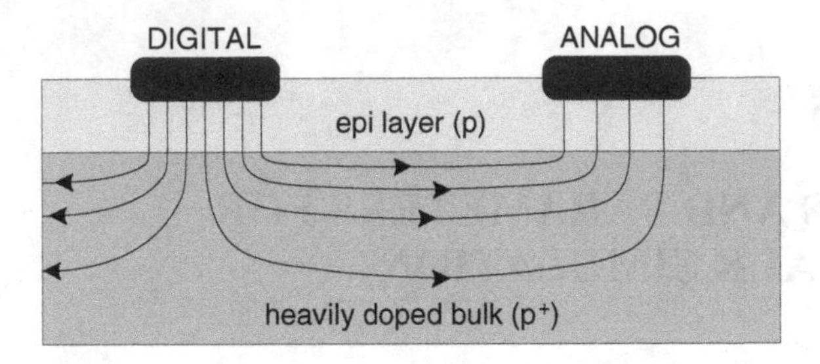

Figure 5-1. Distribution of current lines in the highly-doped substrate and in the epi layer.

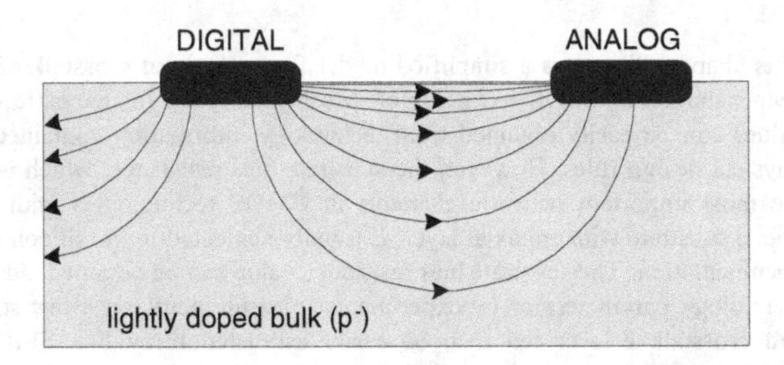

Figure 5-2. Distribution of current lines in the low-doped substrate.

Since the epi layer is 1000 times more resistive than the bulk, any noise current injected by a digital device flows into the deep substrate. *Figure 5-1* shows a qualitative distribution of current lines. On the contrary, in older technologies with lightly doped substrate, the current flows near the silicon surface, as illustrated in *Figure 5-2*.

In CMOS technologies with epi layer, the substrate can be considered as a single equipotential node, due to its very low resistivity. By contrast, the non-negligible resistance due to the light well doping and to the epi layer resistance affects the local voltage inside well regions and inside the epi-layer. This means that the noise collected by an analog device is independent of its distance from the digital switching noise source, at least to a first approximation. *Figure 5-3* illustrates a cross section of an NMOS and a PMOS transistor in a twin-well submicron CMOS technology with epi layer.

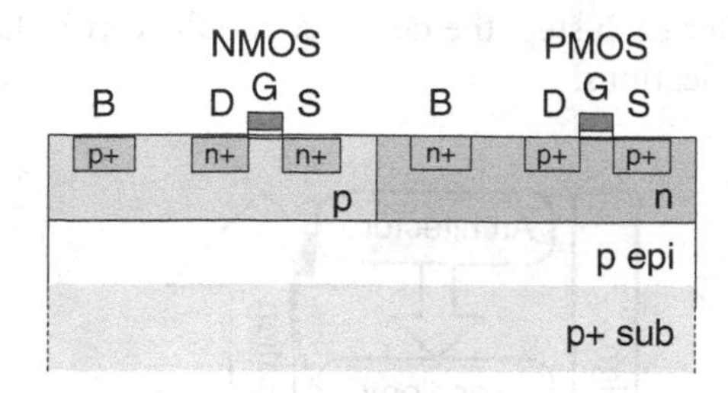

Figure 5-3. Cross section of an NMOS and a PMOS transistor integrated in a twin-well CMOS technology with epi layer.

Mixed-signal design in epi-layer technologies is not straightforward. Indeed, non-ideal interconnections prevent from having quiet power supply and ground voltages [4], the low-resistivity bulk provides a pathway for most of the noise injected by digital gates [6], and coupling between digital and analog parts can cause system failure or severe performance degradation, often requiring a complete redesign [7].

Software tools are available for efficient parasitic extraction from layout, but they can only be used for final verification, when the design process is completed. Analog designers need to evaluate crosstalk effects at early stages of circuit design.

This chapter presents a simplified approach to crosstalk simulation and modeling, suitable for SPICE-level simulation. Most important parameters can be estimated before designing the circuit layout, thus allowing the designer to analyze crosstalk effects on different circuit architectures and topologies. The approach has been validated by comparing results obtained with the simplified model both to simulation results from a back-annotated netlist extracted with a commercial tool, and to experimental measurements from a test structure integrated in a 0.35-μm twin-well CMOS technology.

2. DESIGN METHODOLOGY

From the above considerations, it is apparent that a correct design methodology is essential for a successful design. Fig. 5-4 summarizes the different steps in the design flow.

2.1 Top-Down Design

Essential steps in a top-down design approach are: architecture selection, choice of circuit topology, component sizing, and physical layout design. After each step, the designer should verify that the design conforms to specifications.

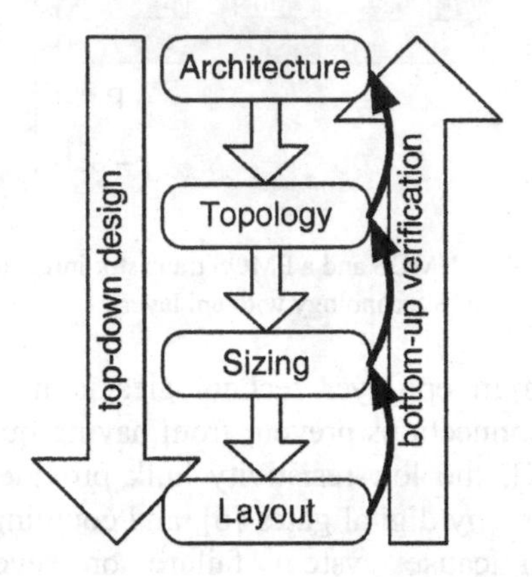

Figure 5-4. Top-down design methodology.

2.2 Bottom-Up Verification

Possible crosstalk problems must be considered from early design stages on, and a careful verification is required after every step. In particular, adequate analysis techniques are required to investigate the effects of digital noise injection and predict the robustness of analog and mixed analog-digital structures.

Post-layout verification is an important design step that is typically performed. In mixed-signal design, verification should consider also substrate coupling problems. An equivalent network of parasitic elements is extracted from the layout, and the whole circuit is simulated to estimate the effect of disturbance propagation [12,13,16].

Software tools are available for efficient parasitic extraction from layout, based on numerical solution either of differential equations (finite difference and finite element methods), or of integral equations (method of moments and boundary element method) [9]. An equivalent network of parasitic elements is extracted from the layout, and the whole circuit is simulated to estimate the effect of disturbance propagation. Several methods (e.g. model order reduction techniques) are available to reduce the complexity of the network and to accelerate the substrate noise analysis [10].

Two examples of commercially available software tools for parasitic extraction are SeismIC by Cadence/CadMOS [11] and SubstrateStorm (formerly known as LAYIN) by Simplex Solutions [12]. Another tool, developed in academia, is SPACE by Technical University Delft [13].

Such tools are useful for design verification, but they have three main disadvantages: (i) they can be used only after completing the layout design, at the end of the design process; (ii) parasitic extraction generates huge netlists and, therefore, time-domain simulations may require very long CPU times; (iii) due to the network complexity, parasitic elements cannot be related to physical and geometrical parameters, thus making it difficult for the designer to improve the design.

Therefore, it is also important that verifications are performed before the layout is completed. Fast simulations at early stages of the design process however need a simple model of crosstalk. The main parasitic elements must be estimated before the layout design, using available parameters both for the fabrication technology and for the package. This is described next.

3. MODELING

Suitable analysis techniques are required to investigate the effect of digital noise injection and to predict the robustness of analog and mixed analog-digital structures.

As disturbances injected by the digital circuitry are strongly correlated with the clock, they cannot be modeled as white noise. Timing characteristics of noise coupling must be accounted for, especially when the analog circuitry is driven by a clocking scheme correlated with the digital part.

Very sophisticated models have been presented in the literature. Although such models can provide a good evaluation of digital noise injection, it is important to start crosstalk analysis with a simple model to be used in conventional SPICE simulations to evaluate design robustness. Two main reasons motivate this approach: a SPICE-level description is still considered the best solution for noise-sensitive circuits [8]; moreover, parasitic values for SPICE simulations can be obtained from technology parameters and are related to both physical quantities and geometrical sizes.

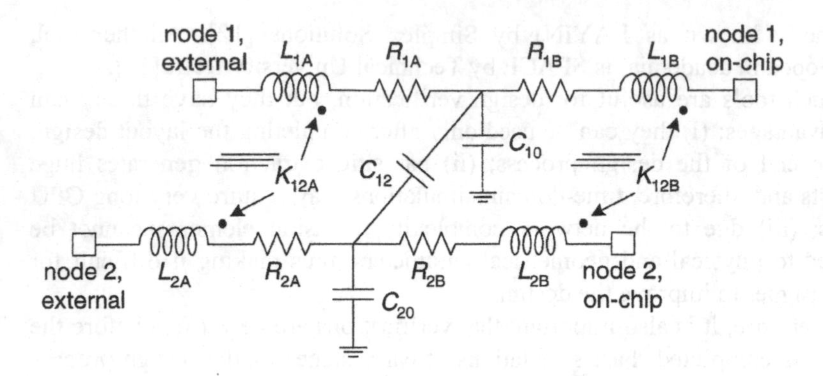

Figure 5-5. Model for package and bonding interconnections.

First of all, a suitable model for the interconnections must be used, accounting for package and bonding inductances, resistances, and capacitances. *Figure 5-5* illustrates the equivalent model for two adjacent interconnections.

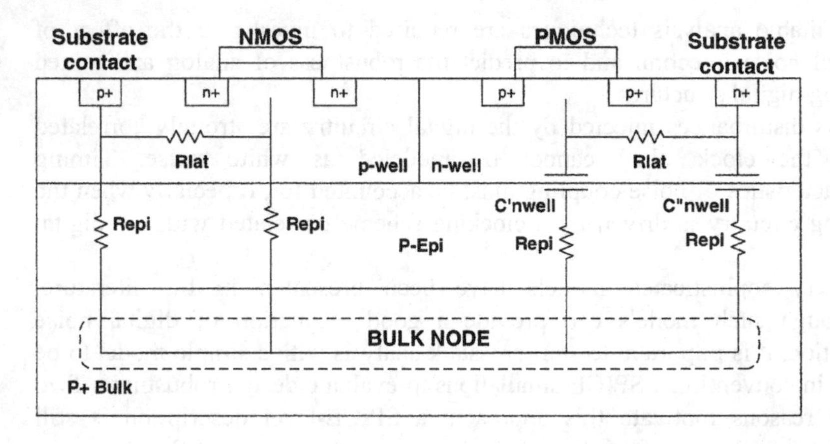

Figure 5-6. Parasitic elements of an NMOS and a PMOS transistor integrated in a twin-well CMOS technology with epi layer.

Then, on-chip parasitics must be considered. A simplified model should account for epi-layer resistance, high substrate conductivity, and different ground biasing in analog and in digital sections. This model is illustrated in *Figure 5-6*. The reader should note that parasitic capacitances related to source and drain areas are already included in SPICE models. High conductivity makes the heavily doped substrate similar to a short circuit between substrate points on the whole chip, thus vanishing the effect of placing analog devices far from digital ones.

Figure 5-7 illustrates a simplified equivalent circuit for digital noise injection in a single transistor amplifier. Any disturbance injected from the digital section into the substrate changes the body bias of the analog transistor, which in turns produces a change in its drain current.

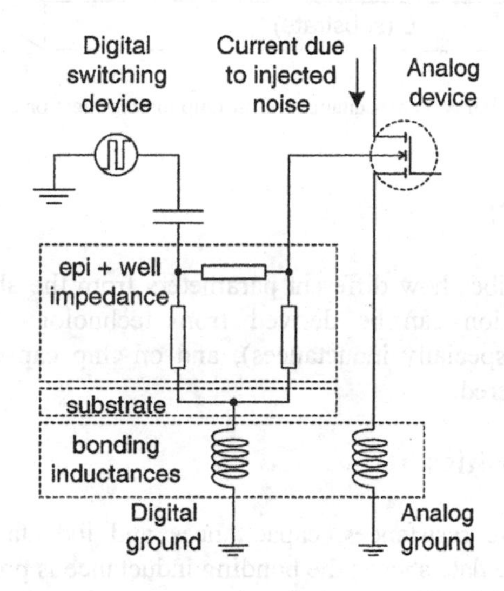

Figure 5-7. Equivalent circuit for simulating digital noise injection.

It is important to point out that the combined effect of bonding inductance and coupling capacitance creates a high-pass network. Hence, any isolation scheme provides good attenuation of crosstalk only at low frequencies. At high frequencies, the capacitive transmission becomes dominant, thus vanishing the isolation benefit [14]. This limitation must be considered carefully, especially when designing RF circuits [15].

Another possible source of coupling is given by interconnection capacitances. The situation depicted in Fig. 5-8 is well studied in digital design, as it may lead to signal propagation delay or spurious transients. Parasitic capacitances between neighboring signal lines (C_{12}, C_{23}) may cause signal integrity problems: opposite logic transitions on adjacent lines may result in glitches or delays in signal propagation. Capacitive coupling with the substrate (C_{10}, C_{20}, C_{30}) may contribute to substrate noise, especially during logic transitions of long interconnections (e.g., the clock net), and therefore mixed-signal crosstalk analysis must account for it.

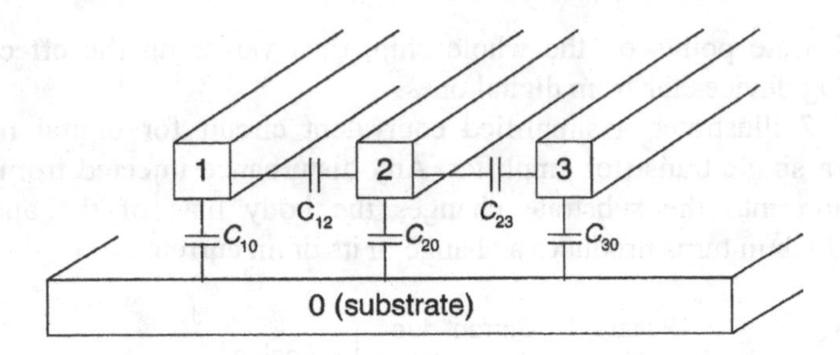

Figure 5-8. Parasitic capacitances of on-chip interconnections.

4. PARAMETERS

This section describes how different parameters from the above models for crosstalk simulation can be derived from technology information. Package parasitics (especially inductances), and on-chip capacitances and resistances are considered.

4.1 Package parasitics

Values of parasitic resistances, capacitances and inductances can be obtained from package data sheets; the bonding inductance is proportional to the bond wire length (~ 1 nH/mm).

4.2 On-chip parasitics: capacitances

The capacitance between the n-well region and the epi-layer, shown in *Figure 5-6*, can be calculated using the surface area data ($A_{n\text{-well}}$) and the specific unit capacitance (C_{Aw}) of the n-well:

$$C_{n\text{-well}} = A_{n\text{-well}} \cdot C_{Aw} \quad (1)$$

For an accurate model, the n-well can be divided into different regions, and each region is associated with a parasitic capacitance towards the substrate.

In a similar way, parasitic capacitances between on-chip interconnect lines and the substrate can be estimated taking into account their surface area data (A_{line}) and the specific unit capacitance (C_{Al}):

$$C_{line} = A_{line} \cdot C_{Al} \quad (2)$$

4.3 On-chip parasitics: resistances

Lateral substrate resistances can be calculated using the sheet resistance of the particular layer (p-well or n-well), which is provided in the technology's design rules manual.

However, in deep submicron technologies a major source of performance degradation and crosstalk sensitivity is the vertical resistance of the epi layer. As vertical parameters, such as doping profiles and layer depth, are not included in the physical design rules for a given technology, a method has been devised for extracting such information from a test chip [16].

A test structure has been integrated, consisting of five test points for microprobes, each of them connected to substrate contacts having the same area and a different perimeter and/or spacing. *Figure 5-9* shows one group of microprobe test points.

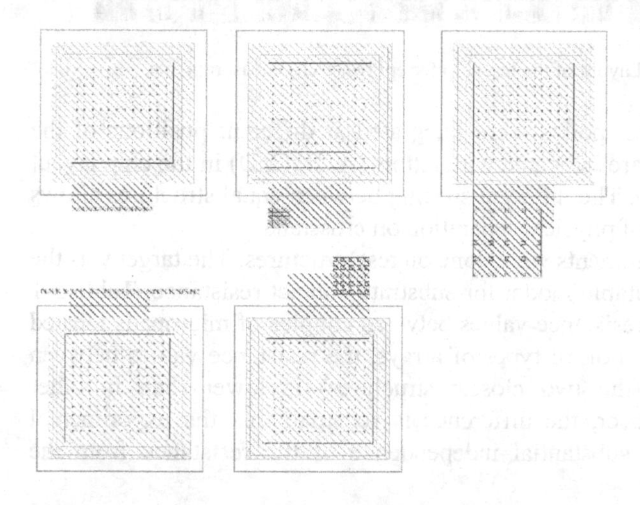

Figure 5-9. **Test points for substrate contact resistance measurements.**

Each group has five arrays of substrate contacts, labeled with numbers (1, 2, 3, 4, 5) in *Figure 5-9*. Each array of contacts is accessible through a 25 μm × 25 μm microprobe pad. Arrays 1 and 2 are linear arrays with 20 contacts, the diffusion area being 1 μm^2 for each contact. In array 1, the diffusion squares are adjacent, thus forming a single diffusion rectangle with area $A_1 = 20\ \mu m^2$ and perimeter $p_1 = 42\ \mu m$. In array 2, the diffusion squares are separated by 1 μm; although the diffusion area is the same, the perimeter is $p_2 = 80\ \mu m$ and the bounding box of the bias diffusion has an area $A_2 = 39\ \mu m^2$. Arrays 3, 4, and 5 are square arrays with 25 contacts and their total diffusion area is 25 μm^2. Perimeters and bounding box areas for these arrays are: $p_3 = 20$ μm and $A_3 = 25$ μm^2 for array 3, $p_4 = 100$ μm and $A_4 = 81$ μm^2 for array 4, $p_5 = 100$ μm and $A_5 = 289$ μm^2 for array 5, respectively.

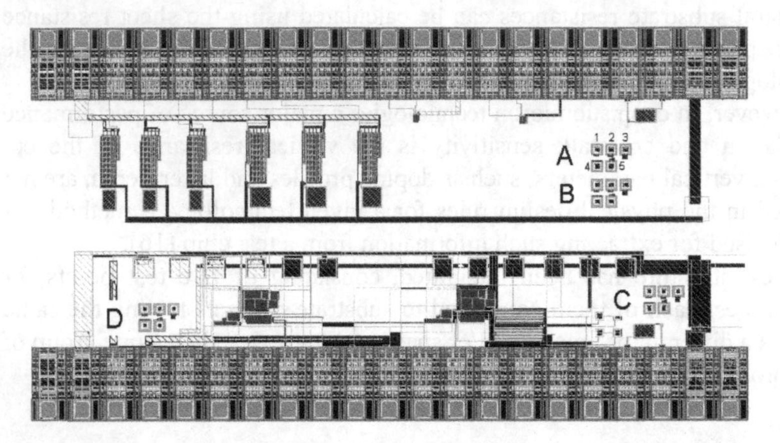

Figure 5-10. Layout of the test chip for epi resistivity measurements.

Four groups of test points were integrated at different positions of the same test chip. They are indicated with letters (A, B, C, D) in the chip layout shown in *Figure 5-10*. The different spacing between equal structures allows evaluating the effect of physical separation on crosstalk.

Resistance measurements were done on test structures. The target was the identification of a suitable model for substrate contact resistance. Table 5-1 shows the measured resistance values between couples of micropads located in different positions. For all types of arrays, the resistance values between pads in A and B (the two closest structures) is lower than in other measurements. However, the difference is so small that the experimental evidence suggests a substantial independence of the resistance from the distance.

Table 5-1. Measured values of substrate bias resistances.

pads	R [kO]	Pads	R [kO]	pads	R [kO]	pads	R [kO]	pads	R [kO]
A1,B1	1.00	A2,B2	0.65	A3,B3	1.28	A4,B4	0.90	A5,B5	0.53
A1,C1	1.07	A2,C2	0.69	A3,C3	1.37	A4,C4	1.05	A5,C5	0.56
A1,D1	1.02	A2,D2	0.65	A3,D3	1.38	A4,D4	0.98	A5,D5	0.54
B1,C1	1.08	B2,C2	0.68	B3,C3	1.35	B4,C4	0.97	B5,C5	0.57
B1,D1	1.10	B2,D2	0.80	B3,D3	1.28	B4,D4	0.84	B5,D5	0.58
C1,D1	1.10	C2,D2	0.70	C3,D3	1.40	C4,D4	1.03	C5,D5	0.59

The substrate resistance can be modeled with lumped elements [17], as shown in *Figure 5-11*. R_{cont} represents the microprobe resistance as well as the contact resistance, R_{lat} is the surface (p-well) lateral resistance, R_{epi} represents the epi layer resistance, and R_{bulk} is the bulk (low-ohmic substrate) resistance. Since R_{cont} and R_{bulk} have a very low value (few ohms) and R_{lat} is very high (tens or hundreds kohms), the measured resistance depends to first order only on R_{epi}, and we can approximate it as $R \approx 2\ R_{epi}$.

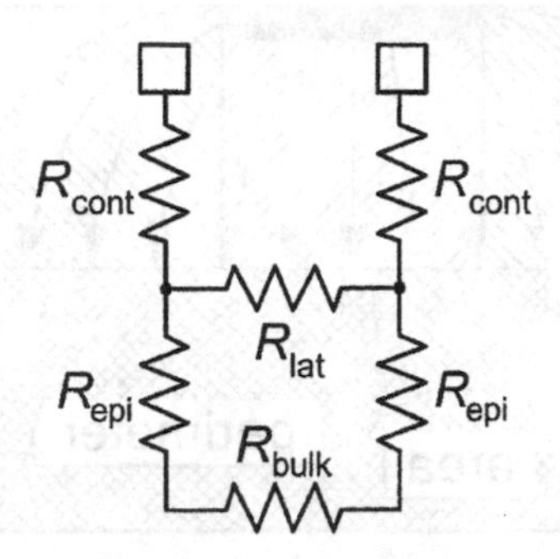

Figure 5-11. Model for the substrate contact resistance.

From Table 5-1, it is apparent that the substrate bias resistance depends on the contact spacing: the larger the substrate diffusion bounding box, the lower the resistance. However, the measurement figures do not completely fit the resistance model published in the literature [18]. Indeed, the epi-layer resistance has been considered to be:

$$R_{epi} = R_{area} \ // \ R_{perimeter}, \tag{3}$$

where the first term $R_{area} = \rho_{epi} \cdot t / A$ is inversely proportional to the diffusion area, and the second term $R_{perimeter} = \rho_{epi} / p$ is inversely proportional to the diffusion perimeter, ρ_{epi} and t being the epi-layer resistivity and thickness, respectively. *Figure 5-12* shows a cross section of a substrate bias contact, illustrating the current paths within the diffusion area and around the perimeter.

However, we can see from measurements that contact arrays 4 and 5 have different resistances, even with the same area and perimeter.

A more accurate model for substrate bias resistance can be obtained by considering that the p-doping concentration is higher in the p-well than in the epi layer. Therefore, the less doped epi layer gives the main contribution to the overall resistance. The thickness t can be considered as the sum of two parts:

$$t = t' + t'', \tag{4}$$

where t" is the thickness of the p-well and t' is the thickness of the epi layer below the p-well.

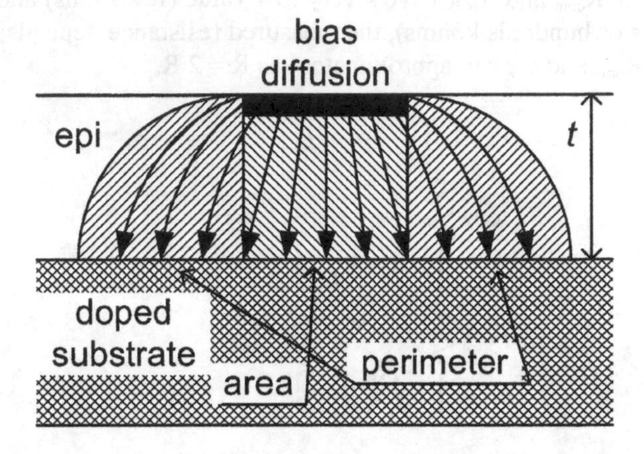

Figure 5-12. Model for the calculation of the substrate bias resistance.

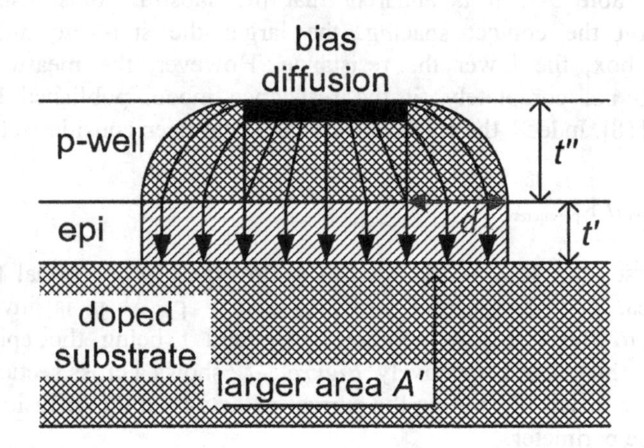

Figure 5-13. New model for the calculation of the substrate bias resistance.

Figure 5-13 illustrates the proposed resistance model. Neglecting the resistance of the p-well, the bias resistance can be modeled as

$$R_{epi} = \rho_{epi} \cdot t' / A', \tag{5}$$

where A' is the "effective" bias area obtained by oversizing the bias diffusion area by a lateral oversized.

Table 5-2. Mean values and standard deviation of substrate bias resistances.

Pads	R [kO]	s_R [kO]
1-1	1.07	0.03
2-2	0.70	0.06
3-3	1.36	0.05
4-4	0.97	0.08
5-5	0.57	0.02

By using the corrected mean values of resistances shown in Table 5-2 (without considering the measured values between A and B arrays), and remembering that

$$R / 2 \approx R_{epi} = \rho_{epi} \cdot t' / A', \tag{6}$$

we get an excellent fitting of experimental data with $\rho_{epi} \cdot t' = 330$ kO · μm^2 and d = 8.5 μm (2d = 17 μm).

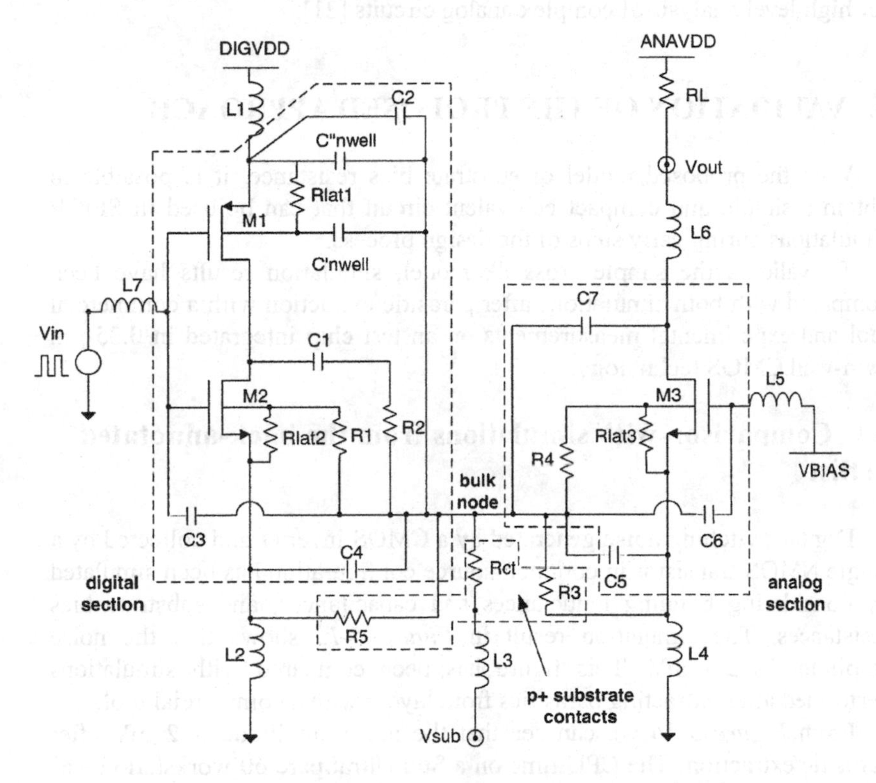

Figure 5-14. Equivalent circuit used in crosstalk simulations; node V_{sub} is an additional test point that can be used to observe substrate noise.

5. SIMULATION

SPICE-level simulation of complex mixed-signal circuits is not a straightforward task. To avoid problems due to the large number of transistors, digital cells can be "lumped" into a small set of transistors, or even into a set of current generators with a time-domain behavior equivalent to the digital subcircuit [18,19]. As an example, it is possible to simulate the effect of digital noise injection from a synchronous digital network by lumping all the digital gates into a single CMOS inverter, which corresponds to the clock driving stage. *Figure 5-14* illustrates the equivalent circuit for substrate noise analysis.

Moreover, if the digital switching pattern is periodic, it may be also represented in the frequency domain [20].

Effects of disturbances can be analyzed with SPICE-level simulations on small analog circuits. Then suitable macromodels can be extracted and used for high-level analysis of complex analog circuits [21].

6. VALIDATION OF THE PROPOSED APPROACH

With the proposed model of substrate bias resistance, it is possible to obtain a simple and compact equivalent circuit that can be used in SPICE simulations during early steps of the design process.

To validate the simple crosstalk model, simulation results have been compared with both simulations after parasitic extraction with a commercial tool and experimental measurements on an test chip integrated in 0.35-μm twin-well CMOS technology.

6.1 Comparison with simulations from the back-annotated netlist

Digital switching noise generated by a CMOS inverter and collected by a single NMOS transistor in common-source configuration has been simulated by considering bonding inductances and capacitances, and substrate bias resistances. The simulation result in *Figure 5-15* shows that the noise amplitude is 2.5 mV. This figure has been compared with simulations performed after extracting parasitics from layout with a commercial tool.

From *Figure 5-16* we can see that the noise amplitude is 2 mV after parasitic extraction. The CPU time on a Sun UltraSparc 60 workstation was 10 s for the simulation in *Figure 5-15* and 10 min for the simulation in *Figure 5-16*.

Another comparison has been done on a mixed-signal circuit including a fully-differential operational amplifier (about 60 MOS transistors). With the simple RLC model, the digital noise amplitude at one output of the operational amplifier was 146 mV and the required CPU time for simulation in the time domain was 2.5 min per clock period of the digital section. After complete parasitic extraction, the digital noise amplitude was 160 mV and the required CPU time was 1 day per clock period.

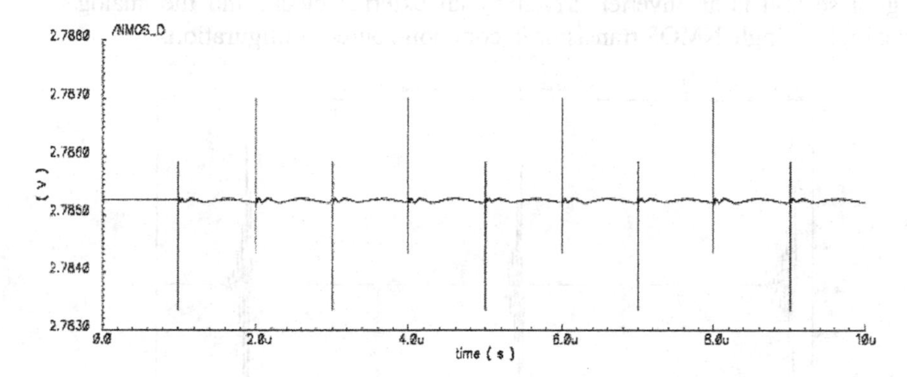

Figure 5-15. Simulation result with the proposed crosstalk model.

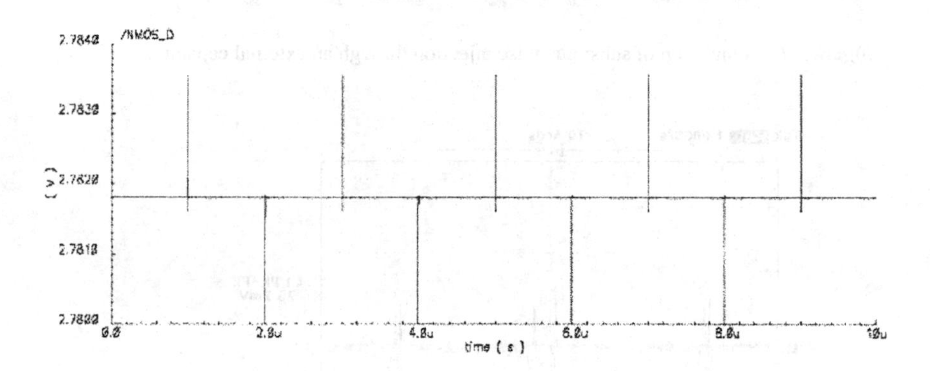

Figure 5-16. Simulation result after parasitic extraction.

By comparing the simulation results, it is apparent that the simple model is fast and accurate, thus being suitable for the evaluation of circuit architectures from the very beginning of the design process.

6.2 Comparison with experimental measurements on a test chip

To validate the proposed simulation methodology, a comparison has been made between SPICE simulations and measurements on integrated prototypes.

A first test was performed on a very simple mixed-signal circuit. The digital section is an inverter driven by an external clock, and the analog section is a single NMOS transistor in common-source configuration.

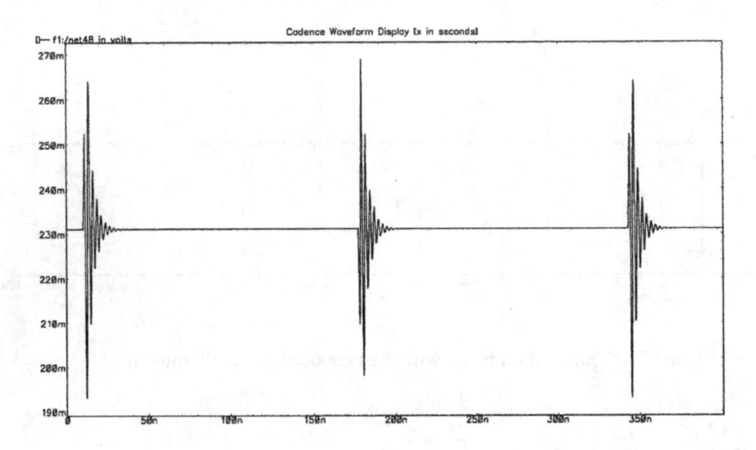

Figure 5-17. Simulation of substrate noise injection through an external capacitor.

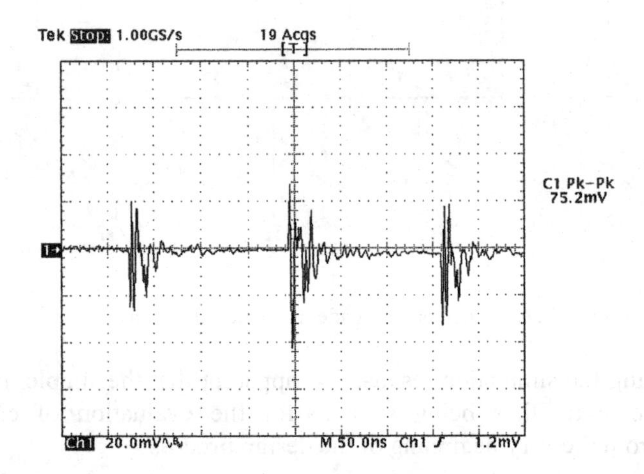

Figure 5-18. Measurement of effects due to substrate noise injection.

The circuit has been simulated with SPICE using the equivalent circuit shown in *Figure 5-14* [22].

Figure 5-17 shows the simulation result of the switching noise effect on the single-transistor amplifier when the injecting digital circuit is driven by an external clock. Transient analysis shows a variation in the drain voltage of the analog MOS transistor, due to the switching of digital circuitry. Oscillations occur after each switching, with an intrinsic frequency depending on bonding parasitics.

Figure 5-18 illustrates the corresponding measurement on the integrated test structure. The simulation results are in good agreement with the measured data. Measurements carried out on several MOS transistors in the test chip (placed at different distances from the digital section, with or without guard rings) confirmed also that in epi technologies the amount of injected noise is independent of the distance and that guard rings are not effective.

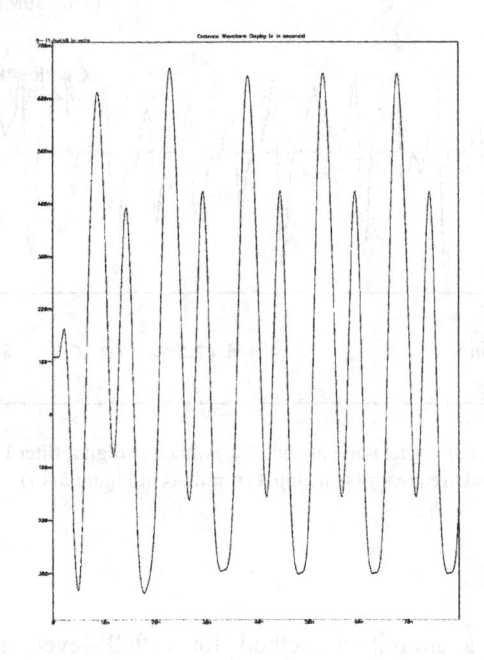

Figure 5-19. Simulation of digital noise: voltage at node V_{sub} when the clock frequency is 66 MHz (time scale: 10 ns/div, voltage scale: 100 mV/div).

Crosstalk noise was also analyzed in a more realistic situation to investigate the impact of digital circuitry on analog blocks located on the same silicon die. For these measurements, a digital decimation filter was used as a source of digital noise. This filter is made up of approximately 130,000 gates. The corresponding simulation was done on the simple crosstalk model using a single switching inverter (*Figure 5-14*). The noise-injecting capacitor (C1) was sized so as to take the high capacitance of the filter clock tree into account. *Figure 5-19* illustrates the SPICE simulation result, using the crosstalk model described above.

Figure 5-20 shows the measurement of the voltage noise collected on the substrate (node V_{sub} in *Figure 5-14*), when the filter is operated at 66 MHz clock frequency. The agreement between simulated and experimental values is apparent.

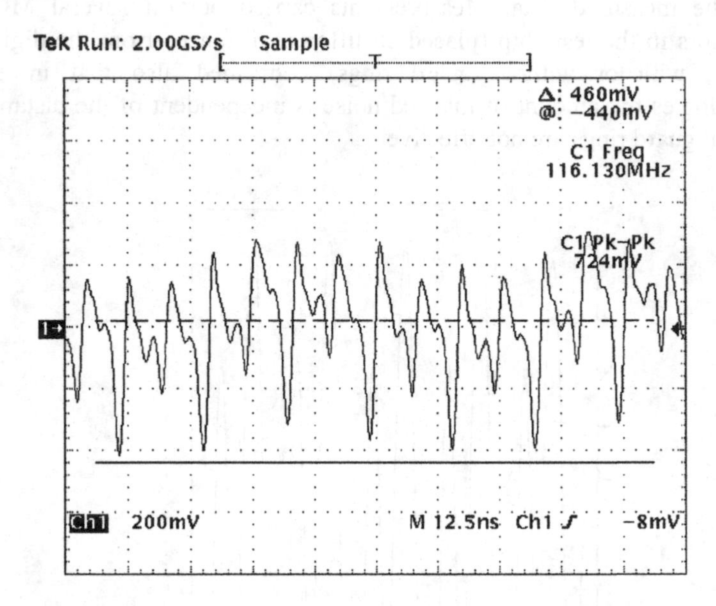

Figure 5-20. Measurement of digital noise at node V_{sub} when the digital filter is operated at 66 MHz clock frequency (same input stimuli as in Figure 5-19).

7. CONCLUSION

In this chapter, a simplified method for SPICE-level simulation of digital-analog crosstalk in mixed-signal integrated circuits has been described.

It has been shown that crosstalk effects can be simulated at early steps of the design process, using fabrication parameters available from physical design manuals and from package datasheets. Epi-layer resistivity, which is one of the most critical parameters in CMOS technologies with heavily doped substrate, can be estimated by measuring the resistance of a simple test structure.

A good modeling of all the non-idealities is essential in speeding up the whole design work. The SPICE macromodel simulations have been shown to be fast and accurate, thus allowing the analog designer to explore different circuit architectures and topologies from the viewpoint of their immunity or insensitivity to the digital switching noise.

ACKNOWLEDGEMENTS

This work has been supported by the European Union through the ESPRIT Project 29261 – MIXMODEST.

REFERENCES

[1] P. M. Zeitzoff and J. E. Chung, "Weighing in on logic scaling trends", *IEEE Circuits and Devices Mag.*, vol. 18, pp. 18-27, Mar. 2002.

[2] G. Badenes and L. Deferm, "Integration challenges in sub-0.25 μm CMOS-based technologies", *Microelectronics Journal*, vol. 31, pp. 861-871, Nov./Dec. 2000.

[3] R. C. Jaeger. *Introduction to Microelectronic Fabrication (2nd ed.)*. Prentice-Hall, Upper Saddle River, NJ, USA, 2002.

[4] J. Twomey, "Noise reduction is crucial to mixed-signal ASIC design success (part I)", *Electronic Design Magazine*, vol. 48, no. 22, Oct. 2000.

[5] J. Twomey, "Noise reduction is crucial to mixed-signal ASIC design success (part II)", *Electronic Design Magazine*, vol. 48, no. 25, Dec. 2000.

[6] M. Felder and J. Ganger, "Analysis of ground-bounce induced substrate noise coupling in a low resistive bulk epitaxial process: Design strategies to minimize noise effects on a mixed-signal chip", *IEEE Trans. Circ. and Syst.-II*, vol. 46, pp. 1427-1436, Nov. 1999.

[7] D. Leenaerts, G. Gielen, and R. A. Rutenbar, "CAD solutions and outstanding challenges for mixed-signal and RF IC design", in *Proc. IEEE/ACM Int. Conf. on Computer Aided Design (ICCAD)*, San Jose, CA, USA, Nov. 2001, pp. 270-277.

[8] H. Nakata and A. Iwata, "A microscopic substrate noise model for full chip mixed-signal design verification", in *Symp. on VLSI Circuits Dig. of Tech. Pap.*, Kyoto, Japan, June 1997, pp. 37-38.

[9] W. H. Kao, C.-Y. Lo, M. Basel, and R. Singh, "Parasitic extraction: Current state of the art and future trends", *Proc. IEEE*, vol. 89, pp. 729-739, May 2001.

[10] E. Charbon, R. Gharpurey, R. G. Meyer, and A. Sangiovanni-Vincentelli, "Substrate optimization based on semi-analytical techniques", *IEEE Trans. Computer-Aided Design of Integr. Circ. and Syst.*, vol. 18, pp. 172-190, Feb. 1999.

[11] S. Ponnapalli, N. Verghese, W. K. Chu, and G. Coram, "Preventing a 'noisequake': Substrate noise analysis identifies potential problems in mixed-signal and RF designs", *IEEE Circuits and Devices Mag.*, vol. 17, pp. 19-28, Nov. 2001.

[12] F. J. R. Clement, E. Zysman, M. Kayal, and M. Declercq, "LAYIN: Toward a global solution for parasitic coupling modeling and visualization", in *Proc. IEEE Custom Integr. Circ. Conf. (CICC)*, San Diego, CA, USA, May 1994, pp. 537-540.

[13] T. Smedes, N. P. van der Meijs, and A. J. van Genderen, "Extraction of circuit models for substrate cross-talk", in *Proc. IEEE/ACM Int. Conf. on Computer Aided Design (ICCAD)*, San Jose, CA, USA, Nov. 1995, pp. 199-206.

[14] K. Joardar, "A simple approach to model crosstalk in integrated circuits", *IEEE J. Solid-State Circ.*, vol. 29, pp. 1212-1219, Oct. 1994.

[15] N. K. Verghese and D. J. Allstot, "Computer-aided design considerations for mixed-signal in RF integrated circuits", *IEEE J. Solid-State Circ.*, vol. 33, pp. 314-323, Mar. 1998.

[16] V. Liberali, "Evaluation of epi layer resistivity effects in mixed-signal submicron CMOS integrated circuits", in *Proc. Intern. Conf. on Microelectronics (MIEL)*, Niš, Yugoslavia, May 2002, pp. 569-572.

[17] V. Liberali, R. Rossi, and G. Torelli, "Crosstalk effects in mixed-signal ICs in deep submicron digital CMOS technology", *Microelectronics Journal*, vol. 31, pp. 893-904, Nov. 2000.

[18] M. van Heijningen, J. Compiet, P. Wambacq, S. Donnay, M. G. E. Engels, and I. Bolsens, "Analysis and experimental verification of digital substrate noise generation for epi-type substrates", *IEEE J. Solid-State Circ.*, vol. 35, pp. 1002-1008, July 2000.

[19] M. van Heijningen, M. Badaroglu, S. Donnay, H. De Man, G. Gielen, M. Engels, and I. Bolsens, "Substrate noise generation in complex digital systems: Efficient modeling and simulation methodology and experimental verification", in *IEEE Int. Solid-State Circ. Conf. Digest of Tech. Pap.*, San Francisco, CA, USA, Feb. 2001, pp. 342-343.

[20] M. Xu, D. K. Su, D. K. Shaeffer, T. H. Lee, and B. A. Wooley, "Measuring and modeling the effects of substrate noise on the LNA for a CMOS GPS receiver", *IEEE J. Solid-State Circ.*, vol. 36, pp. 473-485, Mar. 2001.

[21] M. Nagata and A. Iwata, "Substrate noise simulation techniques for analog-digital mixed LSI design", *Analog Integrated Circuits and Signal Processing*, vol. 25, pp. 209-217, Dec. 2000.

[22] V. Liberali, R. Rossi, and G. Torelli, "A simple model for digital/analog crosstalk simulation in deep submicron CMOS technology", in *Proc. European Conf. on Circuit Theory and Design (ECCTD)*, vol. I, pp. 169-172, Espoo, Finland, Aug. 2001.

Chapter 6

HIGH-LEVEL SIMULATION OF SUBSTRATE NOISE GENERATION IN COMPLEX DIGITAL SYSTEMS

Mustafa Badaroglu, Marc van Heijningen and Stéphane Donnay
IMEC – DESICS, Kapeldreef 75, B-3001 Leuven, Belgium

Abstract: With increasing clock frequencies and resolution requirements in mixed-mode telecom circuits, substrate noise is becoming more and more a major obstacle for single chip integration. To simulate the performance degradation of the sensitive analog circuits the total amount of the generated substrate noise must be known. Existing approaches usually extract the model of the substrate from layout information and then simulate the extracted transistor-level netlist with this substrate model using a transistor-level simulator. For large digital circuits, the substrate simulation is however not feasible with a transistor-level simulator. We have developed a high-level methodology to simulate this substrate noise generation in EPI substrates by taking the noise coupling from the switching gates and also from the supply rails into account. Experimental results show an error of 5% in the RMS value of the substrate noise generation with respect to a full SPICE simulation for a 1Kgate circuit, while maintaining a speedup of 3 orders of magnitude with respect to SPICE simulations. The approach has also been applied to the 86K digital ASIC introduced in chapter 2 and compared to measurements.

1. INTRODUCTION

There is a trend towards single-chip integration of more complex mixed-signal systems, higher speeds and lower supply voltages. In these mixed-signal ICs, the low cost and lower static power consumption of CMOS logic are overshadowed by the larger noise generation due to the large rail-to-rail voltages and the sharp current spikes during switching. Substrate noise coupling is a vital factor in the signal-integrity analysis in mixed-signal ASICs.

The analysis of substrate noise coupling consists of three aspects: generation of substrate noise by the digital circuits, propagation of the noise through the substrate and its impact on analog circuits. Up to now most of the research has concentrated on modeling the propagation and the impact of substrate noise [1][2]. All these techniques model the substrate as a network of resistors and capacitors attached to the nodes of the transistor-level netlist. The simulation of the transistor-level netlist with its substrate model requires large computer resources and is time consuming. Another disadvantage is that these approaches start from the layout of the circuit. However, there is a need for an analysis methodology during the design phase of digital systems, e.g. during gate level design.

There are a few publications on the gate-level simulation of noise generation by a large digital circuit. In [3][4], methodologies are presented to simulate the substrate noise generation using an event-driven simulator. The mathematical functions to formulate the noise for each switching activity are rather simple. Methodologies that make use of real substrate noise waveforms extracted for each standard cell are presented in [5]. The methodology presented in [5] does not include noise coupling from the power supply (which we showed to be the dominant noise source in chapter 2) and is not verified with measurements. In [6] the noise coupling from the supply lines is considered while the substrate coupling from the source/drain regions is ignored. The latter can be important when the package inductance is lower. This methodology uses only the root-mean-square of the power supply current, not the transient behavior of this current.

Up till now, no good methodologies exist to accurately simulate the actual waveform of the substrate noise voltage of a large digital circuit by considering both power supply coupling (Ldi/dt) and capacitive coupling (CdV/dt). This chapter presents a methodology that allows the simulation of the actual time domain waveform of the substrate noise voltage, related to the real circuit operation in EPI-type substrates [7]. The difference between the simulated RMS substrate noise voltage and the measurements is less than 10% [10][9]. The substrate noise simulation time is of the same order of magnitude as a VHDL gate-level simulation.

Our high-level simulation methodology, SWAN, makes it possible to accurately predict the substrate noise generation of a large digital circuit and to combine these results with simulations (e.g. using SPICE) of an embedded analog circuit to study the performance degradation in mixed-signal systems (see chapter 7). The simulation results can also be used to explore low-substrate-noise design techniques and implementations of a digital circuit (see chapter 11).

The substrate noise simulation methodology consists of two main parts: (1) library characterization and (2) substrate noise simulation (switching activity extraction, chip-level substrate model extraction and the substrate noise simulation). The methodology flow is shown in *Figure 6-1*.

The chapter is organized as follows. First in section 2 we will describe the library characterization of substrate macro models. In section 3 we will describe the extraction of the chip-level substrate model and how to simulate it. In section 4 we will present experimental results on two example circuits and comparison to the measurements of an 86Kgate digital ASIC [10][9]. Finally, we draw conclusions in section 5.

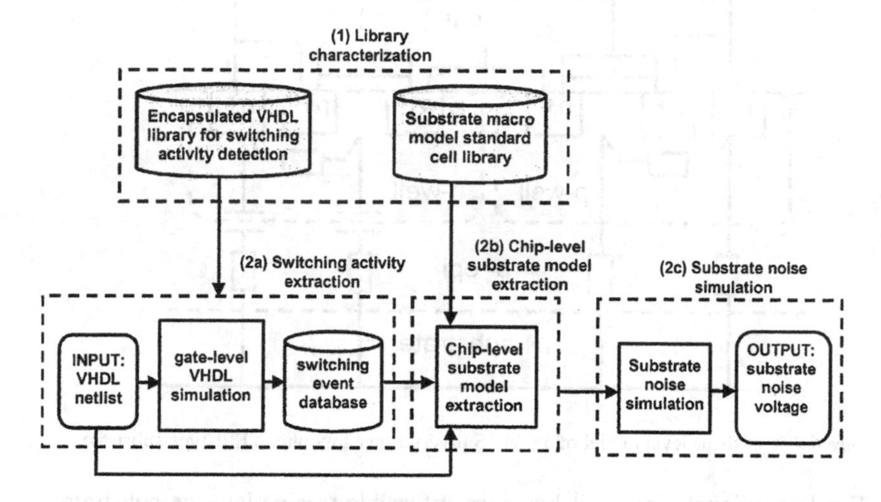

Figure 6-1. High-level substrate noise simulation methodology.

2. LIBRARY CHARACTERIZATION

2.1 Substrate macro model characterization

In large digital circuits, high peaks and fast rise/fall times of the supply current create ringing (Ldi/dt noise) in the supply network due to the damped LC-tank, formed by the on-chip capacitance between VDD and VSS and the package inductance with series resistance in the supply connection. On a typical p-type substrate, this supply ringing couples into the substrate capacitively from VDD via the n-well junction capacitance and resistively via the substrate contacts from VSS. Fast switching of the CMOS gate outputs (CdV/dt noise) couples into the substrate from the drain of the transistors via the diffusion capacitance. A SPICE model of an inverter on a low-ohmic EPI-type substrate taking into account all these noise injection mechanisms is shown in *Figure 6-2*. Experimental verification of the SPICE-level model is presented in chapter 2. The typical resistivity is around 10mΩcm for the conductive p+ substrate under the EPI layer while it is 10Ωcm for the EPI layer. So the conductive p+ substrate can be approximated as one single equipotential node so that a π-circuit representation can be used.

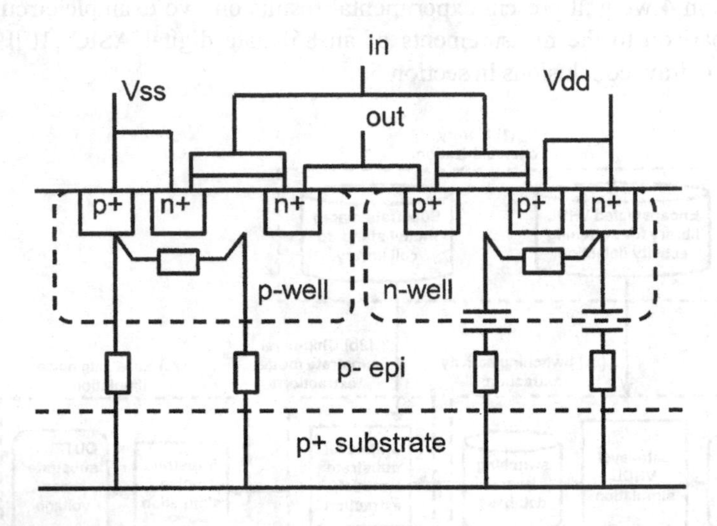

Figure 6-2. Transistor-level model of a CMOS inverter on a low-ohmic EPI-type substrate.

For large digital circuits it becomes infeasible to simulate the substrate noise generation by using the transistor-level models together with detailed substrate models. Therefore a macro modeling approach is necessary. The SPICE-level model shown in *Figure 6-2* will form the basis of the substrate macro model characterization for every standard cell in the library. *Figure 6-3* shows the substrate macro model of an inverter implemented in a 0.35µm CMOS process on an EPI-type (10Ωcm, 4µm EPI) substrate. For every gate, a substrate macro model consists of two current sources, the bulk (Ibulk) current and supply current (Isupply), with the coupling impedances between VDD, VSS and the common node at the conductive p+ substrate, which we will refer to as the substrate node.

The macro models are extracted once for the entire standard-cell library with SPICE simulations that include a detailed substrate model obtained with SubstrateStorm™ from Cadence [17]. The macro model passive element parameters are extracted by performing an AC analysis between Vdd, Vss and the substrate node of the gate by using the SPICE level model shown in *Figure 6-2*. Next, a digital input pattern, which covers all possible combinations of switching inputs and states of the gate, is applied in transient simulation and the corresponding supply current waveform (Isupply) is recorded in a look-up table for every switching combination by measuring the current flowing through the power supply. At the same time, the substrate voltage is recorded from the SPICE level model. Ibulk is then derived in such a way that it generates the same substrate voltage when applied to the macro model. For the macro model shown in *Figure 6-3*, the impedance between VDD and VSS is represented by a capacitance (C_c) in series with a resistance (R_c). The series resistance from the VSS contact to the substrate is represented by R_s. The capacitance due to the reversely biased n-well junction diode and the resistive path underneath the n-well are represented by C_w and R_w respectively. *Figure 6-3* also shows the current waveforms defined for both rising and falling edges at the input. These waveforms are stored in a database. This waveform database is quite small as the number of inputs of a gate is 3-4 in average. For each switching of a gate, the waveforms are stored up to 2ns at a sampling period of 10ps.

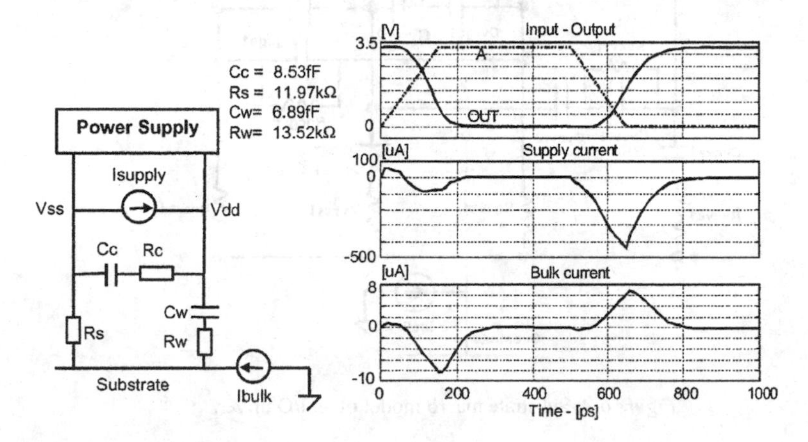

Figure 6-3. Substrate macro model of a minimum-size inverter in a 0.35μm CMOS process on an EPI-type substrate.

The current through the supply line does not only flow through the core cells but also through other chip components such as the I/O and supply pads and off-chip components such as PCB parasitics, off-chip decoupling and external power supply. Therefore it is important to characterize the I/O and supply pads and generate accurate macro models. In *Figure 6-4*, an example is shown of the macro model of an I/O cell with multiple supplies. A typical I/O cell has two power supplies. One (vdd_{ring}, vss_{ring}) is used for the last output stage of the cascaded buffers while the other supply (vdd_{core}, vss_{core}) is used for the remaining buffers and circuits. Such a separation in the supplies is necessary as the last output stage of the I/O cell is usually noisy and the ground of this supply (vss_{ring}) is not connected to the substrate, whereas the core supply ground (vss_{core}) is typically connected to the substrate. For the cells, which have multiple supplies, the coupling between the supply domains is included in the model. This is equivalent to an impedance matrix extraction for an N-port device. The rest of the parameters are computed in a similar way as for a standard core cell with a single power supply.

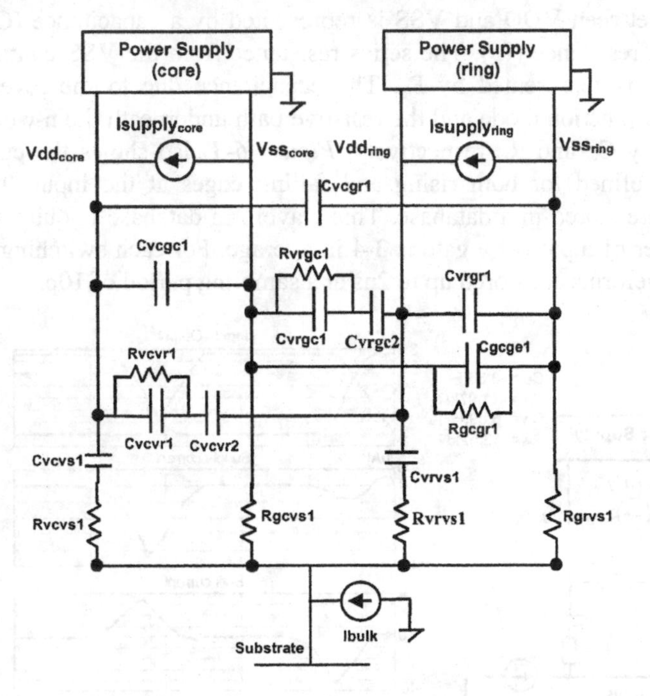

Figure 6-4. Substrate macro model of an I/O driver.

2.2 Effect of load and input transition time

The load of a gate consists of parasitics such as the interconnect capacitance and the fanout. The load represented by the fan-out gates is already modeled by the input capacitance of these gates. Only the interconnect capacitance has to be added to the driver's macro model.

The load also has an important influence on the currents flowing through gate, the supply rails and the load itself during a switching event. Input transition (rise and fall) time also has an important impact on these currents.

For an accurate simulation of the substrate noise, it is therefore important to model load and input transition time effects within the macro cell models.

An example of load and input transition time dependency of the peak-to-peak and the root-mean-square (RMS) values of the supply current for a minimum size inverter gate implemented in a 0.35μ CMOS on an EPI-type substrate is shown in *Figure 6-5*. As the load increases, the RMS value of the supply current also increases linearly. The peak-to-peak value increases with the load for small values, until it saturates at a critical load value, which depends on the transition time.

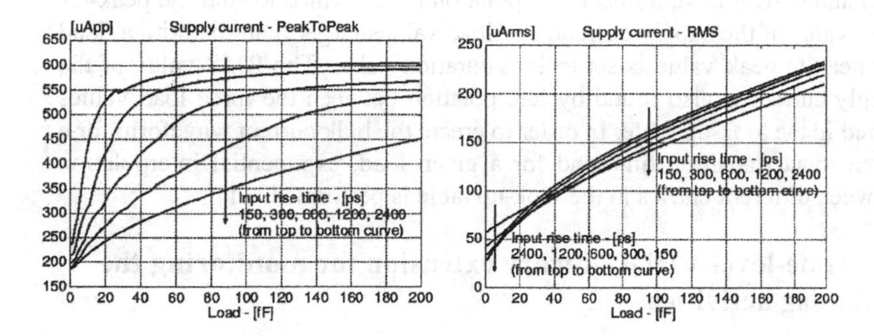

Figure 6-5. Effect of load and input transition time on the supply current for a minimum size inverter in a 0.35μ CMOS process on an EPI-type substrate

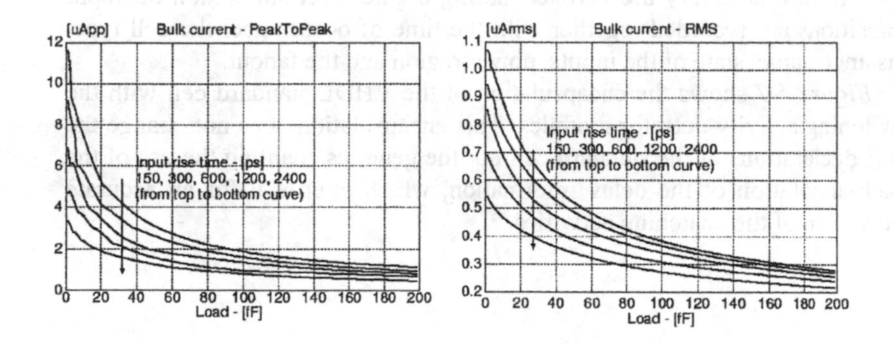

Figure 6-6. Effect of load and input transition time on the bulk current for a minimum size inverter in a 0.35μ CMOS process on an EPI-type substrate.

Figure 6-6 shows the effect of load and input transition time on the bulk current for the same inverter. As the load increases, both peak-to-peak and RMS values of the bulk current decrease and saturate for larger load values.

An increase in input transition time will give us smaller values for the peak-to-peak and the RMS values of the bulk current. The comparison of *Figure 6-5* and *Figure 6-6* clearly shows that the supply current is around two orders of magnitude larger than the bulk current. Therefore the supply current is an important noise contributor in the chips with supply line parasitics (see chapter 2).

To characterize load and input transition time dependency the current waveforms for every standard cell have been recorded into a look-up table for different switching activities and with three load values: zero load, critical load and two times the critical load. For load values between zero load and critical load, a linear interpolation is performed to find the peak-to-peak value of the supply current. For load values larger than the critical load the peak-to-peak value is set to its saturation value. The RMS value of the supply current is also found by interpolation between the three load values stored in the look-up table. In order to create the bulk current waveform for a given input transition time and for a given load, exponential interpolation between different entries in the look-up table is performed [11].

2.3 Gate-level VHDL library extension for monitoring the switching activities

The VHDL standard cell library has to be extended to enable the detection of all input switching events. This library is created by adding switching event detection processes to the original VHDL library. When the cells from this library are invoked during a gate-level simulation all input transitions are recorded together with the time of occurrence, the cell type, instance name, state of the inputs, power region and the fanout.

Figure 6-7 shows the encapsulation of the VHDL standard cell with the switching activity detector modules. This encapsulation does not change the port declarations of the standard cell nor the generics enabling the use of the back-annotation of the delay information, which is crucial for an accurate extraction of the switching activities.

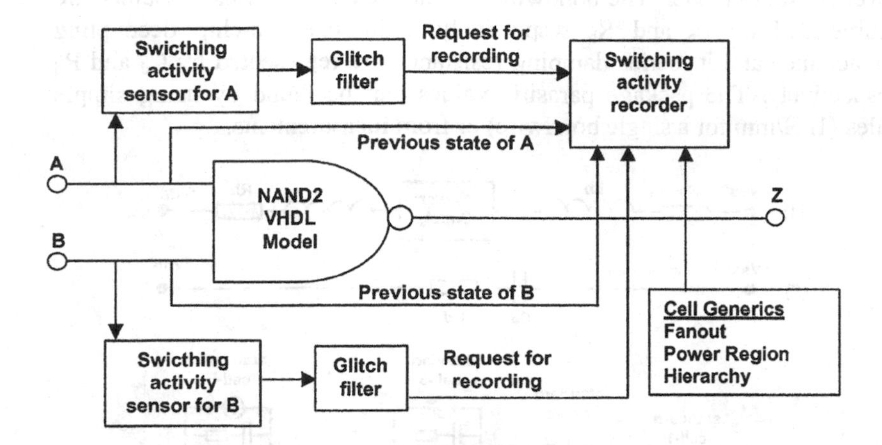

Figure 6-7. Encapsulated VHDL model of a NAND2 cell for recording the switching activities.

3. SUBSTRATE NOISE SIMULATION

3.1 Overview of substrate noise simulation

First a chip-level lumped substrate network for the entire chip is generated using the previously created substrate macro models of the different gates. Next, a gate-level event-driven simulation is performed to extract the switching activity information. Using these switching events and the previously extracted macro model currents taken from the database, the total bulk injection currents and power supply currents are computed. Finally, the chip-level substrate network is combined with the total bulk injection currents and the power supply currents to simulate the substrate noise generation of the entire digital circuit. The sections from 3.2 to 3.4 will describe these steps in more detail.

3.2 Chip-level substrate lumped network

To simulate all these effects for a large system, we construct a chip-level substrate network using substrate macro models for every logic cell as shown in *Figure 6-8*. The chip-level substrate network consists of four layers: (1) package parasitics, (2) on-chip decoupling, (3) the standard cells, and (4) the resistive mesh in the bulk. This mesh can be represented as a single node in low-ohmic EPI substrates, allowing a much simplified chip-level substrate model. The bondwire inductance and its series resistance are represented by L_b and R_b respectively. Additional on-chip decoupling capacitance and its series damping resistance are represented by C_d and R_d respectively. The package parasitic values can be found by using simple rules (1nH/mm for a single bondwire) or from measurements.

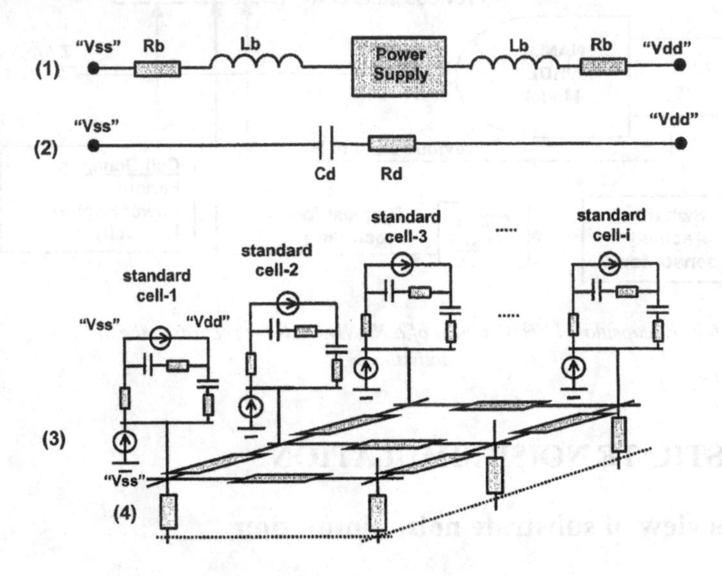

Figure 6-8. Chip-level substrate model network construction for a single supply region.

When the chip is partitioned into different supply regions, they each have their own substrate model. For low-ohmic EPI-type substrates, these substrate macro models for the core, I/O and supply pads in each supply region can be combined in parallel as illustrated in *Figure 6-9*.

Usually the time constants of the standard cells are comparable to each other so that the parallel combination can be done separately for each individual component in a given power region as shown below:

$$\frac{1}{R_s} = \sum_{all\,gates} \frac{1}{R_{s,gate}}, \; C_{w,tot} = \sum_{all\,gates} C_{w,gate}, \; C_{c,tot} = \sum_{all\,gates} C_{c,gate} \qquad (1)$$

As a result of parallel combination of the impedances between VDD, VSS and the single equipotential node on the substrate, a simple equivalent lumped circuit of the entire chip is obtained (see *Figure 6-10*). An accurate chip-level substrate model does not only contain the core cells but should also contain other chip components such as the I/O and supply pads together with off-chip components such as PCB parasitics, PCB and package decoupling and external power supplies. If one assures a good decoupling on the PCB close to the package, the modeling of on-chip components (core cells, pad cells) and package parasitics is sufficient for this high-level substrate model characterization.

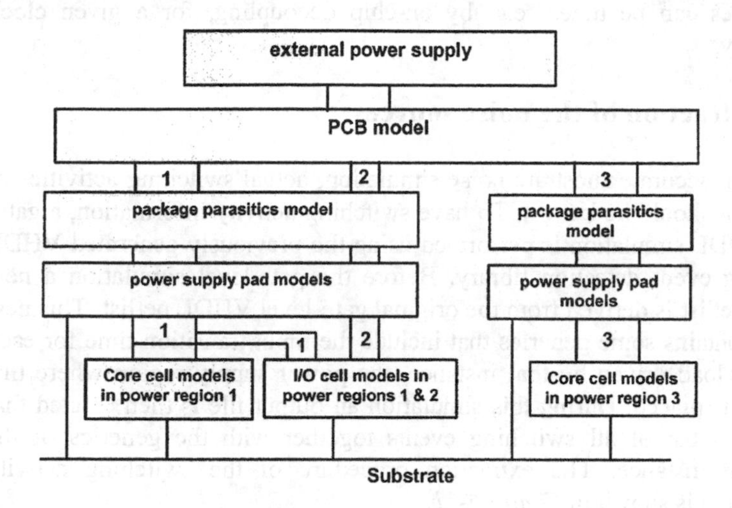

Figure 6-9. Combination of chip-level substrate models in different power supply regions.

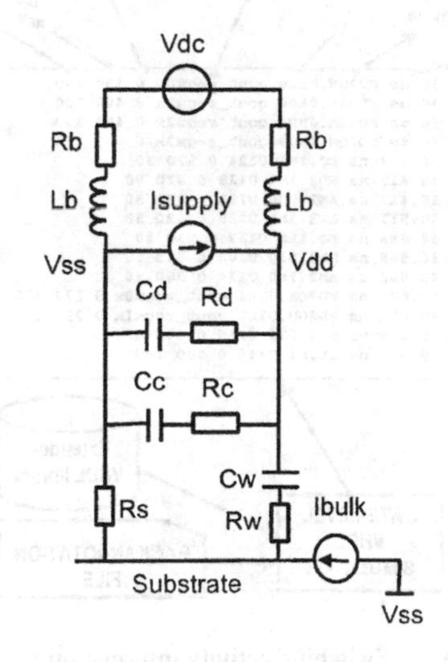

Figure 6-10. Chip-level substrate model.

An accurate extraction of the chip-level substrate model including package parasitics is important to estimate the frequency spectrum envelope of the substrate noise, that can be used for frequency planning of for instance an integrated front-end and also for decreasing the supply bounce problem of a stand-alone digital design. Especially the frequency bands in which the ringing of the digital power supply occurs can be problematic. These resonances can be tuned, e.g. by on-chip decoupling, for a given clock frequency.

3.3 Extraction of the noise sources

For an accurate substrate noise simulation, actual switching activities of the circuit should be known. To have switching activity information, a gate-level VHDL simulation is performed using the previously generated VHDL switching event detection library. Before the gate-level simulation a new VHDL netlist is derived from the original gate-level VHDL netlist. This new netlist contains some generics that include the input transition time for each port, the load driven by that instance, the power supply region where this instance is placed. During this simulation an output file is then created that contains a list of all switching events together with the generics of the switching instance. The extraction procedure of the switching activity information is shown in *Figure 6-11*.

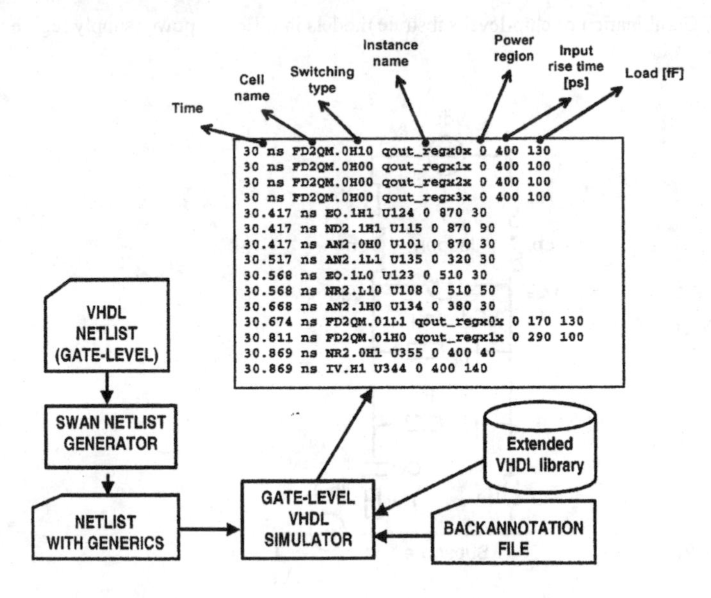

Figure 6-11. Switching activity information.

This switching activity data is used to generate the supply and bulk current noise sources by accumulating all individual waveforms related to the switching events according to the switching timing. For the simplified case of EPI-substrates the superpositioning of the individual waveforms is performed as follows:

$$I_{supply,tot}(t) = \sum_{all\ switching\ events} I_{supply,event}(t - t_{event}) \tag{2}$$

$$I_{bulk,tot}(t) = \sum_{all\ switching\ events} I_{bulk,event}(t - t_{event}) \tag{3}$$

3.4 Substrate noise simulation

The resulting chip-level substrate model with all linear lumped elements and lumped independent current sources can be represented as an s-domain transfer function with multiple inputs. This transfer function is then transformed into a z-domain transfer function to perform the transient simulation with difference equations in order to increase the simulation speed.

4. EXPERIMENTAL RESULTS

In this section the comparison between our high-level simulation methodology SWAN and SPICE simulations will be shown for two circuits: a 4-bit counter and a multiplier. Also simulation results of an 86Kgate digital telecom ASIC are presented and compared to measurements. The last part of this section summarizes the simulation times and circuit details.

4.1 Four Bit Counter

The first test circuit is a 4-bit counter, consisting of 4 flip-flops and a combinatorial feedback circuit. Using our simulation methodology we can simulate the generated substrate noise of this circuit for different inductance values.

Figure 6-12 shows the comparison between the SPICE and the high-level simulations of the substrate voltage produced by this counter, for one clock period (a rising and falling clock edge). At the rising clock edge (at 55ns) the four flip-flops are clocked and 13 combinatorial switching events occur. The negative clock edge (at 60ns) only causes a falling edge on the clock inputs of the flip-flops. Although no output changes occur at the falling clock edge still a significant amount of substrate noise is generated. *Figure 6-12* clearly shows the good correspondence between the SPICE and the high-level simulation. A quantitative comparison is given in *Table 6-1* where the root mean square (RMS) values of the substrate noise are listed for a clock period of 10ns. This RMS value is an indication for the total substrate noise power.

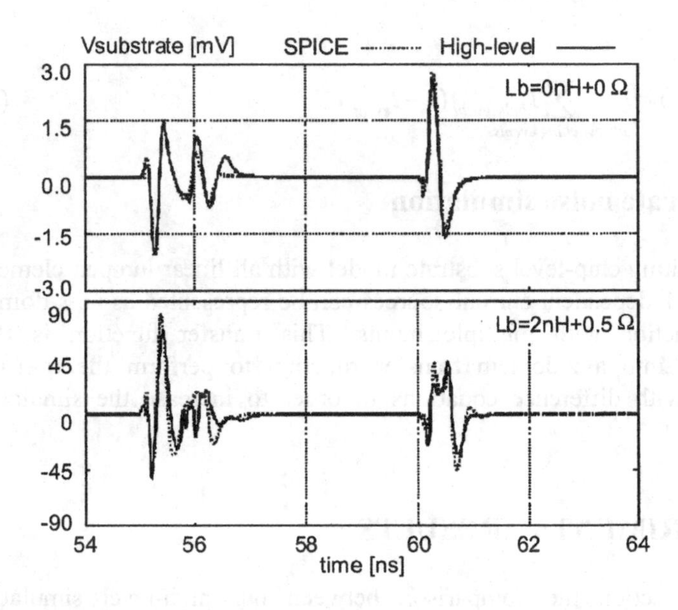

Figure 6-12. SPICE versus high-level simulation of generated substrate noise with (bottom) and without (top) power supply noise coupling for the counter circuit.

Table 6-1. Comparison of SWAN with SPICE in the counter

	Vsub,rms [mV]
SPICE	12.3
High-level (only bulk current)	0.48
High-level (bulk current + supply noise)	12.6

For the high-level simulation with power supply noise coupling, an inductance value of 2nH+0.5Ω has been used. These results show that a high-level substrate noise simulation without power supply noise coupling severely underestimates the amount of generated substrate noise (also see chapter 2).

4.2 Multiplier

The multiplier circuit consists of an 8-bit up counter and an 8-bit down counter followed by a 16-bit Booth multiplier, which multiplies the two counter values.

Figure 6-13 shows the comparison of the SPICE and the high-level simulations for the substrate voltage noise and power supply current without power supply coupling (Lb=2nH+0.5Ω). The noise is generated as a result of approximately 170 switching activities, occurring after one rising clock edge. The simulation is also performed with power supply noise coupling by using the same inductance value of 2nH+0.5Ω as in the inverter. *Table 6-2* gives the comparison of the RMS value of the substrate voltage for a SPICE simulation and a high-level simulation with and without power supply noise coupling for a clock period of 24ns.

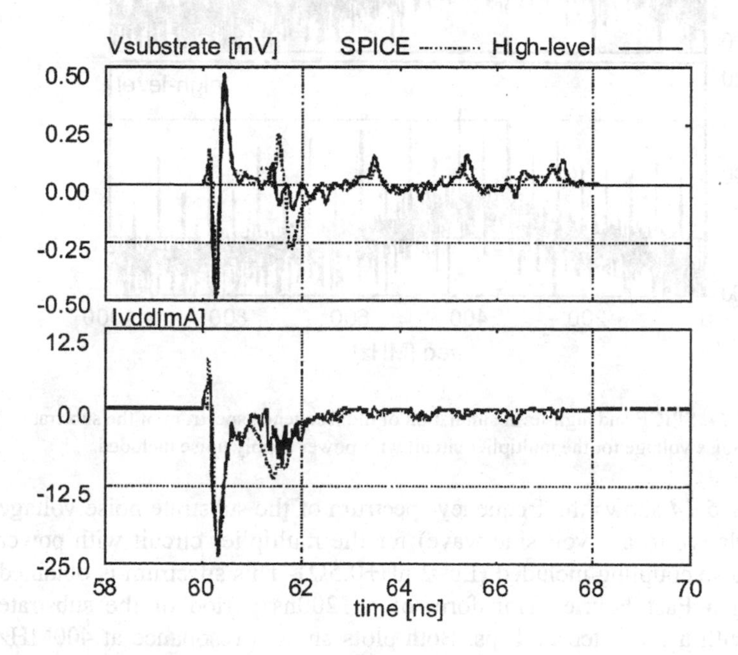

Figure 6-13. SPICE and high-level simulation of substrate noise voltage (top) and power supply current (bottom) for the multiplier circuit without power supply noise.

Again it is clearly visible that the power supply noise coupling is a dominant source of substrate noise. To obtain a good correspondence with the SPICE simulations, the power supply noise coupling must be included in the high-level simulations.

Table 6-2. Comparison of SWAN with SPICE in the multiplier

	Vsub,rms [mV]
SPICE	18.4
High-level (only bulk current)	0.079
High-level (bulk current + supply noise)	19.4

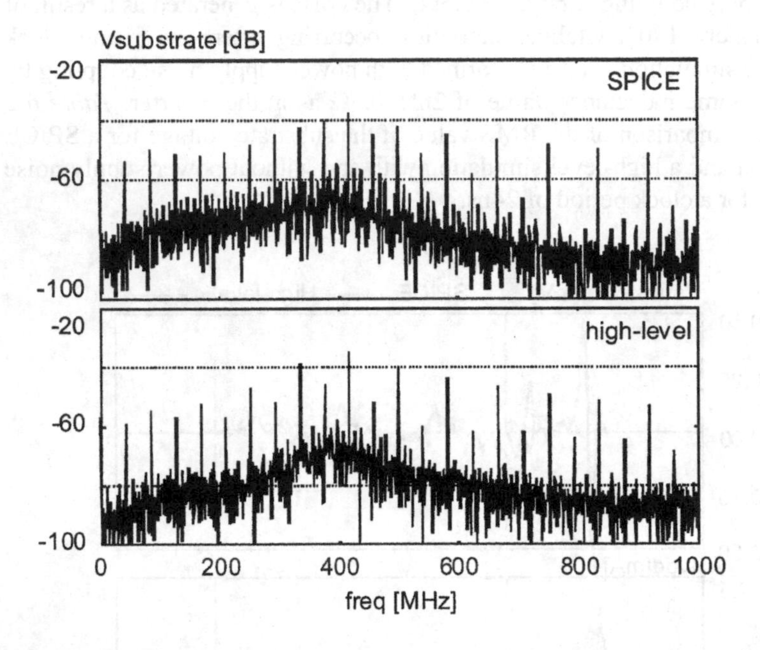

Figure 6-14. SPICE and high-level simulation of the frequency spectrum of the substrate noise voltage for the multiplier circuit with power supply noise included.

Figure 6-14 shows the frequency spectrum of the substrate noise voltage (in dB relative to a 1 volt sine wave) for the multiplier circuit with power supply noise coupling included (Lb=2 nH+0.5Ω). This spectrum is obtained by taking a Fast Fourier Transform of a 1200ns period of the substrate voltage with a time step of 10ps. Both plots show a resonance at 400MHz with the same bandwidth. The spectral peaks around the major resonance are 30-40dB above the noise floor. There is 3-4dB error in the major peaks caused by the clock harmonics. This error is due to the overestimation of the supply current caused by the glitches extracted as a full switching event during the high-level simulation. These glitches in fact do not create a significant supply current when simulated in SPICE. Additionally the piecewise linear approximation of the supply current bring sharp edges which appear as high frequency components in the spectrum.

The increase of the substrate noise around 400 MHz is caused by ringing of the power supply, which is coupled to the substrate. Most substrate noise is concentrated at multiples of the digital clock frequency of 42 MHz. The location and amplitude of the major noise peaks correspond well for the SPICE and high-level simulations.

4.3 Accuracy of SWAN in comparison with measurements for 86K gate digital ASIC

We have designed and measured a number of chips, which allows us to verify the accuracy of SWAN. For an 86Kgate multi-rate channel selection filter (Robo4) ASIC in a 0.5μm CMOS-EPI process (see chapter 2), the measured and simulated substrate noise waveforms are compared. The chip micrograph and its specifications are shown in *Figure 6-15*.

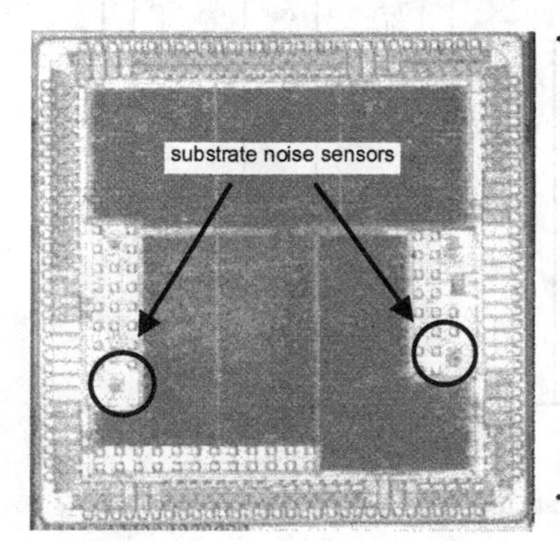

Technology	CMOS 0.5μm,3.3 V
Master Clock	163.84 MHz
Internal Clocks	40.94, 20.48 and 10.24 MHz
IO Word	12 bits
Internal Word	14 bits
Gate Count	86k
Core Area	24.22 mm^2
Chip Area	38.40 mm^2
Package	120 pin CPGA
Substrate Type	EPI (4 μm thick)
EPI Resistivity	10 Ω.cm
Bulk Resistivity	10 mΩ.cm

Figure 6-15. Microphotograph of the test chip (Robo4) and its specifications.

The model, used for the substrate noise simulations, is shown in *Figure 6-16*. Macro models of the core cells, the I/O cells and power supply pads take into account the impedance between the different power supply and substrate nodes. The low-ohmic substrate is modeled as a single node and all core and I/O cell models are placed in parallel [7]. *Figure 6-16* also shows the element values for the parallel combination of all core and I/O cell models. Substrate noise generation is modeled by current sources that represent noise injection by switching gates ($I_{sub,core}$ and $I_{sub,IO}$) and power supply current consumption (I_{vdd}, I_{vss}, I_{vdde} and I_{vsse}), calculated from switching events that are extracted from a VHDL gate-level simulation.

Measured substrate noise is shown in *Figure 6-17*, compared to the simulations from SWAN. The difference between measured and simulated RMS substrate voltage is less than 10%. A frequency domain comparison is shown in *Figure 6-18*. It is seen that the major resonances around 40MHz, 85MHz, 130MHz and 175MHz have been predicted correctly. The error at the fundamental clock frequency is 5dB. The low-frequency peaks in the measurements are due to PCB parasitics outside the chip package, which are not included in the model.

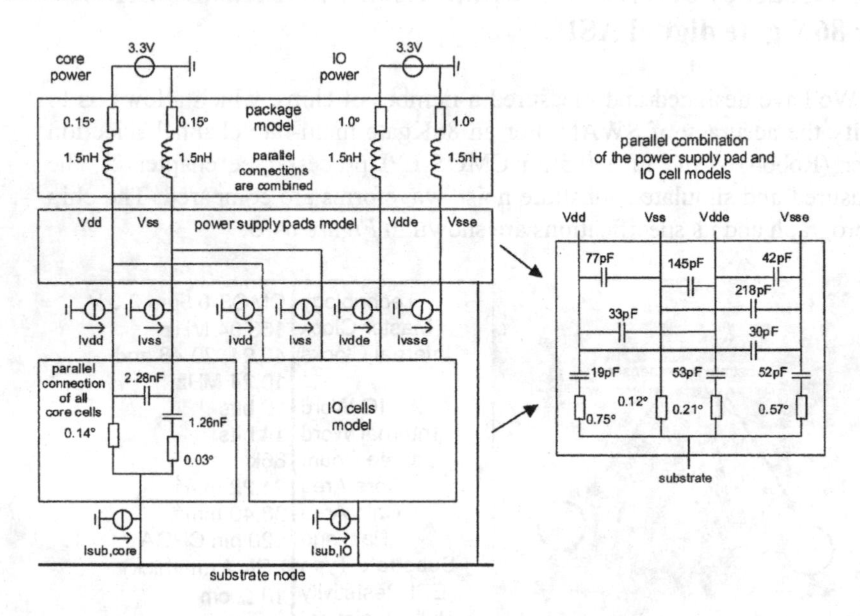

Figure 6-16. Extracted chip-level substrate model of the Robo4 ASIC.

4.4 Speed-up of SWAN in comparison with SPICE simulations

The substrate noise simulation time is of the same order of magnitude as the digital gate-level simulation time for large digital circuits. When this simulation methodology is compared with full SPICE-level simulation, which is of course only possible for very small designs, e.g. the 8x8 bit multiplier, a speedup of 1500 times is achieved. *Table 6-3* lists the details, simulation times of the two example designs and the Robo4 ASIC design, the latter of course no longer simulatable at circuit-level by SPICE. The substrate noise simulation time for the Robo4 design is about the same as the VHDL gate-level simulation time.

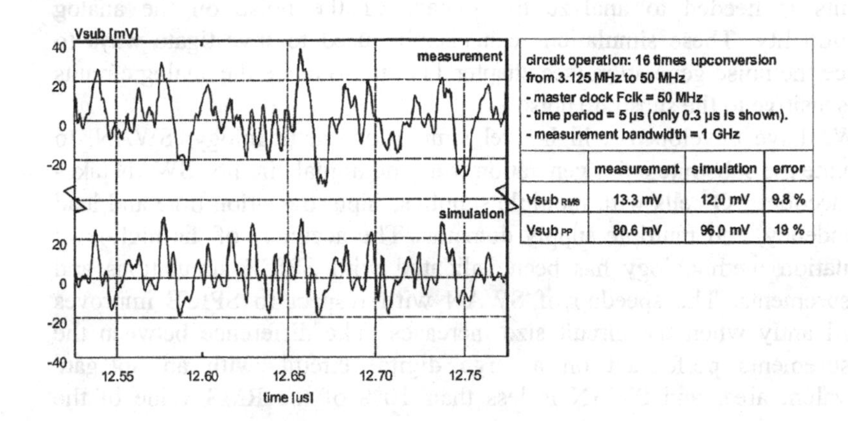

Figure 6-17. Time domain comparison of SWAN simulations with measurements.

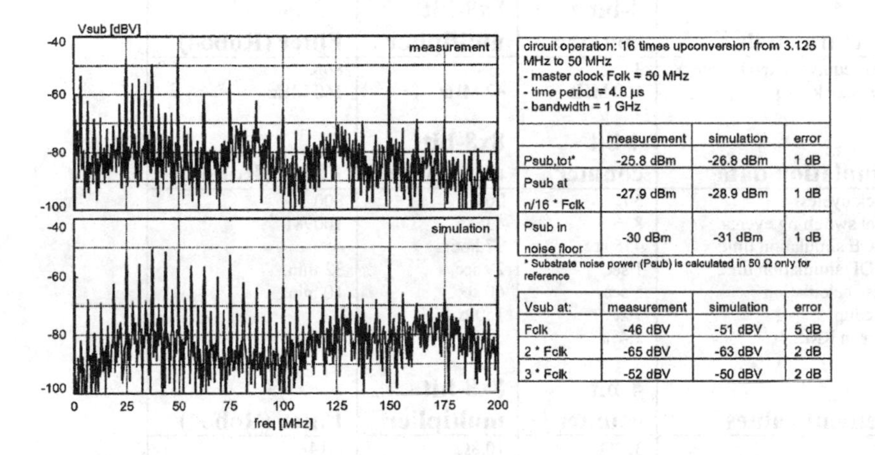

Figure 6-18. Frequency spectrum comparison of SWAN simulations with measurements.

5. CONCLUSIONS

In today's highly integrated mixed-signal ICs, substrate noise coupling from the digital to the analog circuits can severely degrade the performance of the analog circuits. Simulation of the noise generation by the digital circuits is needed to analyze the impact of the noise on the analog functionality. These simulations can also be used to investigate ways to reduce the noise generation (see chapter 11) and to make the analog circuits less sensitive to the substrate noise.

We have developed a high-level simulation methodology, SWAN, to simulate the substrate noise generation from the digital circuits. SWAN takes into account I/O cells with multiple supplies, input transition time and load dependency, and multiple supply domains. The accuracy of the high-level simulation methodology has been validated with SPICE simulations and measurements. The speedup of SWAN with respect to SPICE improves significantly when the circuit size increases. The difference between the measurements performed on a large digital circuit, with an 86Kgate equivalent area, and SWAN is less than 10% of the RMS value of the substrate noise obtained from the measurements.

Table 6-3. SWAN vs. SPICE/measurements

Circuit details	**4-bit counter**	**8x8-bit multiplier**	**Filter (Robo4)**
Area (equiv. NAND2 gates)	34	994	86K
Max. clock freq.	100 MHz	42 MHz	160 MHz
Simulation data	**4-bit counter**	**8x8-bit multiplier**	**Filter (Robo4)**
Clock cycles	50	208	800
Total switching events	856	63545	100981
SPICE simulation time	435 sec.	37 hours	-
VHDL simulation time	1 sec	29 sec.	32 min.
Noise calculation	5 sec	60 sec.	10 min.
Speedup	70x	1500x	-
Error in RMS	2.4%	5.4%	-
Element values	**4-bit counter**	**8x8-bit multiplier**	**Filter (Robo4)**
Rs	327Ω	10.8Ω	0.14
Cw	536fF	15.24pF	1.26nF
Cc	850fF	26.40pF	2.28nF

ACKNOWLEDGMENTS

The authors would like to acknowledge Vincent Gravot and Kris Tiri for their support in this work. This work was supported in part in the frame of the ESPRIT Project-BANDIT, funded by the European Commission.

REFERENCES

[1] T. Blalack and B.A. Wooley, "The effects of switching noise on an oversampling A/D converter," *ISSCC Digest of Tech. Papers*, pp.200-201, 1995.

[2] N. K. Verghese and D. J. Allstot, "Computer-aided design considerations for mixed-signal coupling in RF integrated circuits," *IEEE J. of Solid-State Circuits*, vol. 33, no. 3, pp. 314-323, March 1998.

[3] M. K. Mayes and S. W. Chin, "All verilog mixed-signal simulator with analog behavioral and noise models," *Tech. Digest of the Symposium on VLSI Circuits*, pp. 186-187, 1996.

[4] M. Nagata and A. Iawata, "Substrate noise simulation techniques for analog-digital mixed LSI design, " *IEICE Tr. Fundamentals*, Vol. E82-A, pp. 271-278, February 1999.

[5] E. Charbon, P. Miliozzi, L. P. Carloni, A. Ferrari, and A. Sangiovanni-Vincentelli, "Modeling digital substrate noise injection in mixed-signal IC's," *IEEE Tr. On Computer-Aided Design of Integrated Circuits, Vol. 18, No. 3, March 1999.*

[6] S. Mitra, R. A. Rutenbar, L. R. Carley, and D. J. Allstot, "A methodology for rapid estimation of substrate-coupled switching noise," *Proc. of IEEE Custom Integrated Circuits Conference*, pp. 7.4.1-7.4.4, 1995.

[7] M. van Heijningen, M. Badaroglu, S. Donnay, M. Engels, and I. Bolsens, "High-Level simulation of substrate noise generation including power supply noise coupling," *Proc. of Design Automation Conference*, pp.446-451, June 2000.

[8] M. van Heijningen, M. Badaroglu, S. Donnay, H. De Man, G. Gielen, M. Engels, and I. Bolsens, "Substrate Noise Generation in Complex Digital Systems: Efficient Modeling and Simulation Methodology and Experimental Verification," *ISSCC Digest of Technical Papers*, pp.342-343, 463, February 2001.

[9] M. van Heijningen, M. Badaroglu, S. Donnay, G. Gielen, and H. De Man "Substrate noise generation in complex digital systems: efficient modeling and simulation methodology and experimental verification," *IEEE J. of Solid-State Circuits*, vol. 37, pp. 1065-1072, August 2002.

[10] SubstrateStorm from Cadence: h*ttp://www.cadence.com/products/substrate_noise_analysis.html*

[11] M. Badaroglu, M. van Heijningen, V. Gravot, S. Donnay, H. De Man, G. Gielen, M. Engels, and I. Bolsens, "High-Level Simulation of Substrate Noise Generation from Large Digital Circuits with Multiple Supplies," *Proc. of Design, Automation and Test in Europe*, pp. 326-330, March 2001.

Chapter 7

MODELING THE IMPACT OF DIGITAL SUBSTRATE NOISE ON ANALOG INTEGRATED CIRCUITS

Yann Zinzius, Georges Gielen, Willy Sansen
K.U.Leuven, Dept. Elektrotechniek, ESAT-MICAS, Kasteelpark Arenberg 10, B-3001 Leuven-Heverlee, Belgium

Abstract: This chapter addresses the impact of digital substrate noise on analog circuits embedded in mixed-signal integrated systems. A high-level modeling methodology is presented that allows to simulate in acceptable CPU times the impact of a complex noise signal resulting from a large digital part on the performance of an embedded analog part in a large mixed-signal system. Measurements were performed on an embedded comparator, and show the important impact of the digital noise on this design. The measurement results were used to predict the impact on an embedded analog-to-digital converter.

1. INTRODUCTION

In the coming years the integration of full systems on chip (SoC) will continue to increase, driven by motivations of volume and cost reduction. Most of these systems are mixed-signal in nature, containing a core of digital circuitry and analog (possibly RF) circuits that interface the chip with the outside world. A significant increase of the performance (especially speed) and complexity of the digital circuitry on these integrated systems also means an increase of the amount of digital switching noise generated by this circuitry. This noise is coupled into the substrate, which is shared with the sensitive analog circuits. The supply and substrate connection network play a role here, since the inductances of the bondwires create ringing and this may even be a significant contributor to the substrate noise. At the same time, the performance and precision levels required from the analog circuits will also increase as dictated by today's applications such as emerging communication systems (e.g. WLAN). This goes together with an increase of the sensitivity or the susceptibility of the analog circuits to digital substrate noise. It is therefore important to be able to predict the impact of digital switching noise on the analog circuit performance at the design stage of the integrated system, before the chip is taped out for fabrication.

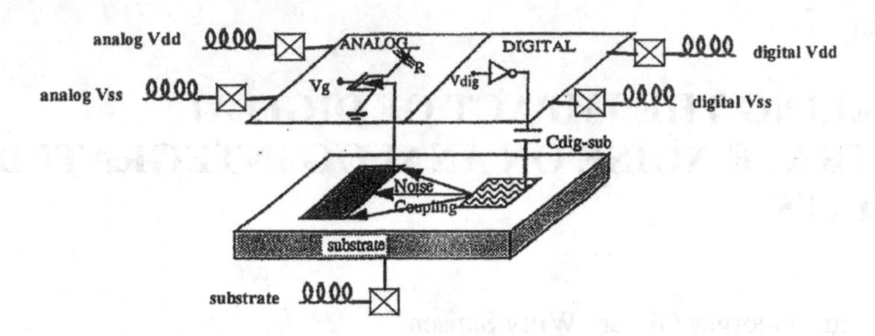

Figure 7-1. Problem of substrate noise coupling between digital and analog circuits.

There are three aspects to such a substrate noise analysis and simulation methodology for mixed-signal integrated systems (see *Figure 7-1*). One is the modeling of the digital switching noise injected in the substrate. Note that this depends on the activity level (the amount of switching) of the digital gates, and therefore depends on the signal patterns. As a result the injected noise is both non-stationary time-varying as well as frequency dependent. An approach to model the switching-dependent digital noise currents injected in the substrate by a complex digital system, based on an event-driven simulator in combination with pre-characterized digital cell library data, was presented in [1] (see also chapter 6). Experimental verification showed good agreements with measurements for technologies with an epi layer [2].

The second part of the analysis methodology is the analysis of the transmission of the noise from the source (the digital circuitry) to the reception point (the analog circuitry embedded in the same substrate) (see *Figure 7-1* as well). This requires a modeling of the substrate, which can be considered as a kind of resistive/capacitive mesh. For CMOS technologies with high-ohmic substrates the resistive nature of the substrate has to be fully taken into account, while for low-ohmic substrates the bulk can be considered as one equipotential node leaving only the epi layer as a resistive layer. *Figure 7-2* shows a model of an extracted network for such a technology with epi layer. In order to extract such a substrate model and to calculate the model parameter values starting from the layout of the chip, and hence to estimate the impact of noise injected at a certain digital injection point on an analog sensing point at another location on the layout, several substrate noise analysis tools were developed, such as

SubstrateStorm [3] or SeismIC [4] (see also chapter 3). Another tool that can extract substrate models is SPACE [5] (see also chapter 4). These tools can provide a good estimation of the impact of digital substrate noise but require a large amount of CPU time and memory, making the simulation of large complex systems with signal-dependent noise injection unfeasible.

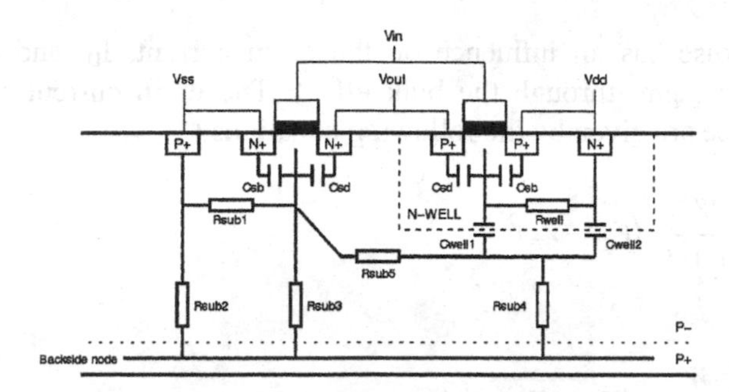

Figure 7-2. Cross-section model of substrate noise coupling between digital and analog circuits.

The third part of the analysis methodology is the modeling of the impact of substrate noise on the analog side. The analog circuitry is not a single noise reception point but has many noise sensing nodes that all have a different sensitivity to the noise. This analysis therefore becomes quite complex and time consuming for large analog circuitry such as entire frontends. Hence it is needed to introduce higher-level (behavioral or macro) modeling for the analog circuits in order to make this analysis tractable. In this chapter such a high-level analog modeling approach will be presented for low-ohmic technologies. We will also present measurement results of the impact of digital substrate noise on a regenerative comparator and extend the results towards the SNR reduction of an analog-to-digital converter embedded within a large digital system.

This chapter is organized as follows. In section 2 a brief overview of the impact of digital substrate noise on analog designs is given. In section 3 the high-level modeling methodology will be presented and illustrated for a regenerative comparator. Section 4 will present measurements results from a test chip that was fabricated. The results of the comparator are extended to a full analog-to-digital converter. Finally, in section 5 conclusions are drawn.

2. OVERVIEW OF SUBSTRATE NOISE IMPACT IN ANALOG CIRCUITS

First we will consider the analog circuits at the transistor level and therefore we analyze the impact of substrate noise on the transistor model. For our calculations we are using the MOS level-3 transistor model equations.

Substrate noise has an influence on the drain current, I_D, and on the transconductance, gm, through the bulk effect. The drain current and the transconductance are given by the following equations :

$$I_D = \frac{K_p \cdot W}{2 \cdot (1+\alpha) \cdot L} \cdot (V_{GS} - V_T)^2 \tag{1}$$

$$gm = \frac{K_p \cdot W}{(1+2 \cdot \alpha) \cdot L} \cdot (V_{GS} - V_T) \tag{2}$$

and $$V_T \approx V_{TO} + \gamma \cdot \left(\sqrt{\phi + V_{SB}} - \sqrt{\phi}\right) \tag{3}$$

where: γ is the body-effect coefficient, ϕ is the surface inversion potential, α is the mobility reduction factor, V_{TO} is the threshold voltage for $V_{SB} = 0V$.

A Taylor expansion of equation (3) shows that V_T to first order depends linearly on V_{SB} :

$$V_T = V_{TO} + \frac{1}{2} \cdot \frac{\gamma}{\sqrt{\phi}} \cdot V_{SB} \tag{4}$$

Since both the drain current and the transconductance depend on the threshold voltage, they are depending on the substrate voltage as well through V_{SB}. If the substrate voltage varies with time due to digital switching noise, instead of being a fixed dc bias voltage, then V_{SB} and V_T are also a function of time according to equation (4), and hence V_{GS}-V_T in equation (1) and (2) can be replaced by the following expression :

$$V_{GS} - V_T = V_{GSTO} - \Delta V_T(\mathrm{t}) \tag{5}$$

$$\text{with}: VGST0 = VGS - VT0 \tag{6}$$

$$\Delta V_T(t) = \gamma \cdot \left(\sqrt{\phi + V_{SB}(t)} - \sqrt{\phi}\right) \tag{7}$$

The variation of V_{SB} due to digital injected switching noise is calculated with a substrate model as shown in *Figure 7-2*, where the values of the model parameters (resistors and capacitors) are extracted from the circuit's layout by means of a tool like SubstrateStorm [3]. (Note that also bondwire inductances are added for external connections). From the previous equations the following conclusions can be drawn about the MOS transistor bulk node sensitivity :

- We can see in equation (1) that the drain current is a function of V_{GS} and V_T, which depends on V_{SB} according to equations (3) and (4). Except for transistors with a fixed gate voltage, e.g. bias transistors acting as current sources, V_{GS} depends on the input voltage of the design. This means that the drain current will be affected at the same time by a variation of the bulk voltage and by a variation of the input voltage.
- The substrate noise sensitivity of a single transistor is reduced for devices with a large V_{GST0}, which is typically the case for transistors used as a current source for example. A transistor with small V_{GST0} on the other hand will be more sensitive to substrate noise, which is typically the case for the input transistors of a circuit that are usually designed for high gm values.

Another coupling mechanism by which the digital switching noise is coupled into the analog signal path is the capacitive feedthrough of noise signals from the bulk node to the source and drain of the transistors via the junction capacitors. This effect however will only play at higher frequencies.

3. MODELING THE DIGITAL SUBSTRATE NOISE IMPACT ON ANALOG CIRCUITS

3.1 Principle of the modeling method

Consider a (possibly clocked) analog system S as shown in *Figure 7-3a*. Assume that this system, S, has D independent inputs, non-differential or differential, $VI_1 \ldots VI_D$. A possible clock signal is not considered as an input since it is a control signal. Since we are working on a low-ohmic substrate, the bulk can be considered as a single node, and in this case the system S has only one single substrate input Vn. We assume also that the system S has a single output Vo.

The system S can be divided into two separate signal paths, or subsystems, as shown in *Figure 7-3b*. The first signal path, shown as the subsystem T, is related to the normal input-output system behavior, and the second signal path, shown as the subsystem F, is related to the substrate noise impact behavior. These two signal paths are of course dependent since the state of the transistors in the system T influences the transfer function from the bulk connection to the output. And at the same time the absolute amplitude of the substrate noise V_n also influences the state of the transistors in the system T, and introduces nonlinearities in the system. Using a linear approximation for small substrate noise signals, we consider the total output signal Vo as a sum of the output signal of the two signal paths. The goal of the modeling approach presented in this chapter is now to model in an efficient way the signal path from the substrate node to the output voltage where the input signals of the system T will be considered as control voltages for this model F, since they determine the "operating point" of the circuit at the moment of the substrate noise.

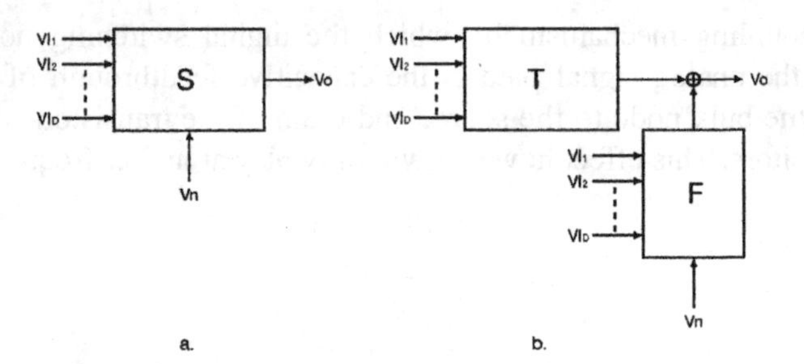

Figure 7-3. Analog system schematic from a high level point of view (a), with the split into two subsystems explicitly showing the impact of the bulk node (b).

From these considerations and assuming small substrate noise voltage excursions, a linear transfer function from the bulk connection to the output voltage can be derived. This transfer function depends on the values of the control input voltages (VI_i) and on the frequency of the substrate noise. To extract these transfer functions used in our model, we sample the voltages VI_i with a fixed sampling step, and for each of this combination of DC input voltage samples, we extract with SPICE AC analysis the corresponding linear transfer function from the substrate node to the output, using a substrate model as shown in *Figure 7-2*. The parameters of each transfer function (poles, zeros and gain) are then stored in a table. The complexity of the obtained table is a function of the number of independent inputs and of the number of sample values for each input. *Figure 7-4* schematically shows the resulting table that contains the different transfer function parameters, poles ($p_{i,j}$), zeros ($z_{i,j}$), and gain ($G_{i,j}$). The independent inputs are numbered from 1 to D, and are sampled with N values each.

$$\begin{bmatrix} (p_{1,1}, z_{1,1}, G_{1,1}) & \cdots & (p_{i,1}, z_{i,1}, G_{i,1}) & \cdots & (p_{N,1}, z_{N,1}, G_{N,1}) \\ \vdots & \ddots & \vdots & \ddots & \vdots \\ (p_{1,j}, z_{1,j}, G_{1,j}) & \cdots & (p_{i,j}, z_{i,j}, G_{i,j}) & \cdots & (p_{N,j}, z_{N,j}, G_{N,j}) \\ \vdots & \vdots & \vdots & \vdots & \vdots \\ (p_{1,D}, z_{1,D}, G_{1,D}) & \cdots & (p_{i,D}, z_{i,D}, G_{i,D}) & \cdots & (p_{N,D}, z_{N,D}, G_{N,D}) \end{bmatrix}$$

Figure 7-4. Table containing the parameters of the substrate transfer functions.

During the extraction of the model, we assume that the substrate voltage can be modeled as a sum of sinusoidal signals at different frequencies and with different amplitudes, spread out over a large range of frequencies, instead of a fixed DC bias voltage. We can then analyze the impact of both the frequency and the amplitude of the substrate noise. In the full verification of complex mixed-signal systems, on the other hand, the instantaneously fluctuating value of the substrate voltage has to be calculated from the signal-dependent switchings of the digital circuitry, using a methodology as for instance described in [1] (see also chapter 6). Hence the substrate noise signal is then not a nice theoretical sinusoid but a real switching signal.

3.2 Description of the model extraction methodology

In this section we will present the model extraction methodology that is based on the previous analysis. Note that the actual model in the end can be implemented in any description language (e.g. VHDL-AMS, or SPICE-like macromodel).

In order to extract the different substrate transfer functions parameters, we sample our input signals and apply the different DC values to the corresponding node VI_i and extract the transfer function from the substrate node to the circuit's output for each of these input combinations. The extracted transfer function parameters are then stored in a matrix and used as a model to replace full transistor-level simulations from then on. But in order to use this approach, we need to make some assumptions :

- The frequency of the input signal should be smaller than the frequency of the substrate noise signal.
- The substrate noise amplitude should be small enough to neglect the nonlinearity introduced in a real system by large V_{BS} voltages. So we make a linearization assumption here.

Figure 7-5 shows the design flow used to extract the noise impact model as presented in our methodology, and its integration in the standard analog design flow. Compared to a standard design flow, the different additional steps required to generate the model are :

- The extraction of the parameters of the transfer function that will be used in the model. These parameters are obtained by running several AC simulations of the original netlist plus the extracted substrate model, with an AC source as the noise source, for different DC values of the input voltages.
- Implement the transfer function parameters in the model, and integration of the generated model in the original simulation netlist. This step will depend of the modeling language used (SPICE-like, VHDL-AMS, etc…), since this approach is not model dependent.
- Simulation and validation of the extracted model by comparing to the results obtained with the original netlist including the extracted substrate model. This step is only needed if a characterization of the model is needed.

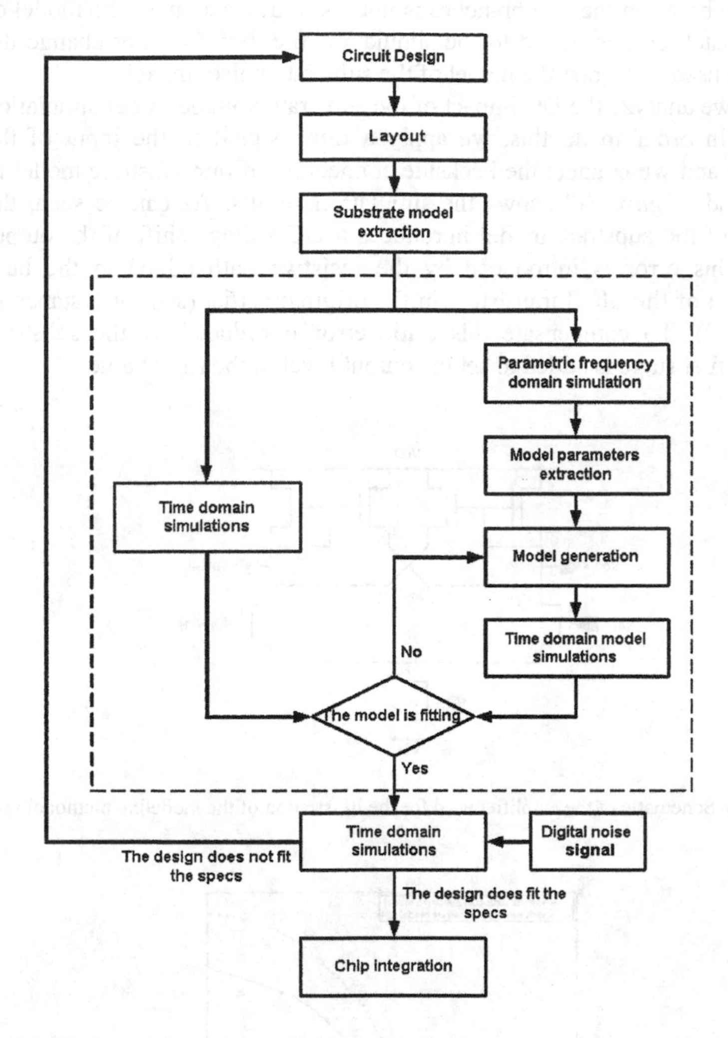

Figure 7-5. Diagram of the analog design flow that takes into account substrate noise impact modeling, with the substrate model extraction methodology shown inside the dashed box.

3.3 Illustration and validation of the modeling methodology

We will now present an example of the simulation results obtained using the developed methodology. As an example of the analysis, we will use the amplifier shown in *Figure 7-6* [6,7]. This amplifier contains a differential pair with cross-coupled active load. It is fully differential. In our work we will consider only one of the outputs, since they are symmetrical if the mismatch between the two branches is not taken into account. The model of the mismatch can be added to the simulation file, but does not change the approach used to extract the model of the substrate noise impact.

First we analyze the DC impact of the substrate voltage to our simulation outputs. In order to do this, we apply a ramp signal to the input of the amplifier and we connect the backside connection of our substrate model to the ground. *Figure 7-7* shows the simulation results. As can be seen, the addition of the substrate model introduces a DC voltage shift of the output signal. This error is introduced by the resistive path added to the bulk connection of the MOS transistors in the original netlist (see for instance in *Figure 7-2*). To compensate this static error introduced by the substrate model, a first stage is added to set the output level to the right value.

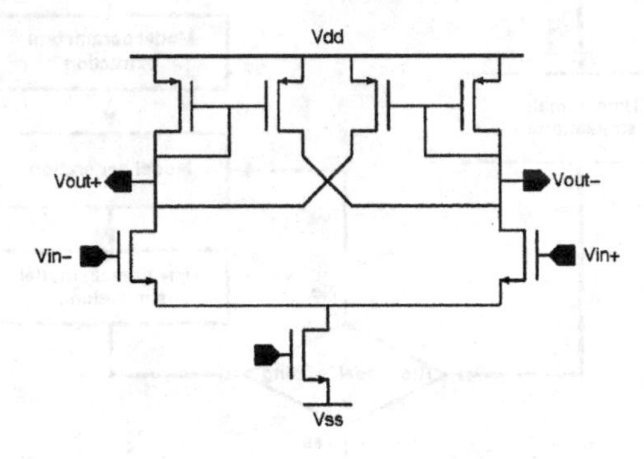

Figure 7-6. Schematic of the amplifier used for the illustration of the modeling methodology.

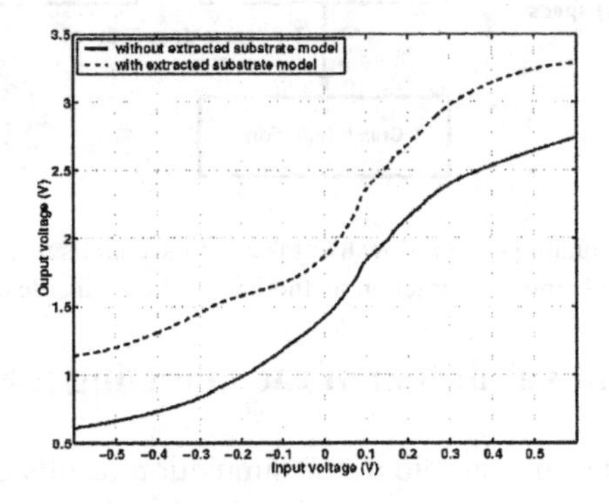

Figure 7-7. Output of the amplifier with and without substrate model versus the input voltage.

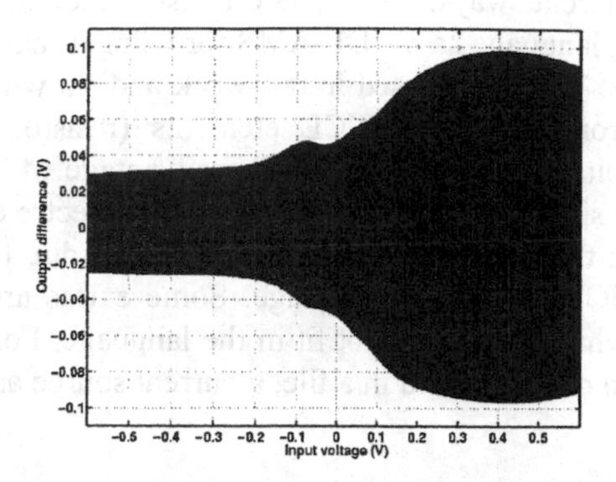

Figure 7-8. Plot of the difference between the amplifier output when a sinusoidal noise signal is applied to the backside node and when there is no noise signal applied, as a function of the DC input voltage in volt.

Let us now analyze the AC effect of the substrate node. Using the same ramp as input signal, we apply a sinusoidal voltage, with an amplitude of 10 mV, as the substrate noise source. Since the noise amplitude at the output is small compared to the output signal of the amplifier, we plot the difference (output difference (V)) between the output signal of an amplifier with the backside node connected to the substrate noise source, and an amplifier with no extracted substrate model added.

Figure 7-8 shows this simulation result for a noise signal with an amplitude of 10 mV and a frequency of 2 MHz. We can clearly see on the plot that the maximum amplitude of the output difference is smaller than the amplitude of the noise signal applied to the system, and that the amplitude is varying as a function of the input voltage DC values.

Now we analyze the same circuit in the frequency domain in order to see the impact of the substrate noise for a noise signal with a large spectrum. In order to perform this analysis, we replace the sinusoidal source by an AC source and we apply to the amplifier inputs a differential DC voltage, varying around the common-mode voltage value. In order to extract the data resulting from a variation of the input voltage over time, we sweep the differential DC input voltage. The resulting simulation plot is shown in *Figure 7-9*, for an input common-mode voltage of 1.9 V and a differential DC sweep between 1.6 V and 2.2 V. We clearly see on the plot of *Figure 7-9* that we have a different transfer function for each input voltage value. This explains the reason for our table-based modeling according to *Figure 7.4*.

The output data of these simulations are then used to generate the model of the substrate impact for this amplifier according to *Figure 7-4*: the filter order, the gain value for each set of DC input values, and the number of poles and zeros, and their frequency positions. The model can then be implemented in different ways: an equation-based model, or using a dedicated description language as VHDL-AMS, or using a macromodel in a simulator like SPICE. The model used in our work and shown in *Figure 7-10* is using a macromodel with SPICE elements (resistors, capacitors, voltage-controlled voltage sources, etc...). The advantage of this is that it can be used with any simulator (for example HSPICE, Spectre or Eldo). The main disadvantage is the precision that can be achieved due to the limited flexibility of the SPICE description language. Some extras are required in order to model the behaviors that do not fit in the language. For example, to model a gain function of data stored in a file, a current source associated to a resistance is used.

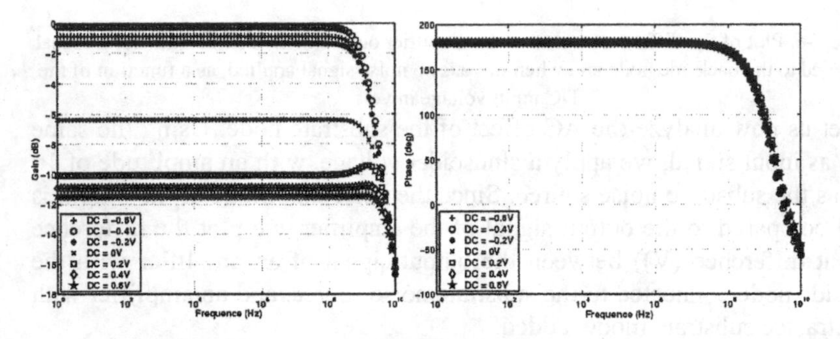

Figure 7-9. Plot of the transfer function (gain and phase) from the backside node to the output node, for a signal with 10 mV AC amplitude at the substrate node, and for different differential DC values as an input signal.

To create this model, an AC analysis is made for several different DC input voltages, over the input range and with a small DC step (a few mV). Since the design shown in *Figure 7-6* is a differential structure, we apply to the input a differential DC voltage, varying around the common-mode voltage. From this analysis the gain and the poles and zeros are extracted and stored in two different tables. As shown in *Figure 7-10*, the macromodel is using two different stages, one containing the information on the gain and the second on the poles and zeros. The table that contains the gain information is called in the Rgain statement, and the table that contains the cut-off frequency is called in the filter-stage definition statement, via Rfilti and Cfilti. The order of the filter is chosen to provide the best complexity to accuracy ratio. In our case a second-order filter was chosen, and this filter is made here by cascading two first-order filter sections.

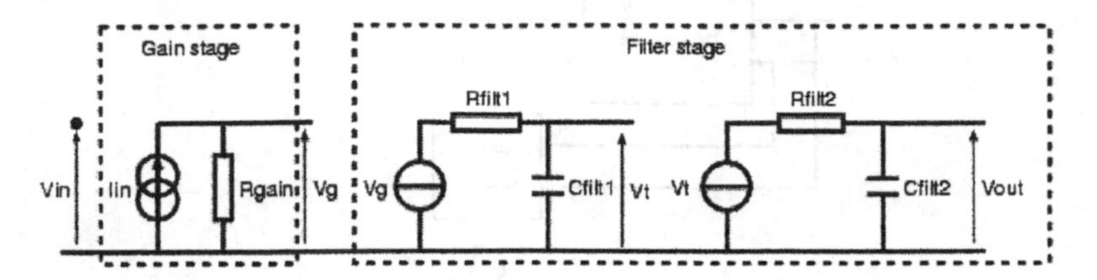

Figure 7-10. Schematic of the macromodel used in our simulations.

Figure 7-11 shows the plot of the gain in dB as a function of a differential voltage applied to the differential input signal. We can see that the variation of the gain is not linearly depending on the input voltage. This is the reason why the gain is described as a table. Since the design is fully symmetrical and the effect of the mismatch is not taken into account, the variation of the gain as a function of the input voltage is symmetrical, with 0 volt as the symmetry axis.

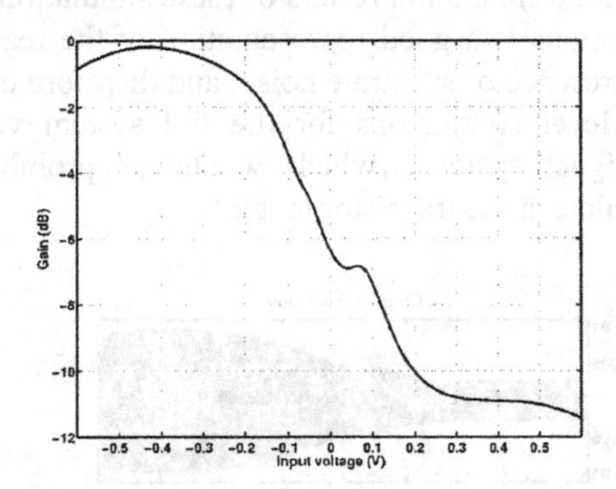

Figure 7-11: Plot of the gain in dB as a function of the input voltage.

The model shown in *Figure 7-10* can now be included in our original netlist and the entire extended circuit can be simulated as shown on *Figure 7-12*. In the figure the substrate transfer function is replaced in the simulations by the macromodel shown before (see *Figure 7-10*).

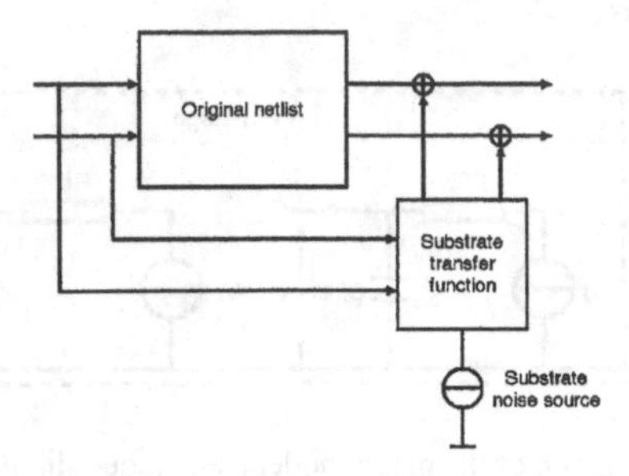

Figure 7-12. Substrate macromodel inserted in the original netlist.

The original transistor-level netlist including the extracted substrate model (see *Figure 7-2*), and the entire circuit including the substrate noise macromodel extracted using our methodology (see *Figure 7-12*), were first simulated using as input signal a ramp varying from –0.6V to 0.6V in 90 μs around the common-mode voltage, and a sinusoidal signal with a fixed frequency and a fixed amplitude as substrate noise voltage. *Figure 7-13* shows the simulation results of these simulations. The model used in our simulations is a good representation of the real behavior of the circuit in the presence of substrate noise, and therefore can be used to replace transistor-level simulations for the full system verification of complex mixed-signal systems, which would be prohibitively time-consuming to simulate at the transistor level.

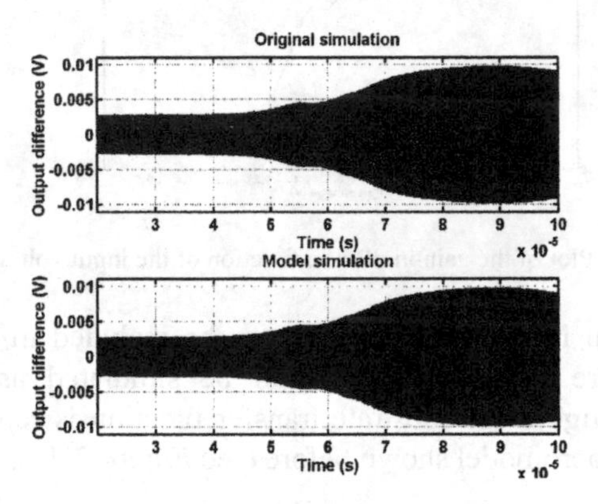

Figure 7-13. Simulation results of the original netlist and of the extracted macromodel for a ramp as signal at the input of the amplifier and for a sinusoidal signal as substrate.

In order to show how accurate the extracted model is versus the transistor-level simulation, we plot on *Figure 7-14* the difference, or the error, between the two signals. Note how small the error signal is.

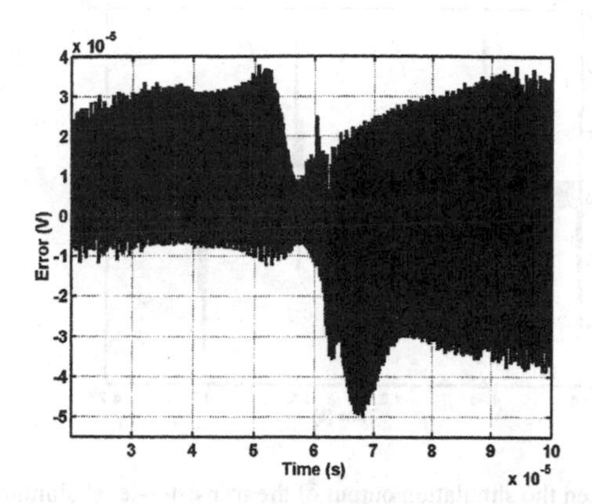

Figure 7-14. Plot of the difference between the simulation result obtains with our extracted macromodel, and the simulation result obtained from transistor-level simulations of the original netlist with the substrate model extracted with SubstrateStorm.

The same model was used for the simulation with a sinusoidal input signal at a frequency of 1 MHz, and with a sinusoidal noise signal at 30 MHz. The simulation results for the original netlist with the extracted substrate model and with the macromodel extracted using our methodology, as well as the error between both results are shown in *Figure 7-15* and *Figure 7-16*, respectively. The simulation results show a good correspondence. Only at the moment when the circuit is switching, some larger error spikes are visible in the plot, which are due to nonlinearities presently not yet taken into account in our model.

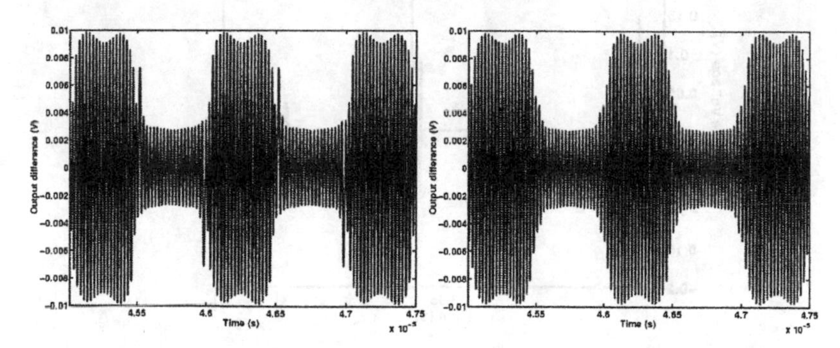

Figure 7-15. Simulation results of the original netlist and the extraction macromodel for a 1 MHz sinusoidal input signal with a substrate noise frequency of 30 MHz.

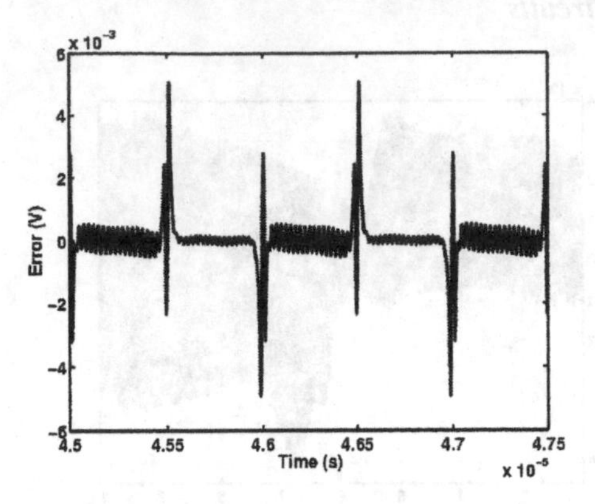

Figure 7-16. Error between the simulation output of the transistor-level simulation and of the macromodel simulation, for a 1 MHz sinusoidal input signal.

Finally, we present the simulation results of our model when not using an artificial sinusoidal substrate noise signal, but when using a more realistic, instantaneously time-varying substrate noise signal generated by real digital circuitry. As a noise generator we are using three inverter lines, with different input frequencies and delays, with the output connected to the substrate by a capacitance. These inverters create the noise injected into the substrate. *Figure 7-17* shows the resulting noise signal present at the backside substrate node (in an epi-type of technology process).

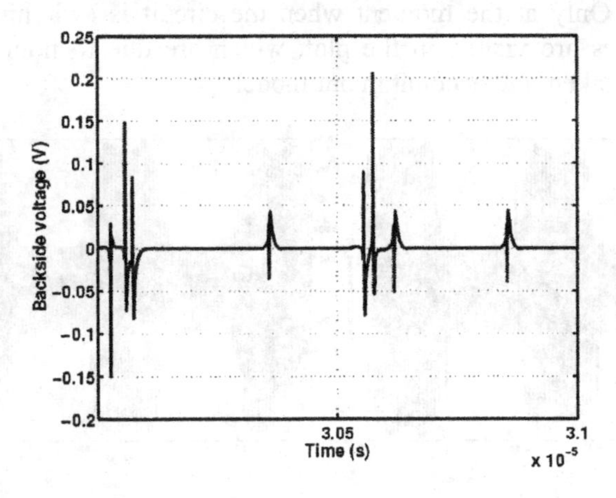

Figure 7-17. Digital noise signal injected at the backside node of the system.

Figure 7-18 shows the simulation results for the original transistor-level netlist and for the macromodel. We can see that the model simulations correspond well to the original transistor-level netlist simulation including the extracted substrate model (see *Figure 7-2*). Only at the switching moments, where we have high-frequency components, we see some peaks in the error plot. This error can be reduced if an extended model is used. Nevertheless, even with the current model the error stays below 20 % of the output signal.

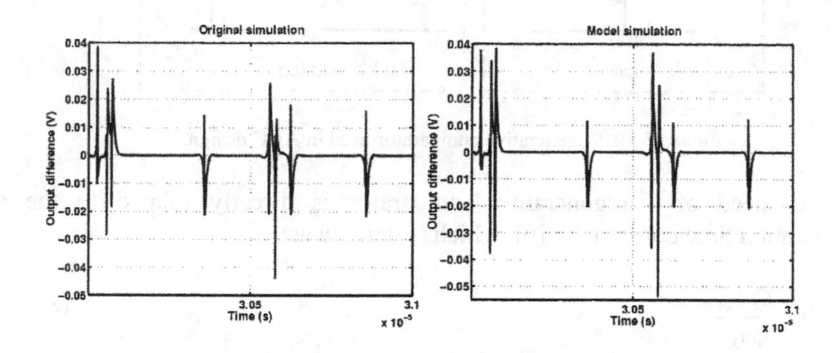

Figure 7-18: Simulation results of the original netlist and of the extracted macromodel, using a real digital noise signal as a source.

4. MEASUREMENTS OF THE IMPACT OF DIGITAL SUBSTRATE NOISE ON ANALOG DESIGNS

In this section we will present the measurements of the digital substrate noise impact on a regenerative comparator, and extend these results to the performance degradation of an analog-to-digital converter (ADC) that would be embedded in a large digital system.

4.1 Digital substrate noise impact on a comparator

A typical regenerative comparator architecture as shown in *Figure 7-19* is composed of two parts: a differential amplification stage, with a low gain

A, followed by a latch [8]. The latch is used to increase the speed when switching from one output state to another.

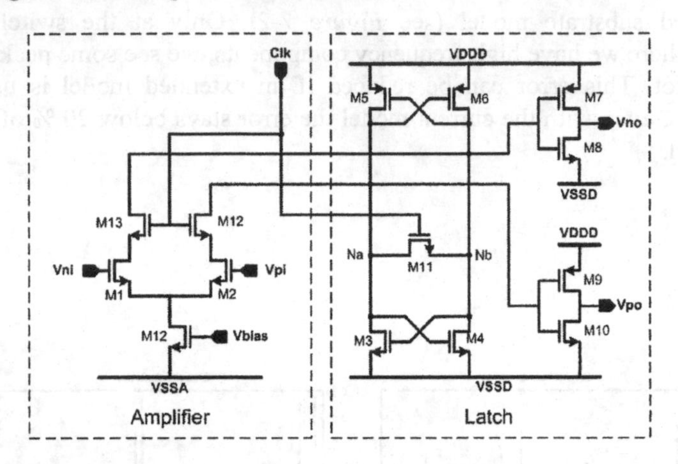

Figure 7-19. Regenerative comparator used in ADC design.

The speed of a regenerative comparator is directly related to the regeneration time constant Tr [9], which is defined as:

$$Tr = \frac{Cgs_{3,4}}{gm_{3,4}} \tag{8}$$

where $gm_{3,4}$ is the initial transconductance of M_3 or M_4, and Cgs the gate-source capacitance of each of these transistors. If we replace $gm_{3,4}$ by equation (2), we can see that Tr is inversely proportional to the source-bulk potential V_{SB}. But as was explained above, the substrate noise voltage is not a fixed value but is time dependent, which makes $gm_{3,4}$ time dependent. Hence the change of the regeneration time constant Tr of the comparator is not constant but depends on the time when you are measuring the output signal of the comparator. This variation can be expressed as an output signal jitter, as shown in *Figure 7-20*.

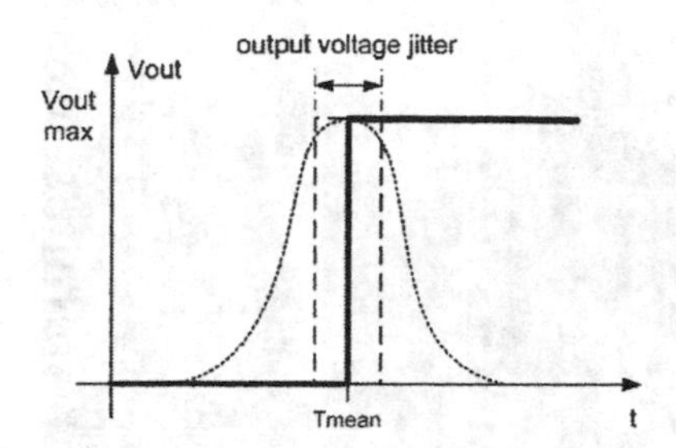

Figure 7-20. Output signal jitter of a comparator.

If this jitter has a Gaussian distribution, then the probability distribution of this jitter equals :

$$P(x) = \frac{1}{\sigma \cdot \sqrt{2 \cdot \pi}} e^{\left(-\frac{(x-\mu)^2}{2 \cdot \sigma^2}\right)} \quad (9)$$

Hence the impact of substrate noise on a comparator can be characterized by the mean value and the standard deviation of the extra jitter introduced due to substrate noise. We will now measure these characteristics on an experimental test chip.

4.2 Experimental test chip and measurement setup

A test chip experiment was set up as part of the BANDIT project [10] to carry out measurements of the impact of digital substrate noise on an analog design embedded in a large mixed-signal chip [11]. The test chip was designed and fabricated in a 0.35 μm standard CMOS process with epi-type substrate, with five metal and two poly layers. A microphotograph of the test chip is shown in *Figure 7-21*. Besides three digital circuits (three different implementation of a digital IQ demodulator, called REF, LN1 and LN2) (see also chapter 11) and a noise sensor, this test chip includes in the bottom right corner an analog comparator array used to measure the impact of substrate noise on embedded analog cells.

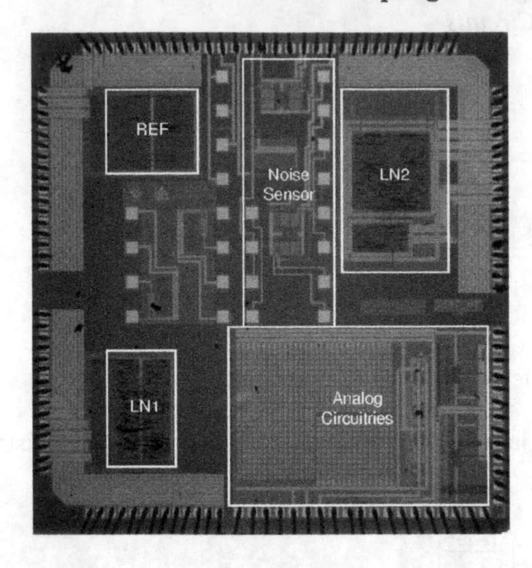

Figure 7-21. Test chip micrograph. The analog circuitries are in the bottom right corner.

The schematic of the comparator is shown in *Figure 7-19. Figure 7-22* shows how the array of comparators is implemented on the test chip. The input signals are common to all 15 comparators, and the output signal of each of them is connected to a digital multiplexing interface which is used to select (by means of six command signals C0 to C5) which signal will be brought to the output. The digital noise is generated using an on-chip digital IQ demodulator [11], and in our measurements the implementation labeled REF in *Figure 7-21* will be used as noise source (see also chapter 11).

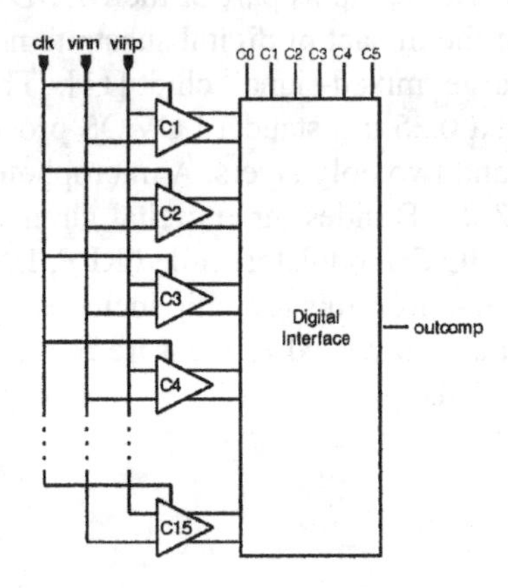

Figure 7-22. Comparator array implementation.

The presented test chip was used to measure the jitter of the comparator output due to the substrate noise. Since this output jitter is randomly varying around a mean value, it is measured using a statistical approach. In this case, the jitter is characterized by two parameters: the mean value, T_{mean}, and the standard deviation, $\sigma_{\Delta t}$. A sampling oscilloscope was used to measure these two parameters.

A noise sensor [12] (see also chapter 2) was also integrated on the test chip to directly measure the substrate noise generated by the digital noise source and hence to provide a reference measurement for our analysis. In our measurements, the input signal to the digital logic was fixed, but the measurements were repeated for different digital clock frequencies.

4.3 Comparator measurement results

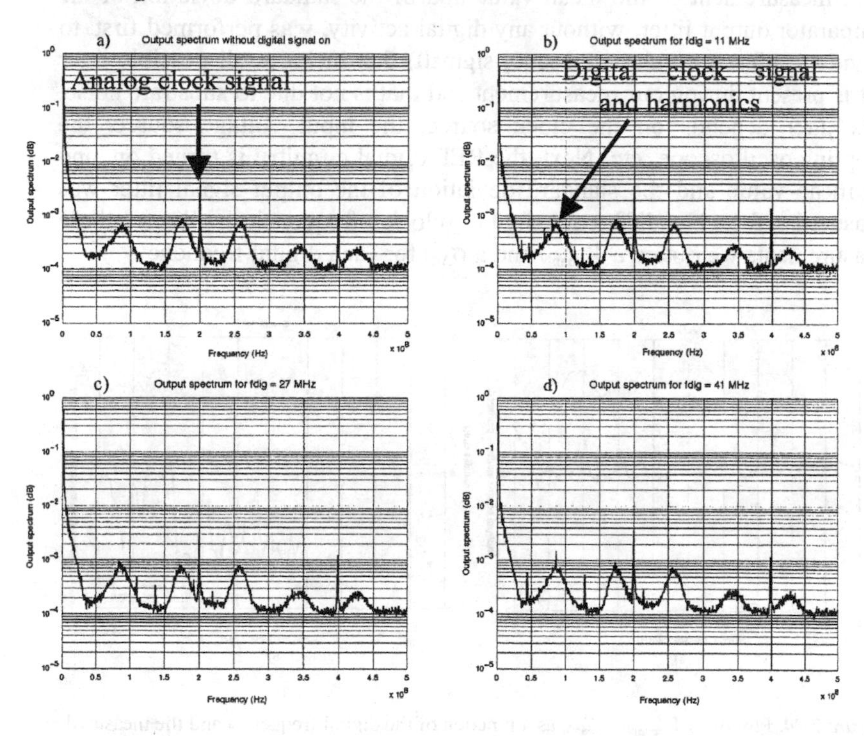

Figure 7-23. Spectrum of the output signal with the digital noise source turned off (a), and with the digital noise source turned on for different digital clock frequencies of fclock = 11 MHz (b), 27 MHz (c), and 41 MHz (d).

The output jitter of the comparator was measured for the presented test chip, with the comparator clock set to a period of 5 ns. An input signal with a frequency of 200 kHz was applied to the comparator input.

In order to see if the output signal of the comparator is affected by the substrate noise, a first measurement was done in the frequency domain. *Figure 7-23* shows the measured spectrum of the output signal when the IQ demodulator is turned off and the digital noise source is therefore not injecting noise in the substrate (*Figure 7-23a*), and when the digital circuit is turned on, with the digital clock set to different frequencies (fclock = 11 MHz, 27 MHz, 41 MHz for *Figure 7-23b, c* and *d* respectively). We can clearly identify the extra spurs on the latter three plots, due to the influence of the digital substrate noise, mainly at the digital clock frequency and its harmonics.

A measurement of the mean value and of the standard deviation of the comparator output jitter, without any digital activity, was performed first, to provide a reference value (Tmean0, sigma0) that includes all possible jitter that is present during our measurement and that is not due to substrate noise. This jitter depends on the clock source, the input voltage source, the sampling oscilloscope, etc. Next, the REF digital circuitry is turned on, and the mean value and the standard deviation of the output signal jitter was measured, for N different digital clock frequencies. From these measurements we obtain a T_{meanN} and a $\sigma_{\Delta tN}$ for each digital frequency.

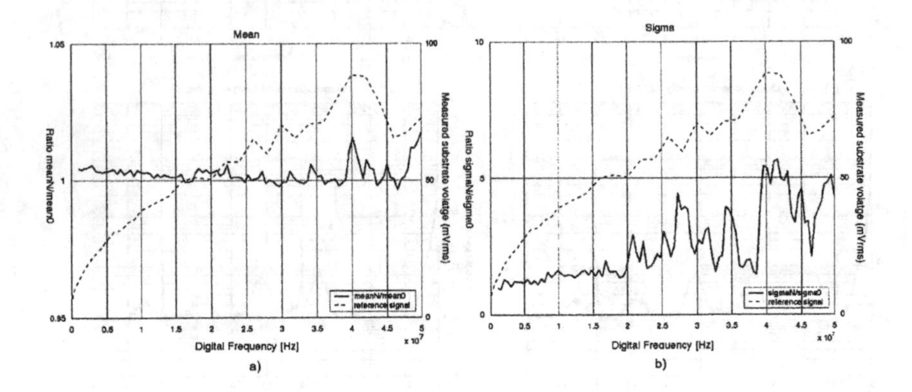

Figure 7-24. Plot of a) T_{meanN}/T_{mean0}, as a function of the digital frequency and the measured substrate noise voltage value in mV_{rms} (dashed line), b) $\sigma_{\Delta tN}/\sigma_{\Delta T0}$, as a function of the digital frequency and the measured substrate noise voltage value in mV_{rms} (dashed line).

In *Figure 7-24* the ratio between T_{meanN} respectively $\sigma_{\Delta tN}$, and the reference T_{mean0} and $\sigma_{\Delta t0}$, are plotted as a function of the digital clock frequency. Plotting the ratio between the measured value with digital activity and the reference value without digital activity shows the amount of increase due to substrate noise. On the same figures in dashed line the substrate voltage is shown as measured with the noise sensor [11].

We can see on *Figure 7-24a* that the mean value is only affected by the substrate noise for high values of substrate noise voltage, but stays very close to T_{mean0}. On the other hand, we can see on *Figure 7-24b* that the standard deviation in the presence of substrate noise is increasing up to more than 5 times the value of $\sigma_{\Delta T0}$ for the peak value of the substrate voltage. We can also observe that the standard deviation is increasing when the substrate noise voltage is increasing, i.e. the straight and the dashed lines more or less track each other in *Figure 7-24b.*

4.4 Impact of substrate noise on an embedded analog-to-digital converter

In the previous paragraphs we saw that the output of a comparator is affected by digital noise, which creates extra jitter uncertainty on the switching moment. In this section we will extend this to derive conclusions about the effect of the substrate noise on an analog-to-digital converter (ADC) that could have been embedded in this same mixed-signal system and that would use these comparators as subblocks, based on the knowledge from the above measurements on the comparator.

One of the main problems in the design of high-speed ADCs is the clock jitter [13, 14] It affects the ADC by changing the time when the input signal is sampled. The jitter on a comparator output due to substrate noise, like we measured above, is different in the sense that it does not modify the sampling time but affects the time characteristics of the comparator, for example the delay, and therefore the time characteristics of the ADC as well.

The impact of clock jitter on the SNR of a sampling system was already discussed in literature [15]. Wakimoto et al. in [16] derived formulas that describe the impact of timing jitter of a sampling system on the effective number of bits (ENOB) and the signal-to-noise ratio (SNR) of an ADC. If the input voltage is represented as $Vin = A \cdot \sin(2 \cdot \pi \cdot f_{in} \cdot t)$, then the rms value of the error due to the jitter and due to the quantization noise, respectively Err_{jrms} and Err_{qrms}, for an n-bit ADC are given by :

$$Err_{jrms} = (\sqrt{8} \cdot \pi \cdot f_{in} \cdot \sigma_T) \cdot S_{rms} \qquad (10)$$

$$Err_{qrms} = \frac{S_{rms}}{2^n \cdot \sqrt{6}/2} \tag{11}$$

where $S_{rms} = A/\sqrt{2}$ is the rms value of the input amplitude and σ_T is the standard deviation of the jitter.

From these expressions the expression for the signal-to-noise ratio (SNR), and the effective number of bits (ENOB or N_{eff}) are derived as follows :

$$SNR = 20 \cdot \log\left(\frac{S_{rms}}{\sqrt{Err_{jrms}^2 + Err_{qrms}^2}}\right) \tag{12}$$

$$N_{eff} = n - \log_2\left(\sqrt{1 + \left(\frac{Err_{jrms}}{Err_{qrms}}\right)^2}\right) \tag{13}$$

Using the numerical jitter results from our comparator measurements and using the above formulas of Wakimoto et al., we can extract the relative variation of the SNR and of the effective number of bits, $Delta_{SNR}$ and $Delta_{Neff}$, due to substrate noise, in the case if the measured comparator would be used in a full n-bit ADC converter embedded in the same mixed-signal system :

$$Delta_{SNR} = \frac{SNR_N}{SNR_0} \tag{14}$$

$$Delta_{Neff} = \frac{N_{effN}}{N_{eff0}} \tag{15}$$

For the 8-bit ADC presented in [7], for an ADC input frequency of 1 MHz and for a digital clock at 20 MHz, we calculate the variation on the SNR and on the effective number of bits as $Delta_{SNR} = 0.8037$ and $Delta_{Neff} = 0.7932$. We see that the SNR is dependent on the injected substrate noise and is reduced by 20%. Similarly, the effective number of bits is reduced by 20% as well, due to the extra distortion introduced by the extra jitter induced by the digital injected substrate noise.

5. CONCLUSIONS

A methodology has been described for the modeling of the impact of substrate noise on analog circuits embedded in large mixed-signal systems. As the analog circuitry is not a single noise reception point but has many noise sensing nodes that all have a different sensitivity to the noise, higher-level (behavioral or macro) modeling for the analog circuits is needed to make the analysis tractable. The accuracy of the derived model has been illustrated for different input signals, including real digital noise signals.

In addition, measurement results have been presented of the impact of digital substrate noise on a regenerative comparator. These results have been extended towards the reduction of the SNR and the ENOB of an analog-to-digital converter embedded within a large digital system.

Future work will concentrate on improving the models and further automating the model extraction methodology.

ACKNOWLEDGEMENTS

This work has been supported in part by the BANDIT project under the Mixed-Signal Initiative of the European Union.

REFERENCES

[1] M. Van Heijningen, M. Badaroglu, S. Donnay, M. Engels, and I. Bolsens, "High-Level Simulation of Substrate Noise Generation Including Power Supply Noise Coupling", *Proc. Design Automation Conference 2000 (DAC'02)*, pp. 446 – 451, 2000.

[2] M. Van Heijningen, M. Badaroglu, S. Donnay, H. De Man, G. Gielen, M. Engels, I. Bolsens, "Substrate Noise Generation in Complex Digital Systems, Efficient Modeling and Simulation Methodology and Experimental Verification", *J. of Solid State Circuits (JSSC)*, pp. 1065 – 1072 , August 2002.

[3] F. J. Clement, E. Zysman, M. Kayal, Prof. M. Declercq, "LAYIN: Toward a Global Solution for Parasitic Coupling Modeling and Visualization", *Proc. IEEE Custom Integrated Circuits Conference (CICC'94)*, pp. 24.4.1 – 24.4.4, 1994.

[4] Online: *http://www.cadmos.com/seismic.htm*

[5] A. J. Van Genderen, N. P. Der Meijs, "SPACE: A Finite Element Based Capacitance Extraction Program for Submicron Integrated Circuits", in *Software Tools for Process, Device and Circuits Modeling*, pp. 45 – 55, Bode Press, Dublin Ireland, July 1989.

[6] K. Bult, and A. Buchwald, "Embedded 240-mW 10-b 50MS/s CMOS ADC in 1 mm2", *IEEE Jour. Of Solid – State Circuits*, Vol. 32, No. 12, pp. 1887 – 1895 , Dec 1997.

[7] J. Vandenbussche, K. Uyttenhove, E. Lauwers, M. Steyaert, G. Gielen "A 8-bit 200MS/s Interpolating/Averaging CMOS A/D Converter", *Proc. IEEE Custom Integrated Circuits Conference (CICC'02)*, pp. 23.3.1 - 23.3.4, May 2002.

[8] B. Razavi, *Principle of Data Conversion System Design*, ed. I.E.E.E. Press, 1995.

[9] B. Razavi, and B. A. Wooley, "Design Techniques for High-Speed, High-Resolution Comparators", *IEEE Jour. Of Solid-State Circuits*, vol. 27 No 12, pp. 1916-1926, Dec. 1992.

[10] Online: *BANDIT Project home page : http://www.imec.be/bandit*

[11] M. Badaroglu, M. Van Heijningen, V. Gravot, J. Compiet, S. Donnay, M. Engels, G. Gielen, and H. De Man, "Methodology and Experimental Verification for Substrate Noise Reduction in CMOS Mixed-Signal ICs with Synchronous Digital Circuits", *Digest of Technical paper IEEE International Solid State Circuit Conference ISSCC 2002,* Session 16, San Francisco – USA, Feb. 3 - 7 2002.

[12] M. Van Heijningen, J. Compiet, P. Wambacq, S. Donnay, M. G. E. Engels, and I. Bolsens, "Analysis and Experimental Verification of Digital Substrate Noise Generation for Epi-Type Substrate", *J. of Solid State Circuits*, July 2000.

[13] D. A. Johns, K. Martin, *Analog Integrated Circuit Design*, ed. I.E.E.E. Press, 1995.

[14] R. Van De Plassche, *Integrated Analog-to-Digital and Digital-to-Analog Converters*, ed. Kluwer Academic Publishers, 1992.

[15] S. Saad Awad, "Analysis of Accumulated Timing-Jitter in the Time Domain", *IEEE Trans. On Instrumentation and Measurements*, Vol. 47, No. 1, pp 69 – 73, Feb. 1998.

[16] T. Wakimoto, Y. Akazawa, and S. Konaka, "Si Bipolar 2-GHz 6-bit Flash A/D Conversion LSI", *IEEE Jour. Of Solid-State Circuits*, vol. 23 No 6, pp. 1345-11350, Dec. 1988.

Chapter 8

MEASURING AND MODELING THE EFFECTS OF SUBSTRATE NOISE ON THE LNA FOR A CMOS GPS RECEIVER

Min Xu* and Bruce A. Wooley
Center for Integrated Systems, Stanford University, Stanford, CA 94305
** Min Xu now works at Big Bear Networks, Inc., Milpitas, CA, 95035*

Abstract: The influence of substrate noise coupling on the performance of a low-noise amplifier (LNA) for a CMOS GPS receiver has been investigated both analytically and experimentally. A frequency domain approach is proposed for modeling both the injection of noise into the substrate by digital circuitry and the mechanisms by which that noise can influence the behavior of sensitive analog circuits. It is shown that substrate noise can affect the analog circuits not only by direct coupling into the signal band but also by intermodulation with analog inputs. A filter bank model is proposed for predicting substrate spectra as a function of digital circuit characteristics. The measured effects of substrate noise on the LNA performance agree well with theoretical predictions.

1. INTRODUCTION

The continued scaling of CMOS technology, together with progress in the design of high frequency analog and mixed-signal CMOS circuits, has enabled the integration of many of the functions needed to implement a broadband communications transceiver [1], with a number of recent papers reporting increasing levels of transceiver integration [2]-[4]. One of the significant challenges in the design of such circuits is the need to implement broadband analog circuits on the same die as the large complex digital signal processing functions that are required for many modern communications applications. Owing to various parasitic coupling mechanisms there is a distinct possibility that the transients in the digital circuitry of such systems will corrupt low-level analog signals and seriously compromise the achievable performance. This may ultimately preclude the practical integration of a complete broadband communication system on a single CMOS chip [5].

Previously reported studies of substrate noise [6]-[14] can be grouped into several categories: 1) modeling of the equivalent substrate impedance using meshes or Green's functions, 2) experimental circuits for measuring substrate noise, 3) the time domain response to the coupling of a single digital transition into the substrate, 4) simplified models for simulating the substrate noise induced by a specific digital circuit, and 5) the experimental measurement of the influence of substrate noise on the performance of some communications ICs. This chapter investigates how the characteristics of substrate noise are affected by some specific characteristics of the digital circuitry and how that noise might affect the frequency-domain behavior of analog circuits. The vehicle used for the investigation is the low-noise amplifier (LNA) for a CMOS GPS receiver [15].

The basic mechanism of noise coupling via a common substrate is illustrated in *Figure 8-1*, wherein the coupling is modeled in three steps. The function F_1 describes the injection of noise into the substrate as a consequence of transients in the digital circuitry. The propagation of this injected noise into regions of the substrate containing analog circuits is represented by the function P. Finally, the coupling of the substrate noise into various nodes of the analog circuitry so as to influence its performance is represented by the function F_2.

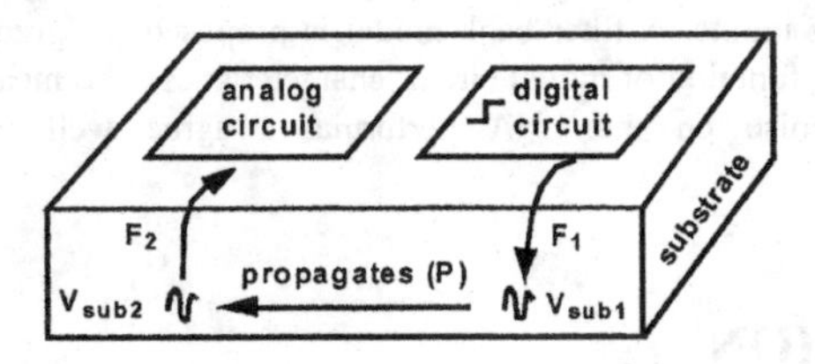

Figure 8-1. Overall picture of the substrate noise effects in the mixed-signal circuit.

In general, the substrate noise propagation function, P, is quite complicated and depends strongly on the substrate doping profiles [6], [7]. However, because the resistivity of the heavily doped p^+ bulk in a typical epitaxial substrate technology is very low, such substrates can generally be modeled as a single node when the distance between where the noise is injected and where it is sensed is a few times larger than the epitaxial layer thickness [8]. Thus, the substrate noise is approximately uniform throughout a heavily doped substrate and $P = 1$. This work focuses on the analysis and experimental characterization of the functions F_1 and F_2 for mixed-signal circuits integrated in an epitaxial substrate. It is generally impractical to formulate closed form expressions for these functions, except in the case of very simple circuits.

The chapter is organized as follows. Section 2 presents a general model of the effects of substrate noise on analog circuits, and reveals the fundamental coupling mechanisms using a frequency domain approach. Section 3 describes the experimental setup for this study and presents measurements of the substrate noise induced by an experimental digital circuit emulator for different operating conditions, as well as an analysis of both the time-domain and frequency-domain characteristics of this noise. Section 4 presents measured effects of substrate noise on the performance of an LNA and considers the mechanisms by which these effects occur, using the models developed in Sections 2 and 3. Finally, Section 5 outlines a statistical approach to a generalized modeling of the substrate noise generated in digital circuits.

2. GENERAL MODEL OF THE EFFECT OF SUBSTRATE NOISE ON ANALOG CIRCUITS

Without loss of generality, an analog circuit can be modeled as a system whose output Y is a function, F, of its input and bias variables. For simplicity in the following analysis, we consider there to be only one input and one bias variable. If there are multiple inputs or bias variables, then the variables in the following equations simply become vectors and the scalar products become matrix products; the physical interpretations of the results remain the same.

The influence of noise on an analog circuit can be modeled as perturbations of its input and its bias. The input is represented as the sum of its dc component, X, the ac input x_0 and the input perturbation δ_x caused by the substrate noise, while the bias is represented as the sum of a dc component B and a perturbation δ_b caused by the substrate noise.

$$Y = F(X + x_0 + \delta_x, B + \delta_b) \tag{1}$$

The function F in (1) can be expanded into a Taylor's series of the second order as:

$$Y = F(X,B) + \frac{\partial F}{\partial x}\Big|_{x=X} x_0 + \frac{\partial F}{\partial x}\Big|_{x=X} \delta_x + \frac{\partial F}{\partial b}\Big|_{b=B} \delta_b + \frac{\partial^2 F}{\partial x^2}\Big|_{x=X} \frac{x_0^2}{2}$$
$$+ \frac{\partial^2 F}{\partial x^2}\Big|_{x=X} x_0 \delta_x + \frac{\partial^2 F}{\partial x \partial b}\Big|_{\substack{x=X \\ b=B}} x_0 \delta_b + \frac{\partial^2 F}{\partial x^2}\Big|_{x=X} \frac{\delta_x^2}{2} + \frac{\partial^2 F}{\partial b^2}\Big|_{b=B} \frac{\delta_b^2}{2}$$
$$+ \frac{\partial^2 F}{\partial x \partial b}\Big|_{\substack{x=X \\ b=B}} \delta_x \delta_b \tag{2}$$

More generally, if the nonlinearity of F is frequency dependent, Y can be expressed as the sum of its Volterra series [16], but since the Taylor's series shows the same basic features as the more complex Volterra series approach, the Taylor expansion is used herein to provide a general description of the noise coupling mechanisms. In (2), the first term *F(X, B)* is the dc output of the circuit. The second term is the ac signal output, where the derivative is the gain in a small-signal circuit analysis. The fifth term is the second harmonic of the signal, while the remaining terms represent the noise resulting from substrate coupling. For simplicity, only the first and second order terms of the expansion have been retained in (2). While the analysis can be extended to higher order terms, as the order increases the amplitudes of the terms diminish exponentially.

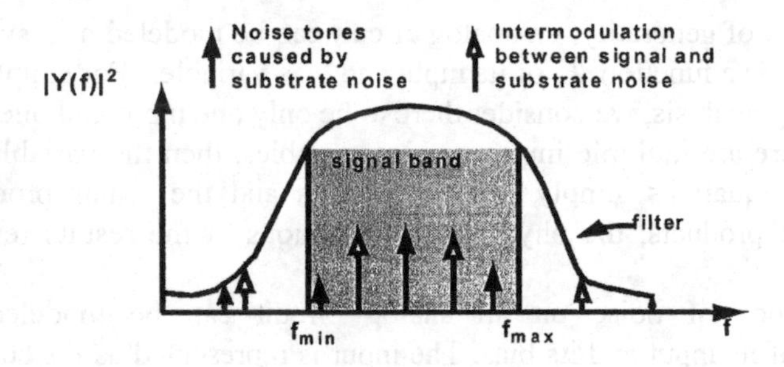

Figure 8-2. Illustration of substrate noise effects in analog circuits.

If f_s is the signal frequency and the substrate noise is concentrated in a single frequency f_n, then the terms in (2) have the frequency components listed in

Table 8-1. In a band-limited mixed-signal system, the effects of the noise terms in (2) depend on their frequency locations. For example, if the noise is located outside the signal band of the analog circuit, it can potentially be filtered out with a subsequent filter. However, if the noise falls into the signal band, as illustrated in *Figure 8-2*, it cannot be removed and will degrade the system's dynamic range.

Table 8-1. Spectral distribution of the terms in equation (2)

Order	Term Number	Frequency Components	Property
0	1st term	0	DC output
1	2nd term	f_s	ac signal
	3rd, 4th term	f_n	1st order noise
2	5th term	0, $2f_s$	2nd order harmonic of the ac signal
	6th, 7th term	$\|f_s-f_n\|, f_s+f_n$	2nd order IM between the signal and the noise
	8th, 9th, 10th term	0, $2f_n$	2nd order harmonics of the noise

If $[f_{smin}, f_{smax}]$ is the signal band in the analog circuits and if the substrate noise occurs in the frequency ranges $[f_{smin}, f_{smax}]$ and $[f_{smin}/2, f_{smax}/2]$, then the frequency components of the substrate noise, as well as their second harmonics, will fall into the signal band. For this reason, the term Direct Coupling Band (DCB) is defined as

$$DCB = [f_{s\min}/2, f_{s\max}/2] \cup [f_{s\min}, f_{s\max}] \quad (3)$$

If instead the substrate noise appears in the ranges $[0, f_{smax}-f_{smin}]$ and $[2f_{smin}, 2f_{smax}]$, then its intermodulation with the signal will fall into the signal band at the output. Thus, the term InterModulation Band (IMB) is defined as

$$IMB = [0, f_{s\max} - f_{s\min}] \cup [2f_{s\min}, 2f_{s\max}] \quad (4)$$

Clearly, the spectral distribution of substrate noise significantly influences its impact on bandlimited analog circuits. Therefore, in this work both theoretical and experimental studies were carried out in the frequency domain.

3. SUBSTRATE NOISE CHARACTERIZATION

A test chip containing the front-end circuits for a CMOS GPS receiver, a digital circuit emulator, and a substrate noise sensor was fabricated in a 0.5-μm, single-poly, triple-metal epitaxial CMOS technology. The analog circuits include the LNA, mixer, IF amplifier, IF filter and PLL for the receiver, all of which are designed to operate from a 2.5-V supply. *Figure 8-3* is a photomicrograph of the test chip. This study focuses on the effects of substrate coupling on the performance of the LNA, a schematic of which is shown in *Figure 8-4* [15]. The test chip was packaged in a 52-pin J-lead chip carrier, and surface-mounted on a two-layer printed circuit board. Special care was taken to minimize package and board-level noise coupling. For example, analog bondwires are positioned perpendicular to the digital bond wires, analog and digital components are physically separated on the board, and separate power supplies are used for various functional blocks on the test chip.

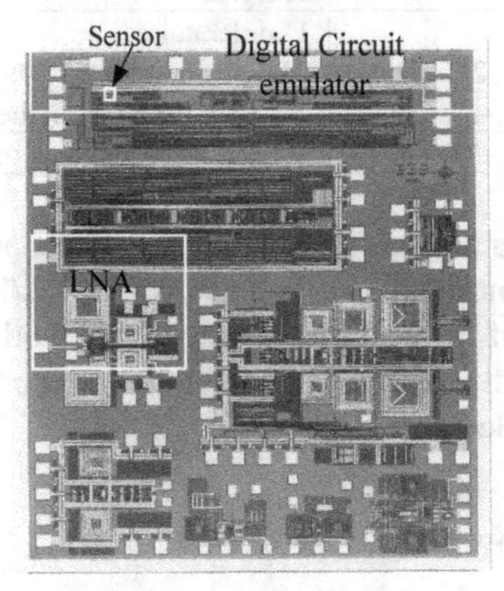

Figure 8-3. Chip micrograph

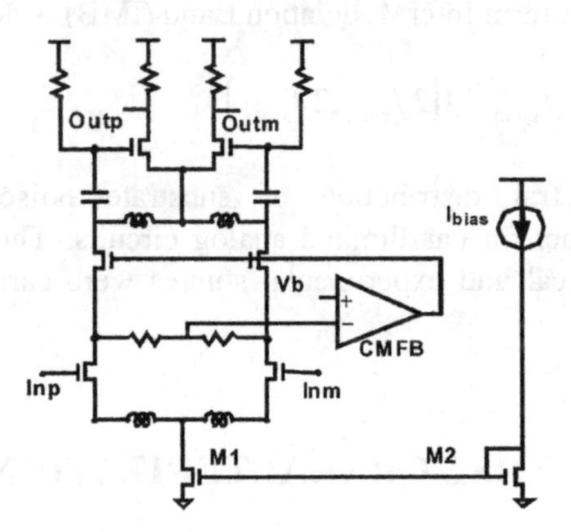

Figure 8-4. Simplified schematic of the fully-differential LNA for a GPS receiver [15].

A schematic of the digital circuit emulator is shown in *Figure 8-5*. It comprises nine tri-state buffers driving capacitive loads that are formed by integer multiples of a unit junction capacitor, C_0 = 0.34 pF. The tri-state buffers are controlled by a shift register so that the total coupling capacitance (C_{couple}) to the substrate can be varied from C_0 to $183C_0$ in increments of C_0. With one exception, the buffers are sized so that the nominal rise and fall times of their outputs, $t_{rise/fall}$, are 0.9 ns. The buffer driving the load capacitance of $56C_0$ actually consists of 3 tri-state buffers in parallel with binary-weighted sizing. This allows the $t_{rise/fall}$ of this buffer to be varied among 0.9, 1.0, 1.2, 1.4, 1.8, 2.7, 5.4 ns. The digital clock driving the inputs of the tri-state buffers can be derived on-chip from a 7.1-MHz ring oscillator or supplied from off chip as a small-swing signal with variable frequency (f_{clock}) and duty cycle.

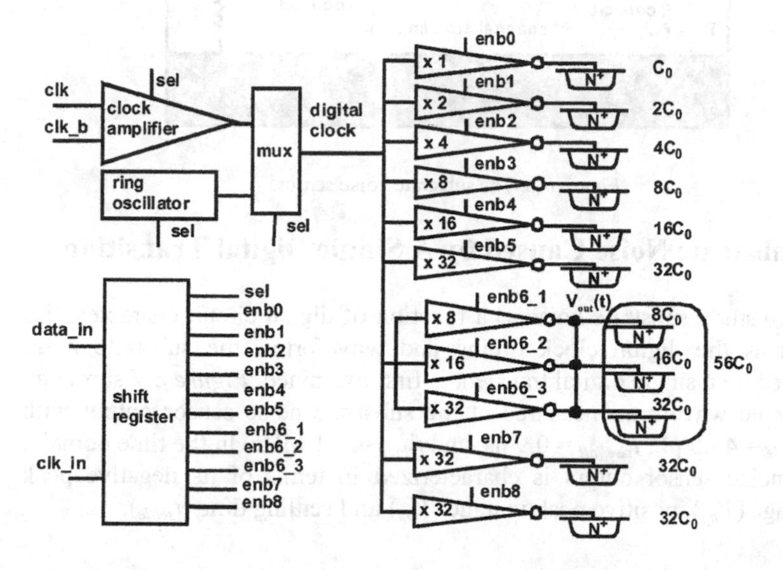

Figure 8-5. Block diagram of the digital circuit emulator.

The substrate noise sensor shown in *Figure 8-6* consists of a single NMOS transistor that senses the substrate noise through the body effect and capacitive coupling to its gate, drain, and source nodes. The source and the gate terminals of the transistor are biased with dedicated power supplies, and the drain is connected through an off-chip 1-kΩ resistor to the 2.5-V supply. The drain is ac coupled to the 50-Ω input of an oscilloscope or spectrum analyzer. To reduce the capacitive coupling to the ac output, the bonding pad connected to the drain is ground shielded. Since the capacitance between the ground-shield pad and the output pad is very small, the substrate noise coupled to the output pad through the ground shield pad is negligible. To reduce the magnetic coupling between the digital bond wires and sensor bond wires, the bond wires for the substrate noise sensor are positioned perpendicular to those for the digital circuits. Simulations indicate that the voltage gain from the "substrate node" to the drain of the sensor transistor is –29 dB.

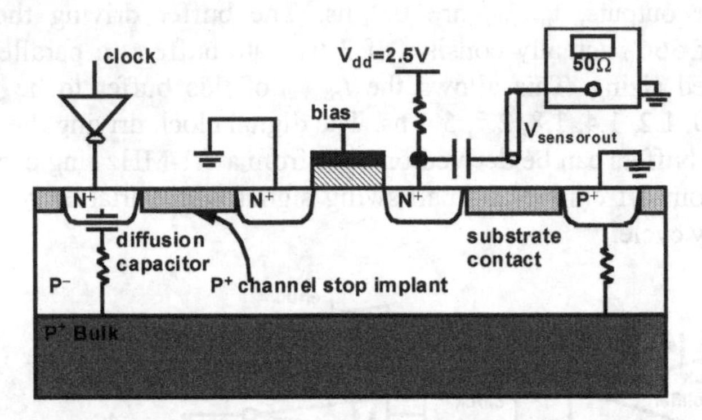

Figure 8-6. The substrate noise sensor.

3.1 Substrate Noise Caused by A Single Digital Transition

To study substrate noise as a function of digital circuit characteristics such as the digital clock timing and waveforms, the substrate noise caused by a single digital transient is first examined. *Figure 8-7* shows an example waveform measured at the substrate noise sensor output with C_{couple} = 43.5 pF, $t_{rise/fall}$ = 0.9 ns, and f_{clock} = 7.1 MHz. In the time domain, the noise sensor output is characterized in terms of its negative peak voltage (V_{np}), positive peak voltage (V_{pp}) and settling time (t_{settle}).

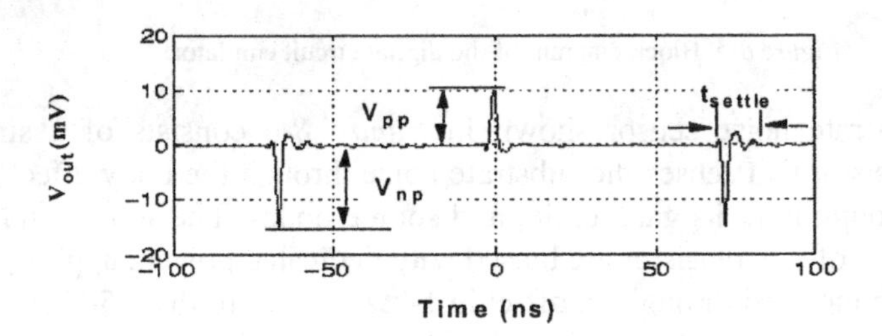

Figure 8-7. Measured substrate noise sensor output waveform when C_{couple} = 43.52 pF, f_{clock} = 7.1 MHz, $t_{rise/fall}$ = 0.9 nsec.

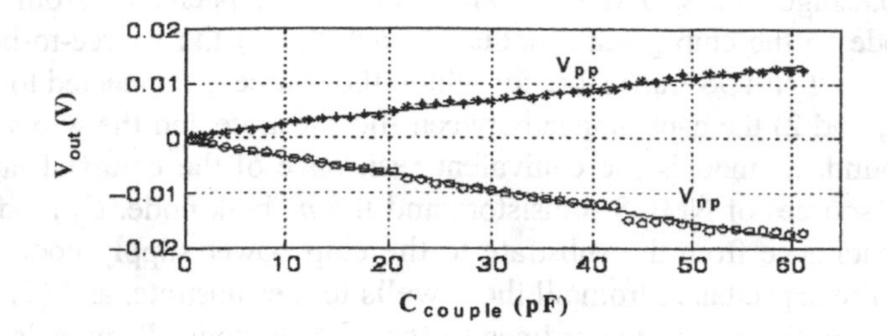

Figure 8-8. Measured V_{np} and V_{pp} as a function of C_{couple}

For the same $t_{rise/fall}$ = 0.9 ns, but different values of C_{couple} the substrate noise sensor output waveforms are all similar to *Figure 8-7*. However, $|V_{pp}|$ and $|V_{np}|$ increase linearly with C_{couple} as shown in *Figure 8-8*, while t_{settle} does not change. These experimental results suggest that the substrate noise is a linear function of the coupling capacitance if $t_{rise/fall}$ is constant.

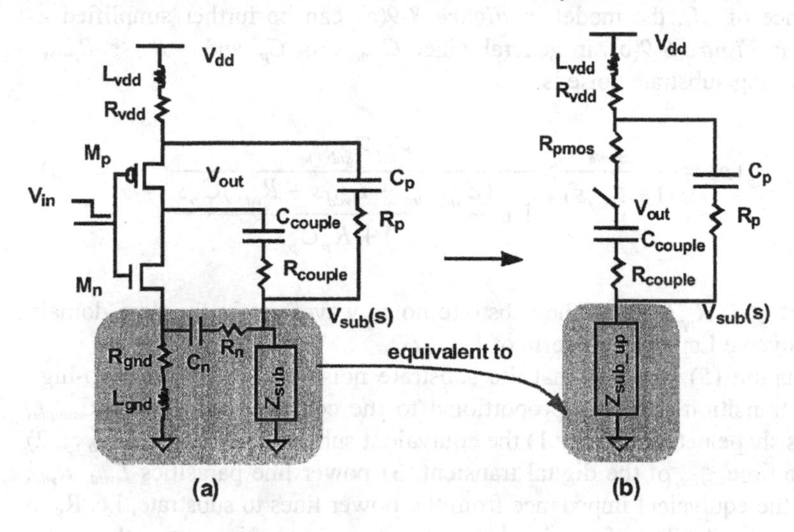

Figure 8-9. Simplified model for the substrate noise injection circuit with single node representation of the heavily doped substrate.

Figure 8-9 (a) is a simplified circuit model for examining substrate noise injection in an epitaxial technology, with the heavily doped p^+ bulk represented by a single node. M_p and M_n model the buffer driving the diffusion capacitor, C_{couple}, which represents the capacitance that couples noise into the substrate. R_{couple} models the epitaxial layer resistance between C_{couple} and the p^+ bulk, L_{vdd} and L_{gnd} model the parasitic inductance of bond wires and package traces, and C_n models the total capacitance from the substrate node to the chip ground node. C_n includes 1) the source-to-bulk capacitance of all NMOS transistors for which the source is connected to the chip ground, and 2) the capacitance between the substrate and the wires for the chip ground. R_n models the equivalent resistance of the epitaxial layer between the sources of NMOS transistors and the p^+ bulk node. C_p models the total capacitance from the substrate to the chip power supply node. C_p includes (1) the capacitance from all the n-wells to the substrate, and (2) the wire capacitance from chip power lines to the p^+ bulk node. R_p models the equivalent epitaxial resistance and the well resistance between the chip power supply node and the p^+ bulk node. Generally, in a digital circuit, C_n and C_p are much bigger than each individual C_{couple}. Z_{sub} represents the lumped substrate contact impedance from the p^+ bulk node to the board ground.

Without loss of generality, suppose at $t = 0$ that V_{in} transitions from high to low, so that M_p is on and M_n is off. Using R_{pmos} to represent the equivalent resistance of M_p, the model in *Figure 8-9*(a) can be further simplified as shown in *Figure 8-9*(b). In general, since $C_{couple} << C_p$ and $R_p << R_{pmos}$, the resulting substrate noise is

$$V_{sub\_up}(s) = \frac{C_{couple}}{(1+\tau_{rise}s)} \times \frac{V_{dd}Z_{sub\_up}}{1+\frac{(Z_{sub\_up}+L_{vdd}s+R_{vdd})C_p s}{1+R_p C_p s}} \tag{5}$$

where $\tau_{rise} = R_{up}C_{couple}$. The substrate noise waveform in the time domain is the inverse Laplace transform of $V_{sub\text{-}up}(s)$.

Equation (5) indicates that the substrate noise caused by a low-to-high digital transition is indeed proportional to the coupling capacitance C_{couple}, with its shape determined by 1) the equivalent substrate impedance Z_{sub_up}, 2) the rise time τ_{rise} of the digital transient, 3) power line parasitics L_{vdd}, R_{vdd}, and 4) the equivalent impedance from the power lines to substrate, i.e. R_p in series with C_p. Therefore, in the s domain, for a given τ_{rise} the noise waveform shape can be characterized using a substrate noise root function ($V_{root\text{-}up}$), which can be viewed as the substrate noise injected from a unit capacitance.

$$V_{root\_up}(s) = \frac{1}{(1+\tau_{rise}s)} \times \frac{V_{dd}Z_{sub\_up}}{1+\frac{(Z_{sub\_up}+L_{vdd}s+R_{vdd})C_p s}{1+R_p C_p s}} \tag{6}$$

A similar root function can be defined for the high-to-low transition of V_{out} with a fall time $\tau_{fall} = R_{down} C_{couple}$.

$$V_{root\_down}(s) = \frac{1}{(1+\tau_{fall}s)} \times \frac{-V_{dd} Z_{sub\_down}}{1+\dfrac{(Z_{sub\_down} + L_{gnd}s + R_{gnd})C_n s}{1+R_n C_n s}} \tag{7}$$

In the time domain substrate noise root functions $u_{up}(t)$ and $u_{down}(t)$ are the inverse Laplace transforms of $V_{root\text{-}up}(s)$ and $V_{root\text{-}down}(s)$, respectively.

The substrate noise root functions reveal the elements that govern the substrate noise characteristics, namely 1) Z_{sub}, the impedance from the substrate contact to the board ground; 2) C_n and C_p, the parasitic capacitances from the substrate to the on-chip digital power and ground, which include the junction capacitors as well as the wiring capacitances from the power lines to the substrate; 3) R_n and R_p, the epitaxial layer resistances under C_n and C_p; 4) L_{vdd} and L_{gnd}, the parasitic inductance from the bondwires and package traces for the power and ground connections; 5) τ_{rise} and τ_{fall}, the transition times of the digital switching.

Due to their large size, C_n and C_p play two contradictory yet critical roles in determining the substrate noise characteristics: 1) coupling the digital power line noise into the substrate, thus introducing substrate noise, and 2) decoupling the noise injected into the substrate by the diffusion capacitances, thus reducing the substrate noise. It is also apparent that the benefits from reducing the bondwire inductance of the power lines are twofold: 1) reducing power supply bounce, and 2) decreasing the equivalent substrate impedance, thus reducing the substrate noise.

3.2 Substrate Noise Spectra Distribution for The Digital Circuit Emulator

With knowledge of the substrate noise caused by a single digital transition, the total substrate noise can be calculated as the sum of noise components resulting from each of the digital transitions. If the digital switching pattern is periodic with a period of T, the corresponding substrate noise spectrum can be derived with the Fourier transform of the time domain noise.

For the convenience of further analysis, some symbols and functions are defined as follows:

$v_{sub}(t)$	total substrate noise at time t.
$S(f)$	power spectral density function of total substrate noise $v_{sub}(t)$.
$U_{up}(f)$	equals $V_{root\text{-}up}(2\pi jf)$.
$U_{down}(f)$	equals $V_{root\text{-}down}(2\pi jf)$.
$III(t)$	shah function [18].

Since $u_{up/down}(t) = 0$ when $t < 0$, the total substrate noise induced by the digital circuit emulator is

$$v_{sub}(t) = \sum_{n=-\infty}^{[t/T]} u_{up}(t - nT)C + \sum_{n=-\infty}^{[t/T-1/2]} u_{down}(t - nT - \frac{T}{2})C$$
$$= \frac{C}{T} \times [u_{up}(t) \otimes III(\frac{t}{T}) + u_{down}(t) \otimes III(\frac{t}{T} - \frac{1}{2})] \tag{8}$$

where C is the total coupling capacitance at each transition, and $\otimes$ denotes convolution. The spectral distribution of the substrate noise is

$$S(f) = \sum (\frac{C}{T})^2 \, |U_{up}(f) + (-1)^n U_{down}(f)|^2 \, \delta(f - \frac{n}{T}) \tag{9}$$

Equation (9) indicates that for periodic digital transitions, the substrate noise spectra consist of discrete tones. The noise frequency components are located at integer multiples of $1/T$, with their magnitude envelope determined by $U_{up}(f)$ and $U_{down}(f)$. If the digital switching transients all have the same transition time ($\tau_{rise} = \tau_{fall} = \tau$), then (9) yields

$$S(f) \propto (\frac{C}{T})^2 \frac{1}{1 + (2\pi f \tau)^2} \delta(f - \frac{n}{T}) \tag{10}$$

Equation (10) indicates that S(f) is inversely proportional to $1+(2\pi f\tau)^2$. Therefore, it predicts that at high frequencies the substrate noise power decreases quadratically with an increase in τ, while at low frequencies the substrate noise power is not significantly influenced by changes in τ. The following experimental measurements validate this prediction.

For a fixed $C = 56C_0$, but different values of τ, time domain responses and their spectra were measured at the sensor output. *Figure 8-10* (a) and (c) show that, in the time domain, when τ is reduced from 5.4 ns to 0.9 ns, the substrate noise magnitude increases by almost a factor of three, while t_{settle}

decreases slightly. *Figure 8-10* (b) and (d) show that in the frequency domain the substrate noise energy is concentrated in tones at integer multiples of f_{clock}. When τ decreases from 5.4 ns to 0.9 ns, the low frequency components of the substrate noise remain almost unchanged, while the magnitude of its high frequency components increase from below the noise floor to about 10 dB above the noise floor.

Experiments were also conducted to examine the effectiveness of distributing the digital switching transitions in time to reduce the substrate noise. To emulate the case of digital switching without such staggering, buffers driving a load of $96C_0$ were switched at a frequency of 20 MHz. To emulate the staggering of the digital transitions, the digital switching was then modified as follows: $48C_0$ was switched during the first half of the 20-MHz clock cycle, while the other $48C_0$ was switched during the latter half of the cycle, which is equivalent to switching $48C_0$ at a frequency of 40 MHz. The measured results shown in *Figure 8-11* indicate that staggering does not reduce the substrate noise in all frequency ranges, although the total noise power and the magnitudes of the noise peaks in the time domain are reduced by a factor of two. This observation is easily explained by (9). These results illustrate that substrate noise reduction in the time domain does not necessarily mean a reduction in noise in a particular frequency band.

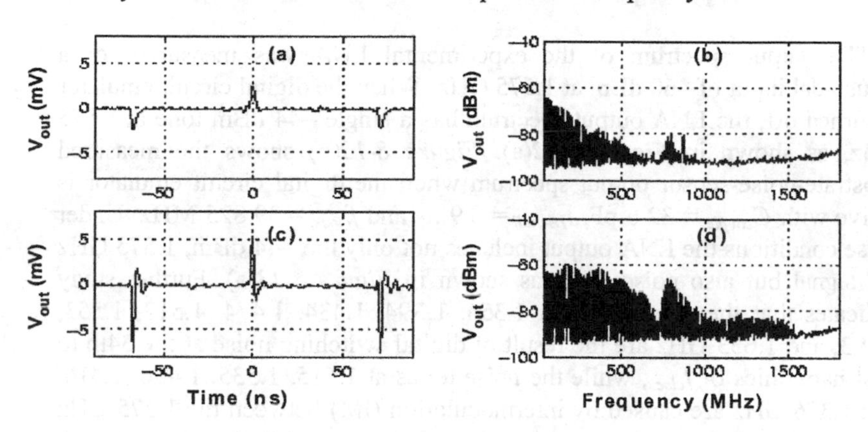

Figure 8-10. Measured substrate noise sensor output waveforms and spectra for (a), (b) τ = 5.4 ns, (c), (d) τ = 0.9 ns.

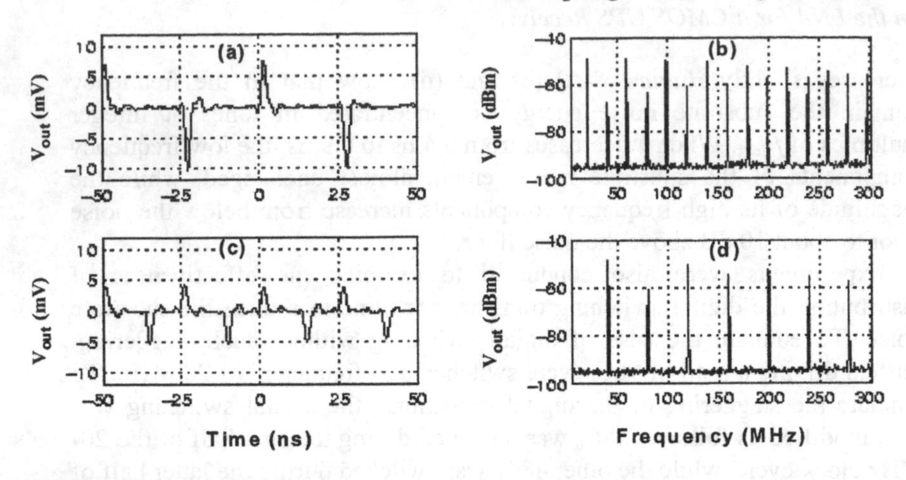

Figure 8-11. Measured waveforms and spectra at the substrate noise sensor output (a), (b) without staggered the digital switching, (c), (d) with staggered the digital switching.

4. NOISE COUPLING INTO THE LNA

4.1 LNA Output Spectrum

The output spectrum of the experimental LNA was measured for a sinusoidal input of –60 dBm at 1.575 GHz. When the digital circuit emulator is turned off, the LNA output spectrum has a single –44 dBm tone at 1.575 GHz, as shown in *Figure 8-12*(a). *Figure 8-12*(b) shows the measured substrate-noise-sensor output spectrum when the digital circuit emulator is active with C_{couple} = 32.6 pF, $t_{rise/fall}$ = 0.9 ns and f_{clock} = 39.825 MHz. Under these conditions the LNA output includes not only the –44 dBm, 1.575 GHz RF signal but also noise tones as shown in *Figure 8-13*(a). Further study indicates that the noise tones at 1.354, 1.394, 1.434, 1.474, 1.513, 1.553, 1.593, and 1.633 GHz are the result of digital switching noise at the 34th to 41st harmonics of f_{clock}, while the noise tones at 1.615, 1.535, 1.456, 1.416, and 1.376 GHz are caused by intermodulation (IM) between the 1.575-GHz RF signal and the substrate noise at f_{clock} and its third, fourth, and fifth harmonics. To confirm this, the RF input frequency was decreased by 5 MHz to 1.570 GHz with the result, shown in *Figure 8-13*(b), that the noise tones at the harmonics of f_{clock} remain unchanged, while the noise tones due to intermodulation are shifted down 5 MHz.

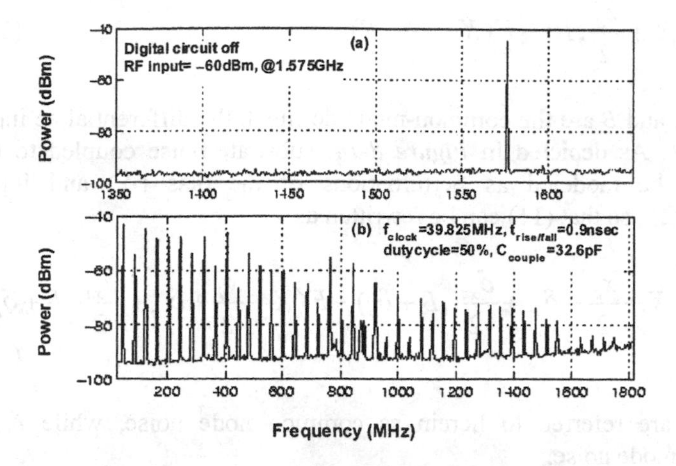

Figure 8-12. Measured (a) LNA output spectrum when the digital emulator is off and (b) spectrum at the substrate noise sensor output when the digital emulator is on.

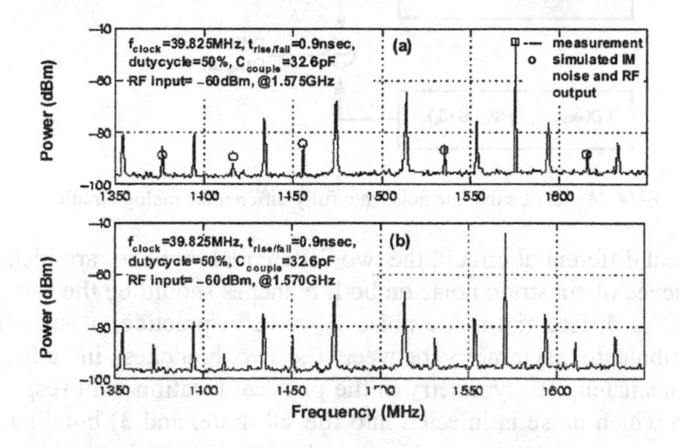

Figure 8-13. (a) Measured LNA output spectrum in the presence of substrate noise. (b) Measured LNA output spectrum when the RF frequency is decreased by 5MHz relative to (a).

4.2 Noise Coupling Mechanism

The approach adopted in Section 2 can be used to analyze the influence of substrate noise on the LNA. However, since the LNA is a fully differential circuit, the output Y in this case is the difference between two nominally identical system functions with different input components,

$$Y = F(X + \frac{x_d}{2}, B) - F(X - \frac{x_d}{2}, B) \tag{11}$$

where X, x_d, and B are the common-mode dc input, the differential ac input, and the bias. As depicted in *Figure 8-14*, substrate noise coupled to this circuit can be modeled as perturbations of the bias (δ_b) and inputs $(\delta_{xc} \pm \delta_{xd}/2)$, so that (11) can be rewritten as

$$Y = F(X + \frac{x_d}{2} + \delta_{xc} + \frac{\delta_{xd}}{2}, B + \delta_b) - F(X - \frac{x_d}{2} + \delta_{xc} - \frac{\delta_{xd}}{2}, B + \delta_b) \tag{12}$$

δ_b and δ_{xc} are referred to herein as common-mode noise, while δ_{xd} is differential-mode noise.

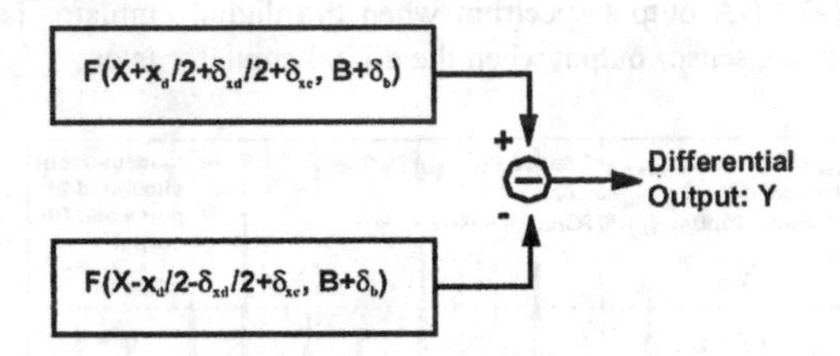

Figure 8-14. Model of substrate noise in a fully differential analog circuit.

For an ideal differential circuit, the two differential branches are identical and the influence of substrate noise on both branches should be the same. In such a case, the differential-mode noise $\delta_{xd} = 0$. In practice, a number of factors contribute to asymmetry between the two branches, including 1) component mismatch, 2) asymmetry in the physical location with respect to the region in which noise is injected into the substrate, and 3) bonding and packaging asymmetry between the two branches, which could cause magnetic field mismatch. Differential-mode noise is difficult to characterize because of its stochastic and distributed characteristics. Generally, it is minimized by designers' efforts to make the two differential branches as symmetric as possible. However, even with two perfectly matched differential branches ($\delta_{xd} = 0$) the common-mode noise components in (12) δ_b and δ_{xc}, will still be present.

The functions $F(X + x_d/2 + \delta_{xc} + \delta_{xd}/2, B + \delta_b)$ and $F(X - x_d/2 + \delta_{xc} - \delta_{xd}/2, B + \delta_b)$ have Taylor series expansions similar to (2). However, the first-order common-mode noise, the second-order harmonics of the signal, the second-order differential-mode noise and the second-order common-mode noise cancel when the outputs of the two branches are combined to form the the differential output *Y*. The remaining terms of *Y* are shown in the following expression:

$$Y = \frac{\partial F}{\partial x}\Big|_{x=X} x_d + \frac{\partial F}{\partial x}\Big|_{x=X} \delta_{xd} + \frac{\partial^2 F}{\partial x^2}\Big|_{x=X} \delta_{xc} x_d + \frac{\partial^2 F}{\partial x \partial b}\Big|_{\substack{x=X \\ b=B}} x_d \delta_b$$
$$+ \frac{\partial^2 F}{\partial x^2}\Big|_{x=X} \delta_{xd} \delta_{xc} + \frac{\partial^2 F}{\partial x \partial b}\Big|_{\substack{x=X \\ b=B}} \delta_{xd} \delta_b \tag{13}$$

Table 8-2 lists the frequency distribution of the terms in(13). Since, as discussed in Section 3, the substrate noise is located at integer multiples of f_{clock}, the fifth and sixth terms are located at the same frequency as the second term. However, the magnitudes of the fifth and sixth terms are much smaller than that of the second term. Thus, the dominant noise contributions in the LNA output are contributed by the second, third, and fourth terms in (13).

Table 8-2. Spectral distribution of the terms in (13)

Order	Term Number	Frequency Compnents	Property
1	1st term	f_s	ac signal
	2nd term	nf_{clock}	1st order differential-mode noise
2	3rd, 4th term	$\lvert f_s - nf_{clock}\rvert$, $f_s + nf_{clock}$	2nd order IM between the signal and common-mode noise
	5th, 6th term	nf_{clock}	2nd order IM between the common-mode noise and differential-mode noise

From *Table 8-2* it is apparent that common-mode noise influences the output through intermodulation with the differential inputs, while differential-mode noise appears at the output directly, scaled by some gain factor. In this example, where the GPS received signal is bandlimited to the range 1.57442 GHz to 1.57642 GHz [20], noise outside of the signal band can be filtered out by the succeeding IF filter in the receiver [15]. Therefore, only high-frequency differential-mode noise and low-frequency common-mode noise mixed with the RF signal will fall into the RF signal band and degrade the LNA's performance.

The effect of common-mode noise on the LNA response can be modeled with a single-node approximation of the heavily doped substrate, and the magnitude of the resulting IM noise tones can be simulated using transient analysis followed by a fast Fourier transform (FFT). As shown in *Figure 8-*

13(a), simulations using the single-node model match the experimental measurements with a maximum difference of 3.4 dB. Simulations indicate that the principal effect of the common-mode noise is to perturb the bias of the LNA, and thereby modulate the RF input.

The differential-mode noise is more difficult to model because a single-node representation of the substrate cannot be used. As a result of asymmetry between the two differential branches in the circuit, digital switching noise can cause differential-mode perturbations through either electrical or magnetic coupling. For the test chip shown in *Figure 8-3*, the digital circuitry is located parallel to the symmetric axis of the LNA, closer to one differential branch than the other. This layout asymmetry is most likely the source of the high-frequency differential-mode noise observed in the LNA output.

The differential-mode noise resulting from the different distances between the two branches in a differential circuit and the physical location where noise is injected into the substrate can be estimated as follows. Referring to *Figure 8-15*, location 3 represents the position of the digital noise source, while locations 1 and 2 correspond to the two differential branches of an noise sensitive circuit. If the distance between the two differential branches 1 and 2 is *l*, then there will be a phase difference between the substrate noise propagated from 3 to 2, and that from 3 to 1. This phase difference can produce differential-mode noise.

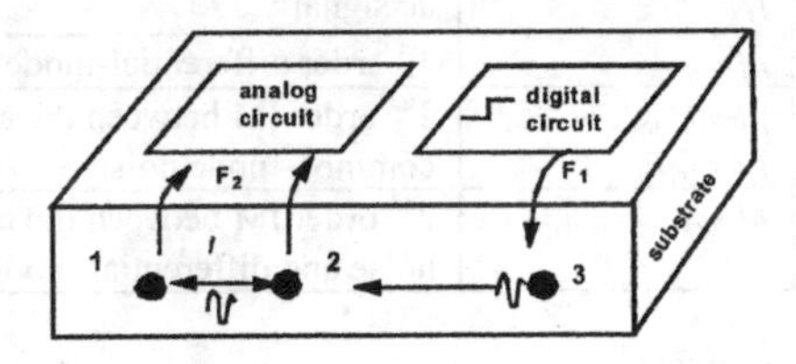

Figure 8-15. Asymmetry in the location of two differential circuit branches with respect to the digital noise source.

Suppose the substrate noise injected at location 3 has a magnitude A. If propagation loss is ignored, the differential mode noise v_{diff} caused by the propagation phase difference will be

$$V_{diff} = |Ae^{j(\omega t+\phi)} - Ae^{j\omega t}| \approx A\phi \qquad (14)$$

where ϕ is the phase difference between locations 1 and 2 resulting from propagation delay,

$$\phi = \frac{2\pi l}{\lambda} = \frac{2\pi l f \sqrt{\mu\varepsilon}}{c} \qquad (15)$$

For silicon, the relative permeability is $\mu = 1$ and the dielectric constant is $\varepsilon = 11.4$.

Equation (15) indicates that $|v_{diff}|$ will increase linearly with the noise frequency and the distance between the two differential branches. For example, if the noise frequency $f = 1.5$ GHz and $l = 100$ μm, then

$$| v_{diff} | = 10^{-2} A \qquad (16)$$

This example suggests that at high frequencies, even a slight difference in distance from two differential branches to the source of substrate noise can induce non-negligible differential-mode noise. Since the differential-mode noise appears at integer multiples of f_{clock}, one way to prevent it degrading the performance of a bandlimited circuit is to choose f_{clock} so that nf_{clock} falls out of the signal band.

4.3 Experimental Verification

The above analysis suggests that the noise in the LNA output can be divided into two categories, IM noise and Directly Coupled Noise (DCN). As predicted by (13), the IM noise is a linear function of the low frequency common-mode noise. Equation (10) suggests that the power of each common-mode noise tone is a quadratic function of the coupling capacitance, C_{couple}, and therefore, it is expected that the power of each IM noise tone will be a quadratic function of the C_{couple}. When plotted on a logarithmic scale, the power of the IM tone, as well as the corresponding substrate noise tone, should then be linear functions of log(C_{couple}) with a slope of 2. Experimental measurements confirm this prediction. *Figure 8-16* (a) plots the measured power of the largest IM tone (at 1.456 GHz) in the LNA output as a function of the coupling capacitance, as well as the power of the substrate noise tone at 119 MHz that causes this noise.

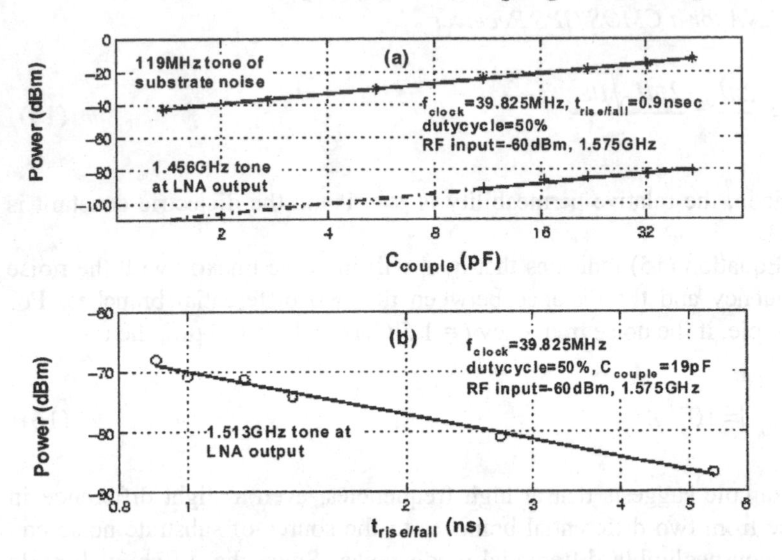

Figure 8-16. In the measured LNA output: (a) the power of the largest IM tone, and the power of the 119-MHz substrate-noise tone that causes this IM tone; (b) the power of the largest harmonic noise tone as a function of $t_{rise/fall}$.

Even though the directly coupled noise (DCN) is difficult to characterize, some useful information about the nature of the DCN can be derived from the earlier analysis. As predicted by (13), DCN is a linear function of the differential-mode noise. Equation (14) and (15) suggest that at high frequencies the differential-mode noise is proportional to the substrate noise, for which the power decreases quadratically with an increase in the digital transition times, $t_{rise/fall}$. Therefore, the power in each of the direct-coupled noise tones in the LNA output should be a quadratic function of $t_{rise/fall}$. Again, if plotted on a log scale, the power of each noise tone should therefore be a linear function of $\log(t_{rise/fall})$, with a slope of –2. Experimental measurements confirm this prediction. *Figure 8-16* (b) plots the 1.513 GHz noise tone, the largest direct-coupled noise tone in the LNA output, as a function of $t_{rise/fall}$.

5. A STATISTICAL APPROACH TO SUBSTRATE NOISE CHARACTERIZATION FOR DIGITAL CIRCUITS

Section 3 considered the nature of the substrate noise resulting from a digital circuit with single periodic switching pattern. In practice, the switching pattern is generally much more complicated and difficult to predict. Statistically, the switching activity is likely to be a combination of periodic and random patterns. For example, in digital circuits employing spread spectrum clocks [17], the switching activity may largely be pseudo-random.

In the case of a digital circuit with multiple transients, all of which introduce substrate noise with the same shape, there is only one root function. As indicated in section 3 , the total substrate noise at time t is then the summation of substrate noise caused by each individual transient.

$$v_{sub}(t) = \int_{-\infty}^{t} u(t-\tau)p(\tau)d\tau \qquad (17)$$

where $p(\tau)d\tau$ is the coupling capacitance $C_{couple}(\tau)$ through which noise coupled in during time $[\tau, \tau+d\tau]$. Therefore, $p(\tau)$ is defined as the *coupling capacitance rate function*. Notice that $u(t-\tau) = 0$ for all $t < \tau$, because substrate noise is zero before the digital transition happens. Equation (17) can be rewritten as the convolution of the substrate noise root function $u(t)$ and the coupling capacitance rate function $p(t)$.

$$v_{sub}(t) = \int_{-\infty}^{\infty} u(t-\tau)p(\tau)d\tau = u(t) \otimes p(t) \qquad (18)$$

Generally, the digital circuit has numerous transitions that induce noise transients in the substrate with different shapes. Multiple substrate noise root functions are then needed to represent the different shapes of the substrate noise. To calculate the total substrate noise, the noise resulting from each individual digital transition can be categorized by its corresponding shape, which is described by the substrate noise root function.

As illustrated in *Figure 8-17* , suppose there are N types of substrate noise root functions for low-to-high transitions, and M types of substrate noise root functions for high-to-low transitions. The substrate noise root functions, $u_{up,k}(t)$ and $u_{down,k}(t)$ are functions of the digital transition times, t_{rise} and t_{fall}, and the subscript k corresponds to the kth distinct value of t_{rise} or t_{fall}. $p_{up,k}(t)$ and $p_{down,k}(t)$ are the capacitance switching rate function corresponding to each root function $u_{up,k}(t)$ and $u_{down,k}(t)$.

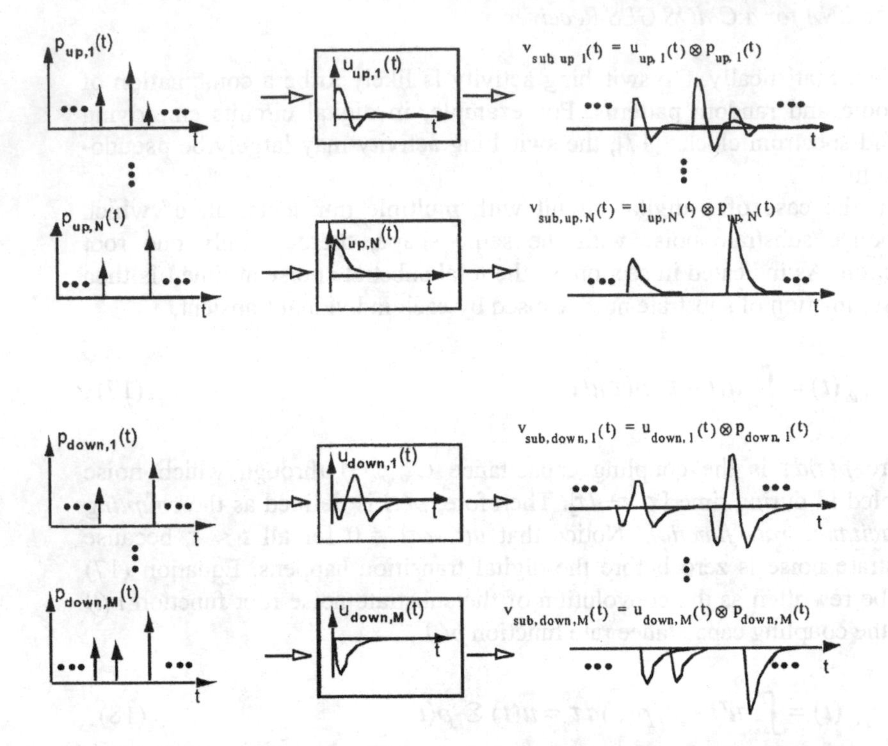

Figure 8-17. Substrate noise when there are M types of substrate noise root functions for high-to-low digital transition and N types of substrate noise root functions for low-to-high digital transition.

The total substrate noise at t is the summation of all the substrate noise components caused by each of the digital transitions occurring prior to t.

$$v_{sub}(t) = \sum_{k=1}^{N} u_{up,k}(t) \otimes p_{up,k}(t) + \sum_{k=1}^{M} u_{down,k}(t) \otimes p_{down,k}(t) \tag{19}$$

Equation (19) decomposes the substrate noise into (N + M) groups of convolutions between $u_{up/down,k}(t)$ and $p_{up/down,k}(t)$. Mathematically, $U_{up/down,k}(f)$, the Fourier transform of $u_{up/down,k}(t)$, can be viewed as a bank of filters, while the digital circuit switching pattern, $p_{up/down,k}(t)$, can be viewed as the input to these filter banks. The substrate noise resulting from a digital circuit can be modeled as the sum of the filter bank outputs, as illustrated in *Figure 8-18*.

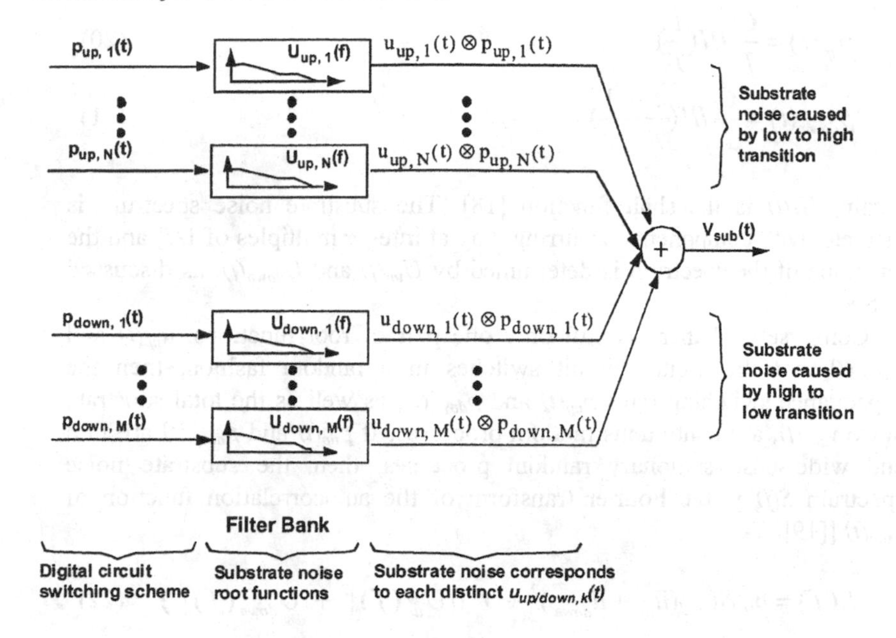

Figure 8-18. Filter bank model of substrate noise as a function of digital circuit characteristics.

This decomposition illustrates that the substrate noise is determined by the digital circuit characteristics that can be separated into 1) physical characteristics and 2) timing characteristics. The filter bank describes the physical characteristics, such as the transistor sizes, digital transition times, substrate impedance and power line parasitics. The information needed to model the filter banks can be acquired through HSPICE simulations. The inputs to the filter bank, $p_{up,k}(t)$ and $p_{down,k}(t)$ represent the timing characteristics, i.e. the digital circuit switching characteristics, such as when the transitions occur and how many occur. An accurate representation of $p_{up,k}(t)$ and $p_{down,k}(t)$ requires accurate statistics on the digital circuit switching activity, which can be very difficult to obtain. Below we consider two extreme cases, periodic switching and random switching, in an effort to provide some preliminary insight.

For periodic digital switching with period T, as in the case of the experimental digital emulator described earlier, there is only one pair of root functions, $u_{up}(t)$ and $u_{down}(t)$. Thus, the capacitance switching rate functions are

$$p_{up}(t) = \frac{C}{T} III(\frac{t}{T}) \tag{20}$$

$$p_{down}(t) = \frac{C}{T} III(\frac{t}{T} - \frac{1}{2}) \tag{21}$$

Again, $III(t)$ is the shah function [18]. The substrate noise spectrum is discrete, with components occurring only at integer multiples of $1/T$, and the envelope of the spectrum is determined by $U_{up}(f)$ and $U_{down}(f)$, as discussed in Section 3 .

Conversely, if there is still only one pair of root functions, $u_{up}(t)$ and $u_{down}(t)$, but the digital circuit switches in a random fashion, then the capacitance switching rates $p_{up}(t)$ and $p_{down}(t)$, as well as the total substrate noise $v_{sub}(t)$, are continuous random processes. If $p_{up}(t)$ and $p_{down}(t)$ are i.i.d and wide-sense stationary random processes, then, the substrate noise spectrum $S(f)$ is the Fourier transform of the autocorrelation function of $v_{sub}(t)$ [[19]],

$$S(f) = m^2 \delta(f)(\overline{u}_{up} + \overline{u}_{down})^2 + \sigma^2(|U_{up}(f)|^2 + |U_{down}(f)|^2) \tag{22}$$

where it is assumed that $p_{up}(t)$ and $p_{down}(t)$ have the same mean m and variance σ^2,

$$m = E(p_{up}(t)) = E(p_{down}(t)) \tag{23}$$

$$\sigma^2 = \text{var}(p_{up}(t)) = \text{var}(p_{down}(t)) \tag{24}$$

and

$$\overline{u}_{up} = \int_{-\infty}^{\infty} u_{up}(t)dt \tag{25}$$

$$\overline{u}_{down} = \int_{-\infty}^{\infty} u_{down}(t)dt \tag{26}$$

This result suggests that if the digital switching becomes random, then the spectral distribution of the substrate noise becomes continuous. The substrate noise spectral contour is determined by $U_{up/down}(f)$ and scaled by a factor of σ^2, as depicted in *Figure 8-19*. If $u_{up}(t) \neq -u_{down}(t)$, then there is a dc offset in the substrate noise. An asymmetry between u_{up}(t) and u_{down}(t) can be caused by differences in the impedances between the substrate and v_{dd}

and the substrate and ground, or differences between t_{rise} and t_{fall}. For a fully symmetric system where $u_{up}(t) = -u_{down}(t)$, (22) simplifies to

$$S(f) = 2\sigma^2 |U(f)|^2 \tag{27}$$

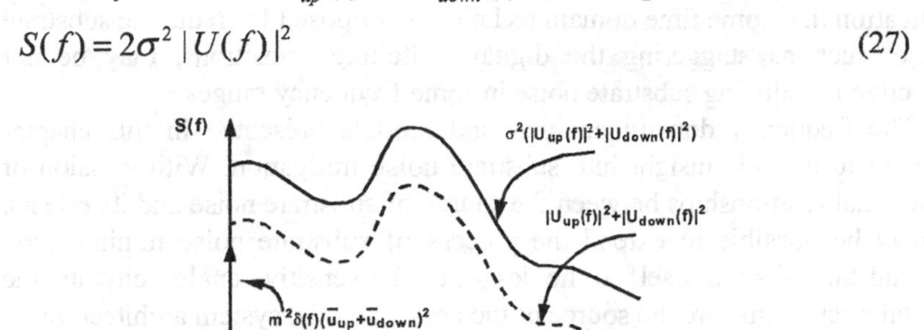

Figure 8-19. Nature of substrate noise spectrum resulting from random digital transitions.

For a digital circuit with random switching, (22) provides some insight on how to minimize the substrate noise in the design of the digital circuitry. The noise injected into the substrate can be reduced by 1) making the substrate noise root function as symmetric (i.e. $u_{up}(t) = -u_{down}(t)$) as possible) so as to reduce $\bar{u}_{up} + \bar{u}_{down}$, 2) reducing m, which means reducing the average switching activity to the extent possible, 3) reducing the variance of the switching activity σ^2, which means trying to make the digital circuit switching as uniform as possible (an extreme is to make $\sigma^2 = 0$, i.e. the digital circuit switching activity is uniformly distributed in the time domain so that the only substrate noise is a dc offset caused by the asymmetry of the up/down root functions), and 4) reducing $U_{up}(f)$ and $U_{down}(f)$ in the analog signal band, which could be done by reducing the equivalent substrate impedance in the direct coupling band and intermodulation band of the analog circuits.

6. CONCLUSION

This chapter has presented experimental and analytical results related to two issues: 1) the relationship between substrate noise and the characteristics of the digital circuitry generating that noise, and 2) the mechanisms by which substrate noise influences the response of a bandlimited analog circuit, in particular the low-noise amplifier (LNA) in a communications receiver. It was shown that the spectral distribution of substrate noise greatly affects its impact in such a system. With the LNA as an example, both analysis and measurements indicate that although the substrate noise power may be several orders of magnitude higher than the received signal power, only those noise components located in specific frequency ranges actually degrade the LNA performance. Among the results of this study are an indication that some time domain techniques proposed for reducing substrate noise, such as staggering the digital switching transitions, may be not effective in reducing substrate noise in some frequency ranges.

The frequency domain analysis and models presented in this chapter attempt to provide insight into substrate noise mitigation. With a vision of functional relationships between the source of substrate noise and its effects, it may be possible to extend the process of substrate noise minimization beyond the substrate itself to the design of the sensitive analog circuits, the digital circuits that are the source of the noise, and the system architecture.

ACKNOWLEDGMENT

The authors gratefully acknowledge the support of Rockwell International, Agilent Technologies, Inc and Big Bear Networks, Inc. They are also indebted to Dr. David K. Su, Dr. Derek K. Shaeffer, Dr. Tomas H. Lee, Jacky Liu, Dr. Won Namgoong and Dr. Arvin Shahani for their contributions to this work and Maurice Tarsia for his support.

REFERENCES

[1] H. Samueli, "Broadband Communications ICs: Enabling High-bandwidth Connectivity in the Home and Office," in *ISSCC Slide Supplement*, 1999, Feb, 1999.

[2] Rofougaran, et al., "A Single-Chip 900MHz Spread-Spectrum Wireless Transceiver in 1-mm CMOS - Part I: Architecture and Transmitter Design," in *IEEE J. of Solid-State Circuits*, vol. 33, No. 4, pp.515-534, Apr. 1998.

[3] M. Steyaert, et al., "A single-Chip CMOS Transceiver for DCS-1800 Wireless Communications," in *ISSCC Digest of Technical Papers*, pp.48-49, Feb. 1998.

[4] T. Cho, et al., "A Single-Chip CMOS Direct-Conversion Transceiver for 900MHz Spread- Spectrum Digital Cordless Phones," in *ISSCC Digest of Technical Papers*, pp.228-229, Feb. 1999.

[5] K. Bult, "Analog Broadband Communication Circuits in Pure Digital Deep Sub-Micron CMOS," *ISSCC Digest of Technical Papers*, pp. 76-77, Feb. 1999.

[6] R. Gharpurey, and R. Meyer, "Modeling and Analysis of Substrate Coupling in Integrated Circuit", in *IEEE J. Solid-State Circuits*, vol. 31, pp.344-353, March 1996.

[7] F. J. R. Clement, E. Zysman, M. Kayal, and M. Declercq, "LAYIN: Toward a Global Solution for Parasitic Coupling Modeling and Visualization," *IEEE Custom Integrated Circuits Conference*. pp.537-540, May 1994.

[8] David K. Su, Marc J. Loinaz, Shoichi Masui, and Bruce A. Wooley, "Experimental Results and Modeling Techniques for Substrate Noise in Mixed-Signal Integrated Circuits," in *IEEE J. Solid-State Circuits*, vol. 28, pp.420-429, April 1993.

[9] S. Mitra, R. A. Rutenbar, L. R. Carley, D. J. Allstot, "A Methodology for Rapid Estimation of Substrate-Coupled Switching Noise," *IEEE Custom Integrated Circuits Conference*, pp. 129-132, May 1995.

[10] P. Miliozzi, L. Carloni, E. Charbon, and A. Sangiovanni-Vincentelli, "SUBWAVE: a Methodology for Modeling Digital Substrate Noise Injection in Mixed-Signal ICs," in *IEEE Custom Integrated Circuits Conference*, pp. 385-388, May 1996.

[11] N. K. Verghese, D. Allstot, "Verification of RF and Mixed-Signal Integrated Circuits for Substrate Coupling Effects," in *IEEE Custom Integrated Circuits Conference*, pp. 363- 370, May 1997.

[12] M. V. Heijningen, J. Compiet, P. Wambacq, S. Donnay, "Modeling of Digital Substrate Noise Generation and Experimental Verification Using a Novel Substrate Noise Sensor," in *ESSCIRC*, pp. 186-189, Sept. 1999.

[13] K. Makie-Fukuda, "Voltage-Comparator-Based Measurement of Equivalently Sampled Substrate Noise Waveforms in Mixed-Signal Integrated Circuits," in *IEEE J. Solid-State Circuits*, vol. 31, pp. 726-731, May, 1996.

[14] R. Gharpurey, "A Methodology for Measurement and Characterization of Substrate Noise in High Frequency Circuits," in *IEEE Custom Integrated Circuits Conference*, pp. 487- 490, May 1999.

[15] D. K. Shaeffer, et al., "A 115-mW, 0.5mm CMOS GPS Receiver with Wide Dynamic-range Active Filters," *IEEE J. Solid-State Circuits*, vol. 33, pp. 2219-2231, Dec. 1998

[16] D. A. Weiner and J. F. Spina, *Sinusoidal Analysis and Modeling of Weakly Nonlinear Circuits*. New York: Van Nostrand Reinhold, 1980.

[17] H. Li, etc, "Dual-Loop Spread-Spectrum Clock Generator," in *ISSCC Digest of Technical Papers*, pp.184-185, Feb. 1999.

[18] R. N. Bracewell, *The Fourier Transform and Its Applications*, second edition, chapter 5, McGraw-Hill, Inc, 1986.

[19] Alberto Leon-Garcia, *Probability and Random Processes for Electrical Engineering*, second edition, chapter 7, Addison-Wesley Publishing Company,1994

[20] J. J. Spilker Jr, and E. D. Natali, "Interference Effects and Mitigation Techniques" in Bradford W. Parkinson and J.J Spilker, Jr., Eds., *Global Positioning System: Theory and Applications, vol. I*, American Institute of Aeronautics an Astronautics, pp. 717-771, 1996.

Chapter 9

A PRACTICAL APPROACH TO MODELING SILICON-CROSSTALK IN SYSTEMS-ON-SILICON

Paul T.M. van Zeijl
Ericsson, Emmen, The Netherlands, Paul.van.Zeijl@eln.ericsson.se

Abstract: This chapter will demonstrate a simple approach in modeling crosstalk on silicon. By splitting the problem into three parts (the digital interference caused by the digital circuitry or source, the transfer of interference in the substrate, and the (undesired) reception of the interference by the analog part) and modeling these three parts in a simple, yet effective manner, simulations for the complete system can easily be done. A comparison of measured data and simulation results shows the effectiveness of the approach for a low-ohmic substrate. A second application, a single-chip Bluetooth ASIC, demonstrates our approach in a system-on-silicon.

1. INTRODUCTION

Mixed-mode ASICs and the realization of a system-on-silicon are becoming more and more important. A bottleneck in a system-on-silicon is the disturbance of sensitive analog blocks by interfering signals from the large amount of digital circuitry on the same die. The complexity of a system-on-silicon in combination with the difficult task of modeling the silicon-crosstalk problem and the size of the problem (for instance more than 1 M digital gates on 10 ... 100 mm^2 silicon in combination with a complete analog Bluetooth transceiver) leaves designers with the immense task of realizing functional and up to specification performing silicon. This chapter will propose several techniques in order to solve the modeling issues and show how this complexity can be reduced significantly to be able to do practical simulations.

The next section will explain the silicon crosstalk problem. An example of how to isolate a (very sensitive) VCO from other (analog) circuitry on the same die will support the basic idea that very good isolation can be obtained on silicon. Section 3 will shortly summarize the well known bottlenecks in trying to analyze the silicon-crosstalk problem. Section 4 will present our approach in the modeling of the digital circuitry, the modeling of the analog circuitry and the modeling of various substrate types (i.e. low resistivity substrates as commonly used in pure CMOS processes and high resistivity substrates as commonly used in BiCMOS/RFCMOS like processes). An example of a simulation-testbench will also be presented, together with a comparison between measured and simulated results. Finally, Section 5 will conclude this chapter with a summary of ways to reduce silicon-crosstalk and some conclusions.

2. PROBLEM STATEMENT

A problem with the design of a system-on-silicon is crosstalk. Crosstalk may be caused by crosstalk on the PCB, crosstalk via the bonding wires and package, crosstalk via ground and supply lines and crosstalk due to the silicon-substrate. PCBs, packages, ground and supply-line series resistances can be modeled, thus increasing the complexity of the design and thus increasing simulation time. The main issue of this silicon-crosstalk problem is that we don't have ways to easily predict the effects (magnitude) of the crosstalk. In this chapter we will mainly focus on this silicon-crosstalk problem.

Figure 9-1 shows a simplified overview of the silicon-crosstalk problem. On the digital side we see an inverter consisting of an NMOS and a PMOS transistor. The power supply (Vcc_dig) and ground (Vgnd_dig) are coming from a battery, so via bonding wires. Input (dig_in) and output (dig_out) are also connected from the PCB to the inverter. Each connection has parasitic capacitances to the substrate. The substrate connections on the digital part of the chip are represented by the Dsub_gnd connection. The backgate connection of the PMOS transistor is connected to the positive power supply. The backgate of the NMOS transistor is directly connected into the substrate. The substrate is represented here by a matrix of resistors. An LNA with a resonator load represents a typical analog circuit.

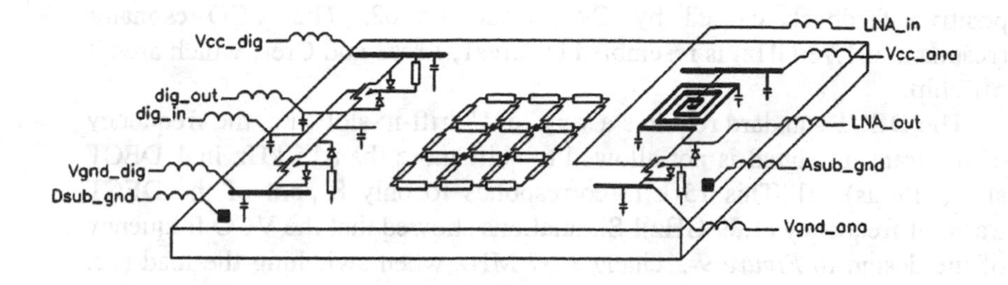

Figure 9-1. An overview of the silicon-crosstalk problem.

The interfering signals can have consequences dominated by either linear (i.e. addition of signals) or non-linear behavior. Non-linear behavior may create modulation of signals (FM/AM, and thus create extra spurious components), or shifts bias points and gives pushing and/or pulling effects on oscillators like a VCO or a crystal oscillator.

An example of how good a VCO can be designed and isolated is given in Reference [1]. We designed a bipolar VCO for the DECT standard, see *Figure 9-2*.

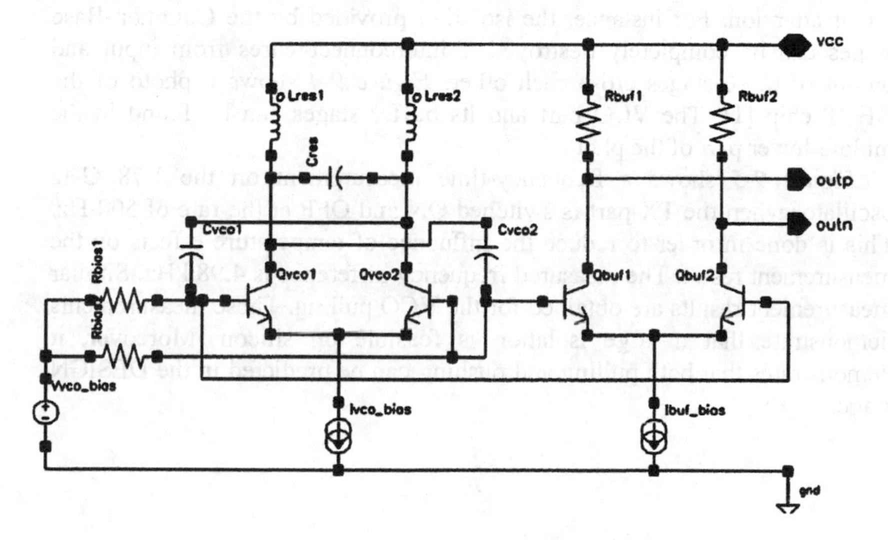

Figure 9-2. The original schematic for the DECT VCO.

The VCO consists of a bipolar differential pair (Qvco1 and Qvco2) with positive feedback caused by Cvco1 and Cvco2. The VCO-resonator (resonant at 1.78 GHz) is resembled by Lres1, Lres2 and Cres, which are all off-chip.

The DECT standard requires a very low "drift-in-slot", i.e. the frequency of the transmit signal is not allowed to shift more than 15 kHz in 1 DECT slot (417 μs) [2]. This 15 kHz corresponds to only 8 ppm of the DECT transmit frequency (1.89 GHz)! Simulations showed that the VCO frequency of the design in *Figure 9-2* changes 57 MHz when switching the load (i.e. the RX- and TX-mixers) from ON to OFF and vice-versa. This is caused by the change in load-impedance seen by the resonator (i.e. the input capacitance of the buffer stage (Qbuf1 and Qbuf2) changes when switching the RX- or TX-mixer from ON to OFF). An isolation of more then 78 dB is required to lower the frequency shift to far below 7.5 kHz. Two extra buffer stages, each containing a Common-Base stage with its own biasing, are added to realize this isolation, see *Figure 9-3*. The first CB-stage is formed by transistors Qcb1a and Qcb1b, which are connected on top of the already present differential pair Qbuf1a and Qbuf1b. An extra differential pair (Qbuf2a-Qbuf2d) hosts the second CB-stage Qcb2a and Qcb2b.

The layout for reaching the 78dB isolation is very critical and requires a lot of attention. For instance, the isolation provided by the Common-Base stages can be completely destroyed if interconnect wires from input and output of these stages cross each other. *Figure 9-4* shows a photo of the DECT chip [1]. The VCO part and its buffer stages can be found in the middle-lower part of the photo.

Figure 9-5 shows a frequency-time measurement on the 1.78 GHz oscillator when the TX part is switched ON and OFF at the rate of 500 Hz. This is done in order to reduce the influence of temperature effects on the measurement result. The measured frequency difference is 4.98 kHz. Similar measurement results are obtained for the VCO pulling. These measurements demonstrate that a large isolation is feasible on silicon. Moreover, it demonstrates that both pulling and pushing can be predicted in the DESIGN stage.

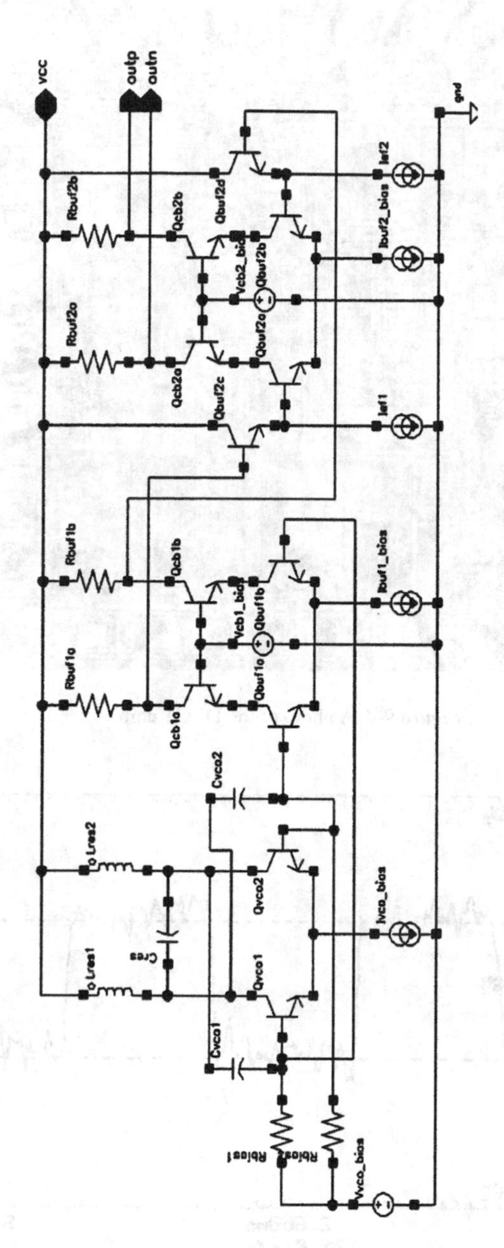

Figure 9-3. The VCO schematic including two extra CB-buffer stages.

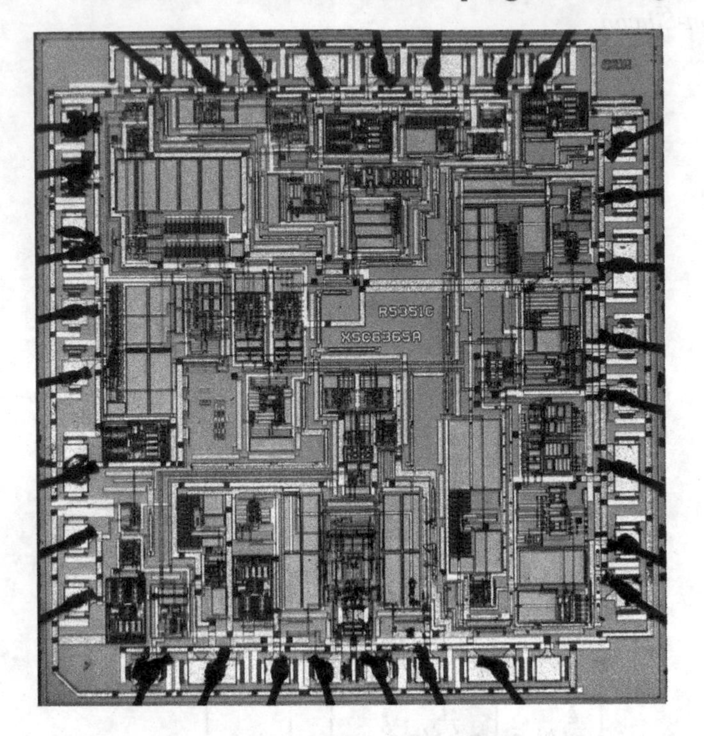

Figure 9-4. A photo of the DECT chip.

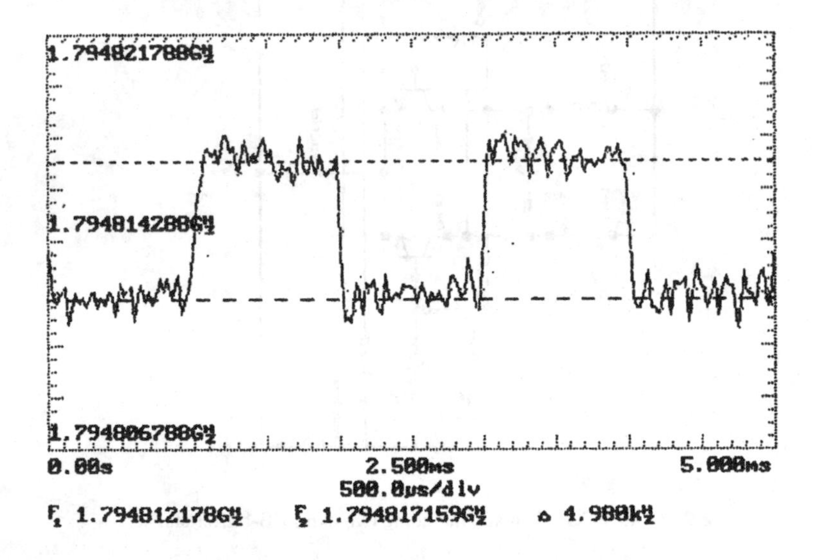

Figure 9-5. The measured pulling of a 1.8 GHz VCO.

In general there are several ways to minimize silicon-crosstalk:

- By separation of the desired signal and the interfering signal in the frequency domain (this may prove difficult due to high data-rates).
- By separation of the desired signal and the interfering signal in the time domain: no digital activity during reception and/or transmission of signals.
- By lowering the amplitude of the interference source (e.g. introducing jitter on clocks).
- By isolating the blocks that generate interference and the blocks that are sensitive to interference (increase separation distance, use extra layout measures such as shielding, triple-well, P- and/or N-rings), or by applying compensation and balancing.

But how much does this improve our design? The answer to this question should be found at the start of the design-phase of the project, otherwise the complete design may fail.

3. LIMITATIONS IN STATE-OF-THE-ART APPROACHES TO SILICON-CROSSTALK

One of the first problems one encounters in trying to model the digital circuitry in a system-on-silicon is the large number of digital gates. An effective approach for modeling digital circuitry has been presented in Reference [3], but even this approach seems to reach its limits at 3 kgates or 1000 substrate contacts. Thus for more then 1M gates it is not really practical to have this kind of complexity in analog simulations. The fast slopes in digital circuits in state-of-the-art CMOS processes give rise to harmonics in the GHz region, even if the circuitry is only clocked at 10-20 MHz. In the case that there is SW in ROM on the silicon, then it even becomes harder to analyze the problem, as SW will "never" be stable. An educated guess or estimation is needed to do something practical.

The complexity in the extraction of a substrate model is so huge that you will not reach a practical size and simulation time for any simulation with a substrate model that has substrate contacts on each device [3]. Moreover, this kind of analysis can only be done after the complete layout of the analog and digital circuitry has been finalized. Analyzing the problem on such a finalized layout is not easy and will take a long time, maybe even longer than silicon processing. It may then even be better to wait for the silicon to come back from processing.

Usually, silicon-crosstalk analysis is done after 1st silicon has been received, because there are "problems". Many designers are also very unfamiliar with the issue, it is like "black-magic" for them.

In the next section our approach to all these problems will be presented.

4. OUR STRATEGY

Our strategy is to start taking care of the silicon-crosstalk issue at the start of the design-phase of the project. Therefore we try to model the three contributing factors from the start:

1. Model the digital circuitry in a simple yet effective manner.
2. Model the analog circuitry.
3. Model the substrate.

Then, by using the simple model for the digital circuitry, the model for the analog circuitry and the model for the substrate, we can run analog simulations on the overall performance, compare the simulation results with the specifications and/or re-specify the analog and RF circuitry. Modeling the digital circuitry will be discussed next.

4.1 Modeling the digital circuitry

The modeling of the digital circuitry requires some educated guesses to lower the complexity, while still realizing a sufficiently accurate model. An important question is whether the behavior of a digital block is dominated by clocked signals or dominated by a pseudo-random-bit-sequence (PRBS). Before we discuss the model, we will first have a look at the spectrum of various digital signals.

Figure 9-6 shows the spectrum of clock signals (f_{clock}=10 MHz, the rise- and fall-times are 1 ns (upper plot) or 100 ps (lower plot), the duty-cycle is 45/55 %).

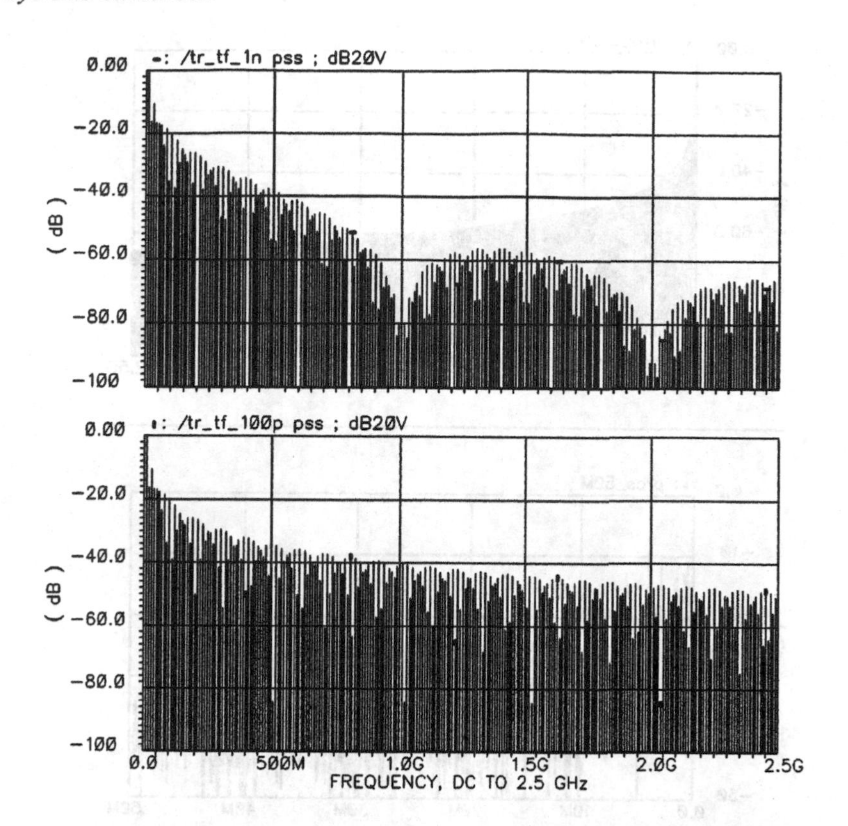

Figure 9-6. The spectrum of clock signal with various rise- and fall-times.

The upper plot shows dips in the spectrum at multiples of 1 GHz, corresponding to the rise- and fall-times. This effect can and is usually exploited in drivers for off-chip circuitry to limit radiation of high-frequency harmonics. It could also be used for all on-chip gates, but this does not fit in the digital design flow. Anyway, harmonics up to 2.5 GHz and higher are not that much lower in amplitude compared to harmonics at 0.5 GHz as the attenuation falls with 1/f, i.e. 20*LOG(2.5/0.5)=14 dB (assuming rise- and fall-times of 100 ps). Comparing the spectra for rise- and fall-times of 100 ps versus 1 ns, the difference in amplitude for the harmonics around 2.5 GHz is only some 15 dB, which may, however, just be enough in some applications.

Figure 9-7 shows the spectrum of a PRBS signal (upper plot, f_{clock}=10 MHz, rise- and fall-times are 100 ps, the duty-cycle is 45/55 %). The lower plot shows a zoom-in to the spectrum (DC to 50 MHz only).

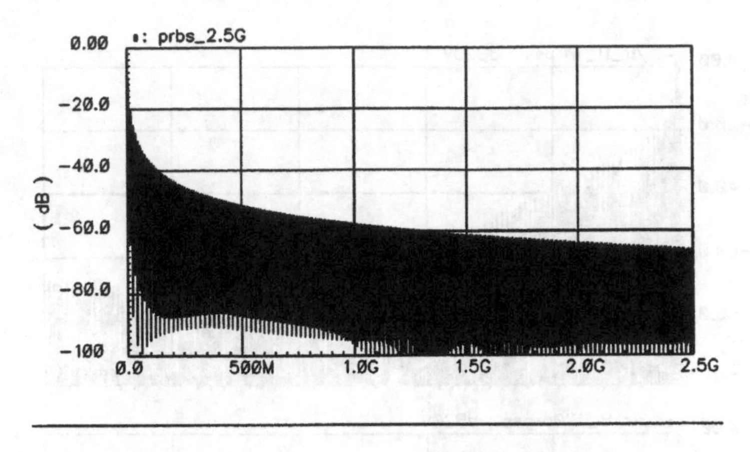

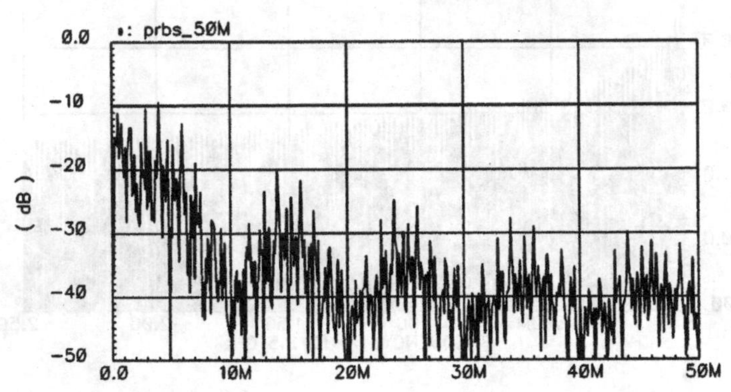

Figure 9-7. The spectrum of a pseudo-random-bit-sequence.

Also the harmonics of a PRBS-signal extend into the GHz region and are somewhat lower then the spectra of clocked signals (15 dB at 2.5 GHz for rise- and fall-times of 100 ps). From the lower plot in *Figure 9-7*, one can see dips in the spectrum at multiples of the bitrate. Such amplitude dips can be exploited, especially in small bandwidth systems.

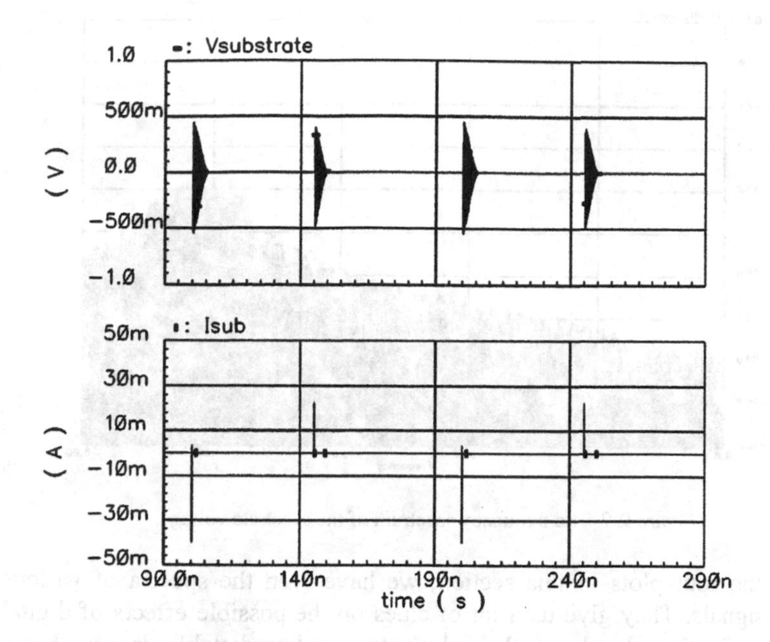

Figure 9-8. The current injected into the substrate and the corresponding substrate voltage.

Figure 9-8 shows the plot of the current injected into the substrate of a 1 kgate large (effective) inverter (0.35 μm technology). The upper plot shows the substrate voltage whereas the package is modeled by a 1 nH series inductance to the PCB-ground (f_{clock}=10 MHz clock on a 1 V supply). The substrate voltage consists of a series of spikes corresponding to the transitions of the 10 MHz clock. Each spike is in fact a damped resonance. The lower plot shows the substrate short-circuit current. The peak current at the transitions are +20 mA_p and -40 mA_p, respectively.

Figure 9-9 shows the spectrum of the substrate voltage as fourier-transformed from the upper plot in *Figure 9-8*. The spectrum looks quite flat up to 3 GHz, and then increases. The flat part of the spectrum is caused by the bonding-wire inductance, whose impedance increases with frequency, while the harmonics fall with 1/f. At 4 GHz, a resonance occurs caused by the bonding wire inductance (1 nH) and the capacitance of the inverter itself (1.3 pF). This resonance frequency may shift to much higher or lower frequencies depending on the specific application. When designing a system-on-silicon, one should be aware of this behavior, as it still may be changed in the design phase of the project.

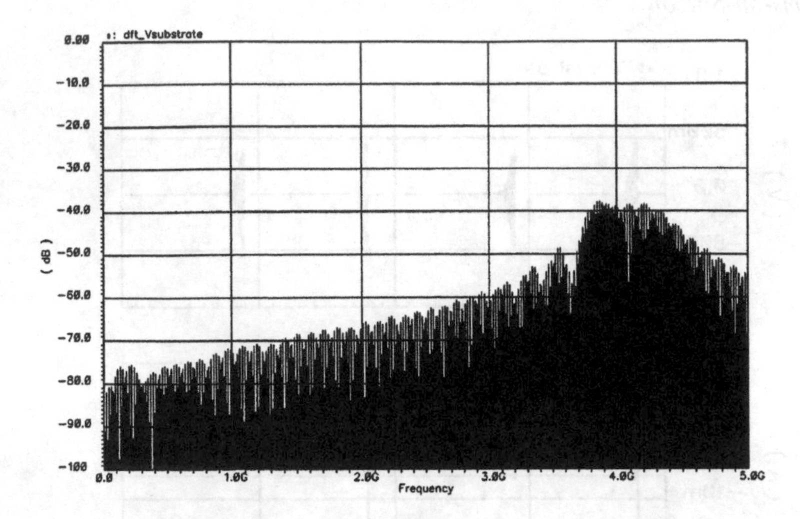

Figure 9-9. The frequency spectrum of the substrate voltage.

On the four plots in this section, we have seen the spectra of various digital signals. They give us a lot of clues on the possible effects of digital circuitry. On the other hand, the only known and predictable data we have, previous to the realization of the digital circuitry in a real layout, are the amount of gates, the power supply voltage, the supply current (or at least a prediction of this) and the clock frequency. We should be able to estimate whether the digital block will behave like a purely clocked block (like a microprocessor) or behave more like a pseudo-random-bit-sequence (like memory access). We also know that the current consumption of digital blocks scales with the number of transitions of the gates in this digital block. From *Figure 9-8* it is clear that current injected into the substrate is caused by current pulses in the backgate of the NMOS transistor, the pulses in the V_{cc} and ground lines and from the interconnect at the input and output of the digital gates.

It is now proposed to model each digital block (of maybe up to 100k gates or more) by its current consumption (I_{cc}) at a given power supply voltage V_{cc}, at a given frequency f_{clock}. Then we replace this complete digital block by this single LARGE inverter. Depending on the behavior, such a block is driven from a clock signal or from a PRBS-signal. This model is very simple and is not as accurate as the modeling-approach in Reference [3], buts its simplicity will give us a lot of insight. It will also speed up simulation-time enormously: instead of having to wait thousands of clock-cycles in order to completely have the digital block reach each digital state, we only have to run one clock cycle for clocked blocks or some 100 clock cycles for blocks that have pseudo-random bit sequences. In practice, this modeling may be pessimistic, but such pessimistic situations will occur in practice, and our system has to be robust for these cases. If needed, the model can be enhanced with the modeling of the capacitances of the V_{cc} and ground lines to the substrate. Of course, all different clock domains have to be modeled separately.

As an example we model two digital blocks on our Bluetooth ASIC (also see Section 4.5): the micro-processor and the RAM. The micro-processor consist of 75 kgates, runs at 10 MHz and consumes 5 mA at Vcc=1.5V. The micro-processor is similar to a purely clocked block. We replace this complete block by a single large inverter which is 70.000 times larger than the minimum sized inverter in our 0.18μ CMOS process. The large inverter is driven by a clock signal of 10 MHz.

The RAM consist of 256 kbit, is accessed with a clock of 5 MHz and consumes 6 mA at Vcc=1.5V. The RAM is accessed in a pseudo-random manner. We replace this RAM by a single large inverter which is 600.000 times larger than the minimum sized inverter in our 0.18μ CMOS process. The large inverter is driven by a PRBS signal with 5 MHz frequency. The combination of both blocks (or even more blocks) can now be used in a silicon-crosstalk simulation.

4.2 Modeling the analog circuitry

Modeling the analog circuitry is relatively simple, see *Figure 9-1*. The modeling of all components requires inclusion of all device-parasitics to the substrate: models of MOS devices shall include the Drain-Bulk and Source-Bulk diodes, and the models of the passive components (resistors, inductors and capacitors) shall include all parasitics (mostly capacitances or reverse-biased diodes) to the substrate. Thus an LNA-block or VCO-block will get an extra node (apart from the power supply, ground, input and output nodes), representing the connection to all parasitic components in the LNA-block or VCO-block to the substrate.

4.3 Substrate modeling for low-impedance substrate (0.35μ pure CMOS), and overall simulations

Some years ago we designed a test-chip in 0.35μm pure CMOS process with an epi-substrate. *Figure 9-10* shows the test-chip. The ASIC contains several analog blocks: an LNA, a VCO, a receiver front-end and a frequency divider.

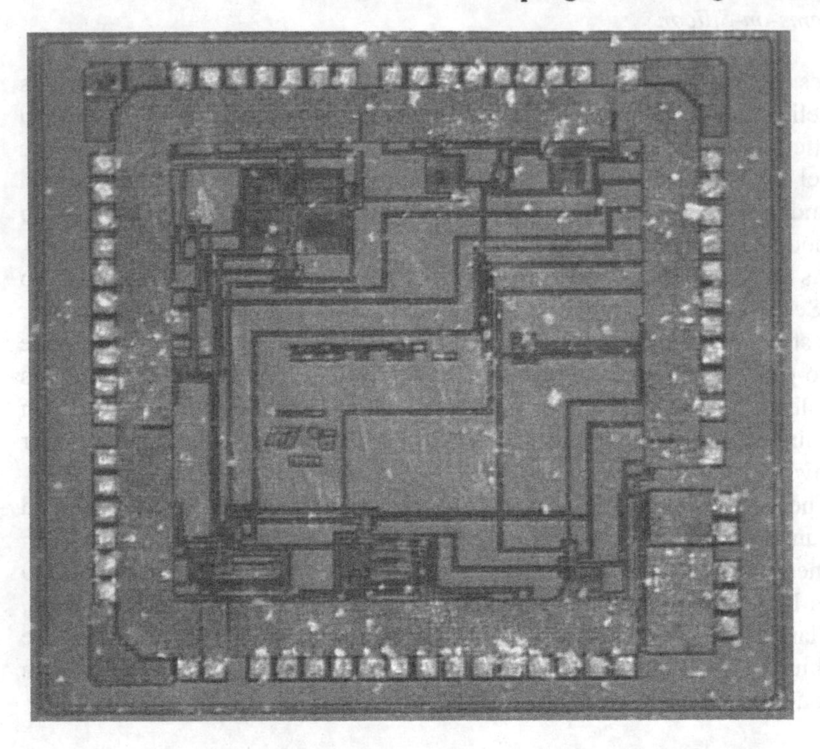

Figure 9-10. The test chip in 0.35um CMOS.

Large inverters serve as digital blocks that can generate interference into the substrate. These blocks are placed at various places so that distance dependence of the silicon-crosstalk can be checked. In order to minimize the crosstalk due to the bonding wires and/or the package, these inverters are driven with a small sinusoidal signal. A second measure in minimizing measurement disturbance is that we measure at the third harmonic of the sinusoidal signal that drives the inverter: the third harmonic can only be generated on-silicon by the large inverter and not by the signal source, nor by the crosstalk on the PCB nor by the crosstalk via the bonding wire.

Figure 9-11 shows the model used in the simulations. The PCB-part of the model consist of the power supply and the sine-wave voltage for perturbing the inverter. The package section models series inductances and resistances of the package [4]. Transistors M5 and M7 together with feedback resistor R0 form an self-biased inverter that converts the sinusoidal voltage source on the PCB to a square-wave signal. The inverters formed by transistors M4 and M6 is 50 times larger than a minimum sized inverter. The output of the inverter is unloaded so that only the devices themselves inject interference into the substrate. Capacitors Cvcc and Cgnd model Vcc and ground-wire capacitance to the substrate. The backgates of the NMOS devices are connected to the substrate. According to literature, the pure CMOS substrate can be assumed to be a short [5,6], also see Chapter 12. All substrate connections of the sensitive analog block are directly connected to this ground node.

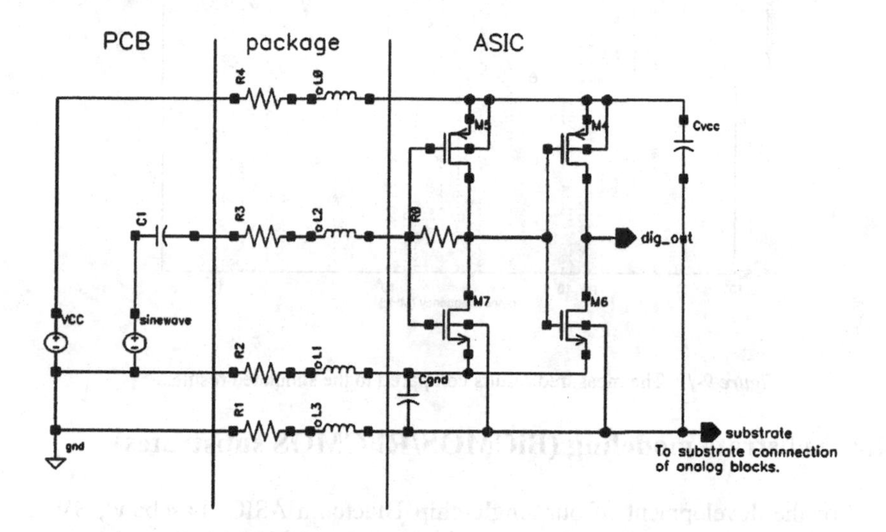

Figure 9-11. The model of the digital circuit and the substrate.

Figure 9-12 shows a comparison between the measurement results and the simulated results for one of the blocks. Along the X-axis we have the frequency of the interference (the third harmonic of the frequency with which the inverters switch), and along the Y-axis we have the measured (denoted by a "*") and simulated (denoted by the "o") level of interference (in dBμV) at the output of the front-end. As can be seen from this figure, there is a very good correlation between measured results and simulated results, the maximum difference is 15 dB at 60 MHz, while at all other frequencies, the difference is below 5 dB. The distance between digital inverter and sensitive analog block proved to be insignificant, as confirmed by literature [6,7,8].

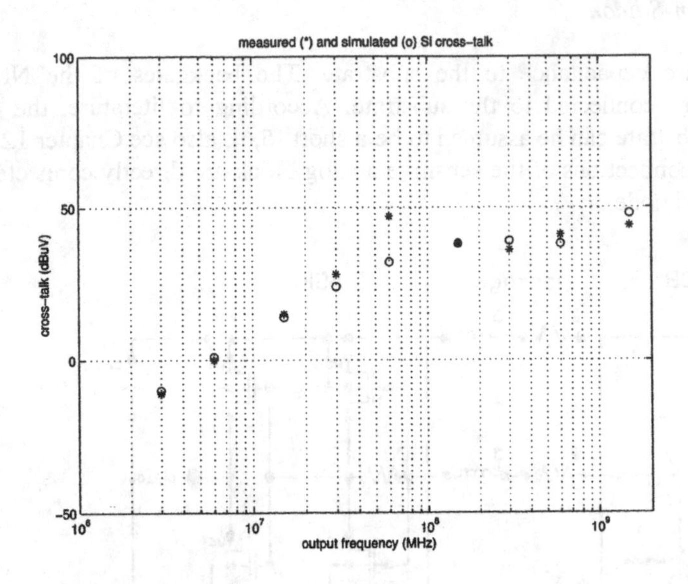

Figure 9-12. The measured results compared to the simulated results.

4.4 Substrate modeling (BiCMOS/RFCMOS substrates)

For the development of our single-chip Bluetooth ASIC (baseband, SW and radio are integrated on the same single die), a BiCMOS-like substrate was chosen. We estimated the silicon area of all blocks and the number of substrate contacts per block. *Figure 9-13* shows the floorplan of the ASIC. This floorplan was then used, together with the program Substratestorm [9], to generate an RC-netlist as model for our substrate.

Even this simplified substrate model proved to be so large that simulations were hardly possible: the first generated netlist used 225 nodes, was 2.4 Mbyte large and needed 9 seconds for a DC Operating Point analysis and 9 seconds/point for an AC-analysis. This first netlist was too large, both in physical size and the number of nodes, and in terms of simulation time. Also the range in element values i.e. resistors with values of milli-ohms and mega-ohms, is very large. A special routine reduced this very large range. After another iteration, the third and final netlist had 30 nodes, was 1 Mbyte large and needed 4 seconds for a DC Operating Point and 2 seconds/point for an AC-analysis.

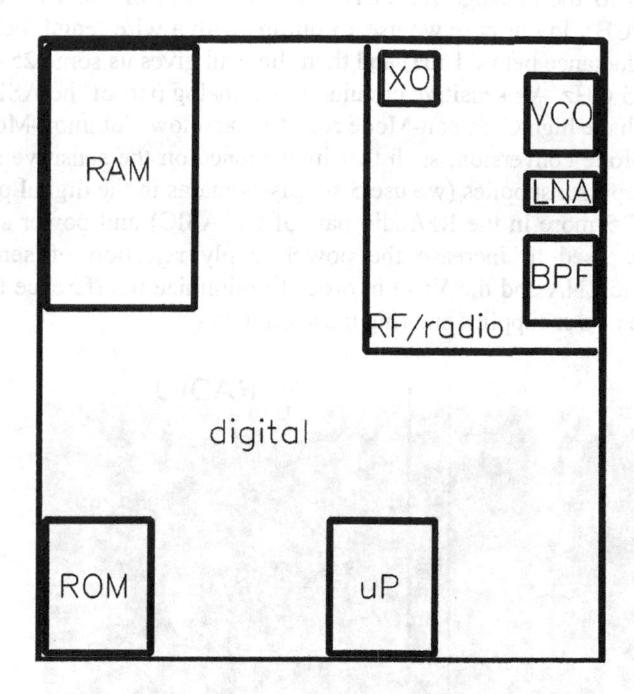

Figure 9-13. The floorplan of the single-chip Bluetooth ASIC.

4.5 Overall simulations

The digital part of the ASIC can now be modeled as 5 large inverters (one for each clock domain: ROM, RAM, uP, standard-cells and the IO's), in total modeling 1.5 Mgates with just 5 NMOS and 5 PMOS transistors. See Section 4.1 for exampes of the size of such blocks. The combination of the 5 large inverters, the substrate netlist and the sensitive analog circuitry, like the LNA and VCO, can now be simulated. *Figure 9-1* and *Figure 9-11* show test-benches that can be used for this purpose. Effects of the digital interference on the analog circuitry could be minimized or gave rise to additional specifications on the analog blocks.

In the case of our single-chip Bluetooth ASIC, several measures were taken for minimizing the crosstalk problem: in the layout, a specially designed P-type wall isolates the radio from the baseband, also see [10]. This wall is approximately 300 μm wide and has multiple bumps to the BGA package such that all digital interference travelling to the analog part is picked up and shorted to the ground connection in the package. The isolation provided by this wall is dependent on the number of ground connections from the wall to the package (or PCB) and the length of the wire to the package (or PCB). In our case we use 13 bumps with a wire-length below 1 mm (or an inductance below 1 nH) and then the wall gives us some 25 dB of isolation at 2.5 GHz. All sensitive circuits in the analog part of the ASIC are balanced and have high Common-Mode rejection and low Common-Mode to Differential-Mode conversion, such that interference on the sensitive nodes is minimal. Separate supplies (we use 5 supply domains in the digital part of the ASIC and 5 more in the RF/radio part of the ASIC) and power supply regulators are used to increase the power-supply rejection of sensitive circuits like the LNA and the VCO in order to minimize interference that is present on the power supplies entering these circuits.

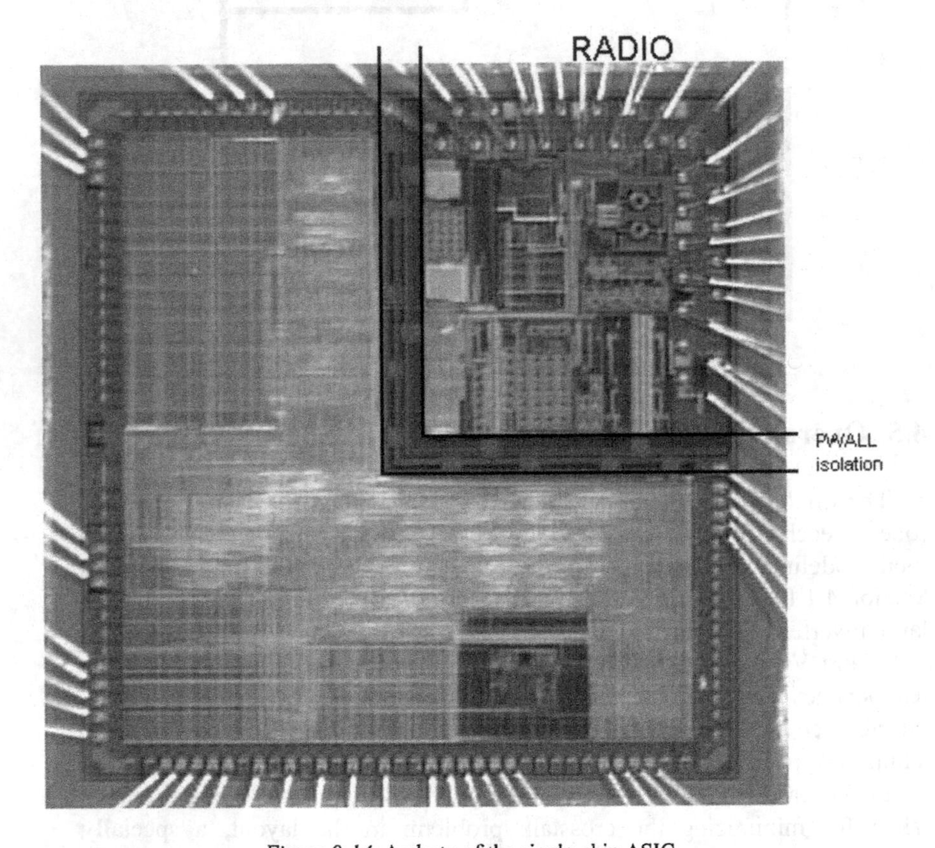

Figure 9-14. A photo of the single-chip ASIC.

Simulations using a test-bench as presented in *Figure 9-1* and *Figure 9-11* show that the interference is more than 20 dB below the sensitivity level of the receiver in the case of bumping the die in a package. A small degradation of some dBs can be expected when the isolation wall is connected via long bonding wires. Measurements on the ASIC, both in bonded and bumped version, show a sensitivity of the receiver of -82 dBm, while Bluetooth transmission and reception is taking place (so the digital part is running). *Figure 9-14* shows a photograph of the Bluetooth single-chip ASIC [11].

5. CONCLUSIONS

When you consider the silicon-crosstalk issue seriously you should do pre-layout simulations, create problem awareness in the design-team, get input from digital designers and start with the silicon-crosstalk problem from the start.

We found the following issues important in reducing silicon-crosstalk in our system-on-silicon:

- Specify power-supply-rejection. Design and implement voltage regulators if the specification cannot be reached.
- Use balanced circuits, specify Common-Mode rejection, and Common-Mode to Differential-Mode conversion. Be aware that simulations incorporating mismatch data on devices is required.
- Specify pushing and pulling for (voltage-controlled-) oscillators and design your oscillators such that the pushing and pulling specifications are fulfilled.
- Use shielded bond-pads and bump-pads for all pads including the digital ones.
- Use triple-well to isolate your NMOS devices from the substrate.
- Use bumping to realize a low impedance to the PCB ground.
- Separate power supplies and grounds, and isolate the interconnect from analog parts and digital parts of the layout.
- A BiCMOS/RFCMOS substrate in general increases the isolation of circuits.

However, all these issues increase the design complexity of the analog blocks and consequently RF design becomes (even more) multidimensional as we get more compromises in (RF) design!

And finally, do not forget to solve all other crosstalk issues (PCB, package, power decoupling) as well!

ACKNOWLEDGEMENT

The author acknowledges STMicroelectronics for processing the 0.35 μm CMOS test-chip, Didier Belot from STMicroelectronics and Francois Clement from Simplex for generating the RC-netlist from our floorplan for the Bluetooth ASIC. Futher acknowledgement goes to my colleagues Jan-Wim Eikenbroek, Peter-Paul Vervoort, Jurjen Tangenberg, Ids Keekstra for help and discussions.

REFERENCES

[1] P.T.M. van Zeijl et all, "Rembrandt: an RF ASIC for DECT TDMA applications", *ESSCIRC'97*, September 1997, Southampton, UK.

[2] prTBR6: *General Terminal Attachment Requirements for the Digital European Cordless Telecommunications standard (DECT),* European Telecommunications Standards Institute, June 18, 1996.

[3] E. Charbon, P. Milliozzi, L.P. Carloni, A. Ferrari, A. Sangiovanni-Vincentelli, "Modeling Digital Substrate Noise Injection in Mixed-Signal IC's", *IEEE Transactions on Computer-Aided Design of Integrated Circuits and Systems*, Vol.18, No.3, March 1999.

[4] M. Felder and J. Ganger, "Analysis of Ground-Bounce Induced Substrate Noise Coupling in a Low Resistive Bulk Epitaxial Process: Design Strategies to Minimize Noise Effects on a Mixed-Signal Chip", *IEEE Tranactions on Circuits and Systems-II: analog and digital signal processing*, Vol.46, No.11, November 1999.

[5] B. Nauta and G. Hoogzaad, "Substrate Bounce in Mixed-Mode CMOS ICs", *ETTCD 1997*, August 1997, Budapest, Hungary.

[6] D.K. Su, M.J. Loinaz, S. Masui and B.A. Wooley, "Experimental results and modeling techniques for substrate noise in mixed-signal integrated circuits," *IEEE Journal of Solid State Circuits*, Vol.28, No.4, April 1993 .

[7] B.R. Stanisic, N.K. Verghese, R.A. Rutenbar, L.R. Carley and D.J. Allstot, "Addressing substrate coupling in mixed-mode IC's: simulation and power distribution synthesis", *IEEE Journal of Solid State Circuits*, Vol.29, No.3, March 1994.

[8] R. Gharpurey and R.G. Meyer, "Modeling and analysis of substrate coupling in integrated circuits", *IEEE Journal of Solid State Circuits*, Vol.31, No.3, march 1996.

[9] http://www.simplex.com.

[10] K. Joardar, "A simple approach to modeling cross-talk in integrated circuits", *IEEE Journal of Solid State Circuits*, Vol.29, No.10, October 1994.

[11] P.T.M. van Zeijl et all, "A 0.18um CMOS Bluetooth ASIC", *ISSCC2002*, February 2002, San Francisco.

Chapter 10

THE REDUCTION OF SWITCHING NOISE USING CMOS CURRENT STEERING LOGIC

Maher Kayal, Richard Lara Saez and Marc Pastre
Swiss Federal Institute of Technology-EPF-Lausanne- STI-IMM-LEG

Abstract: The main advantage of the current steering technique is the small amount of noise generated during state commutations of logic gates. However, it presents a steady state consumption, which is considered as a limitation for low power applications when compared to the conventional static logic.

1. INTRODUCTION

The current steering technique has been used in bipolar technology, basically with the objective of speeding up logic gates. Recently, this strategy has been exploited in bulk CMOS technology [1]. The objective is to reduce the generation of noise by digital circuits, in order to preserve the integrity of analog signal processing in mixed-mode circuits.

This chapter presents the CMOS Current Steering Logic (CSL) and CMOS Folded Source Coupled Logic (FSCL).

The CSL is a single-ended logic family, whereas the FSCL is a fully differential logic.

A thorough analysis is developed for CSL and FSCL inverters. Theoretical expressions are given for static and dynamic characteristics. Special attention is given to the current spikes generated during switching of logic gates.

The relationship between gate performance and technological parameters, transistor dimensions, power consumption is addressed in this chapter.

2. DEFINITIONS

Figure 10-1 presents the Voltage Transfer Curve (VTC) of a generic inverter along with the definitions of the following nominal voltages:

- V_{OH} Output-high voltage,
- V_{OL} Output-low voltage,
- V_{IH} Input-high voltage,
- V_{IL} Input-low voltage,
- V_{TH} Gate threshold voltage, i. e., the point where the output voltage is equal to the input voltage.
- NM_H and NM_L are the high and low noise margins, respectively.

Additional information about static characteristics can be found in [4] and [5]. Throughout this text, the current factor $[\mu_0 \cdot Cox \cdot (W/L)_a]$ of transistor M_a is denoted by β_a, and the threshold voltage of transistor M_a, type n or p, is denoted by $V_{TOn,p}^{\alpha}$. In the transient analysis, the propagation delay tp_{HL} (tp_{LH}) is defined as the time required for the output to swing from the high (low) voltage to the gate threshold voltage.

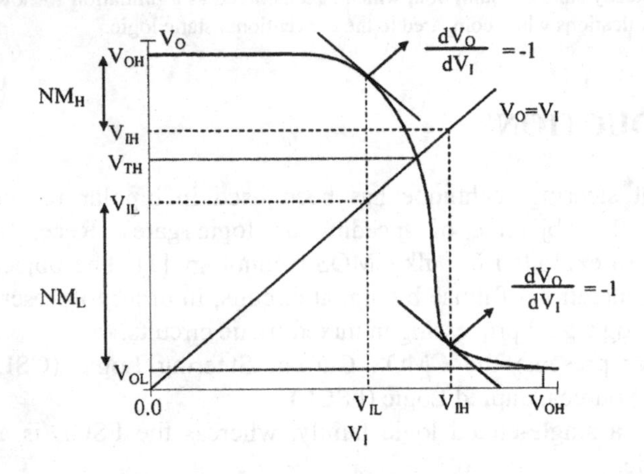

Figure 10-1. Definitions of static characteristics for a generic inverter.

3. CSL INVERTER

Figure 10-2 shows the NMOS version of the CSL inverter together with a bias circuit. The CSL inverter is composed of a p-channel current source (M_P) that is connected to the drain of an NMOS logic transistor (M_L) and to the drain of an NMOS diode-connected transistor (M_D). The inverter input is connected to the gate of M_L.

The bias circuit is composed of a PMOS diode-connected transistor biased by a current source. The output of the bias circuit is shared with the CSL gates.

In the analysis that follows, we assume that the current through M_P is constant and equal to I, whatever happens to the output voltage. In fact M_P is assumed to operate in the saturation region and second order effects such as channel length modulation are neglected. When V_I is logic high, the current through M_L defines the low logic level. It is designed to be less than V_{TOn}^D, consequently cutting off M_D. When V_I is logic low, M_L is off and the current through M_D defines the output high voltage.

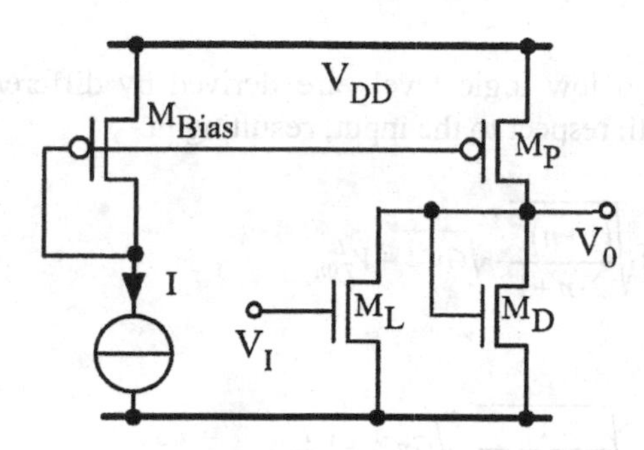

Figure 10-2. Complete CSL inverter schematic, with bias structure.

3.1 Static Characteristics

The CSL inverter static and dynamic characteristics are here defined as a function of G_v, the ratio of β_L to β_D (1). G_v can be adjusted by changing the length of M_D, and keeping M_L unchanged.

$$Gv = \frac{\beta_L}{\beta_D} \tag{1}$$

Using the EKV MOSFET model in strong inversion [2], V_{OH} is determined by the current flowing through the diode-connected transistor:

$$V_{OH} - V_{TOn}^D = \sqrt{\frac{2 \cdot n \cdot I \cdot Gv}{\beta_L}} \tag{2}$$

The derivation of the expression for V_{OL} is straightforward from the expression of the current that flows through M_L when V_I is V_{OH}. V_{OL} is given by:

$$V_{OL} = \left(V_{OH} - V_{TOn}^{L}\right)\cdot\left(1-\sqrt{1-Gv^{-1}}\right) \tag{3}$$

To find the gate threshold voltage, V_{TH}, we set V_I=V_O=V_{TH}, that leads to:

$$V_{TH} = V_{TOn}^{L} + \left(\frac{V_{OH} - V_{TOn}^{D}}{\sqrt{1+Gv}}\right) \tag{4}$$

The input high and low logic levels are derived by differentiating the output of the VTC with respect to the input, resulting in:

$$V_{IH} = \left(V_{OH} - V_{TOn}^{D}\right)\sqrt{\frac{(1+n)^2}{2\cdot n+1}}\cdot\sqrt{Gv^{-1}} + V_{TOn}^{L} \tag{5}$$

$$V_{IL} = (V_{OH} - V_{TOn}^{D})\cdot\sqrt{\frac{1}{Gv\cdot n^2+1}}\cdot\sqrt{Gv^{-1}} + V_{TOn}^{L} \tag{6}$$

The CSL inverter characteristics are robust to technological fluctuations because its static characteristics are given by two MOS transistors of the same type, i. e., NMOS transistors that can be physically laid out very close to each other. In the previous analysis, the slope factor (n), has been assumed to be constant.

3.2 Noise Margins

To complete the static characterization of the CSL inverter, it is necessary to determine the parameters that affect noise margins. They are deduced directly from the static characteristics previously derived. High noise margin (NM_H) and low noise margin (NM_L) are given in (7) and (8).

$$NM_H = \left(V_{OH} - V_{TOn}^{D}\right)\cdot\left(1-\sqrt{\frac{(1+n)^2}{2\cdot n+1}}\cdot\sqrt{\frac{1}{Gv}}\right) \tag{7}$$

$$NM_L = \left(V_{OH} - V_{TOn}^D\right) \cdot \left(A - B\right) + V_{TOn}^L \quad (8)$$

$$\text{Where: } A = \sqrt{\frac{1}{G_V}} \cdot \sqrt{\frac{1}{n^2 . G_V + 1}} \text{ and } B = \frac{1}{n} . (1 - \sqrt{1 - G_v^{-1}}) \quad (9)$$

NM_L has a V_{TOn}^L - dependent term that changes slightly with temperature and over the chip due to mismatch. The first term in (8) is similar to the term that gives the high noise margin. This term depends on temperature by the mobility factor, and depends on mismatch as well.

These expressions show that noise margins are not symmetric, and that NM_H is the smallest noise margin.

3.3 Dynamic Characteristics

To derive the rise and fall propagation delays for the CSL inverter, the simplified scheme shown in *Figure 10-3* is used. The load capacitance C_L includes wiring; fan out, as well as the inverter's capacitance associated with the output node. In order to simplify the analysis, the input is assumed to be a step voltage.

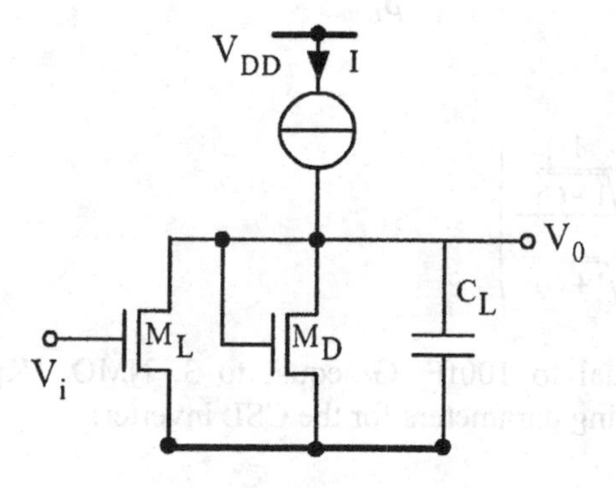

Figure 10-3. Circuit used to derive the rise and fall propagation times

For a step input, applied to M_L, ranging from V_{OL} to V_{OH} and with the output voltage at V_{OH}, this transistor pulls the output voltage down, discharging C_L. The diode-connected device conducts until the output voltage reaches V_{TOn}^{D}, and finally M_L brings the output voltage to V_{OL}. The resultant expression for the fall propagation time (tp_{HL}) is given by:

$$tp_{HL} = -\frac{\frac{2\cdot n\cdot CL}{\beta_D}}{\sqrt{\frac{2\cdot n\cdot I}{\beta_D}}\cdot\sqrt{Gv-1}}\cdot(A-B) \qquad (10)$$

$$\text{Where: } A = arctg\left[\frac{1}{\sqrt{(Gv)^2-1}}\right] \text{ and } B = arctg\left[\frac{1}{\sqrt{Gv-1}}\right] \qquad (11)$$

For a step input varying from V_{OH} to V_{OL}, transistor M_L is assumed to be instantaneously cut off, and the bias current charges capacitance C_L. Transistor M_D is off until the output voltage reaches V_{TOn}^{D}, then it starts conducting. After that, the output voltage increases until it reaches V_{OH}. The resultant expression for the rising propagation time (tp_{LH}) is:

$$tp_{LH} = \frac{CL}{I}\cdot\left(V_{TOn}^{D} - V_{OL}\right) + \frac{CL}{\beta_D\cdot\sqrt{\frac{2\cdot n\cdot I}{\beta_D}}}\cdot\ln(C) \qquad (12)$$

$$\text{Where: } C = \left(\frac{1-\frac{1}{\sqrt{1+Gv}}}{1+\frac{1}{\sqrt{1+Gv}}}\right)$$

Assuming C_L equal to 100fF, G_v equal to 3, NMOS Kp=127µA/V^2, n=1.4, and the following parameters for the CSL inverter:

- $I=20.10^{-6}$ A,
- $(W/L)_D$=2.5µm/3µm,
- $(W/L)_L$=2.5µm/1µm,
- $(W/L)_P$=8µm/1µm

This results in: tp_{HL}=0.71ns

This result is independent of V_{DD}, as long as the power supply voltage is high enough to saturate the p-channel current source. On the other hand, tp_{LH} has two terms, one due to the charging of a capacitive load by a constant current source, the second one referring to the situation when the diode-connected transistor starts to conduct. Considering the same parameters as before, and a V_{TOn} of 0.75V, the low to high propagation time is: tp_{LH}=5.27ns.

The propagation time of an inverter with 100fF of capacitive load, calculated by the average of the two propagation time gives: tp ˜ 3ns.

3.4 Current Spikes

Ideally, the CSL structure does not present current spikes. However, in practical implementations, there exists a parasitic capacitance connecting the inverter output to the power supply. When the output switches between logic levels, a current spike flows into the power supply bus due to the parasitic capacitive coupling of the output with the power supply.

This section presents analytical expressions that give insight about factors that affect the current spikes.

Figure 10-4(a) shows a CSL inverter together with its capacitances. *Figure 10-4(b)* shows the equivalent circuit used in the analysis of the current spikes. This circuit is composed of the CSL inverter and an equivalent capacitance that has been split into C_{CS} and C_{EQ}. C_{CS} is the drain-to-bulk parasitic capacitance of the PMOS transistor. C_{EQ} is the sum of the load capacitance C_L, the drain-to-bulk capacitance of M_L, the drain-to-bulk, gate-to-bulk and gate-to-source capacitances of M_D.

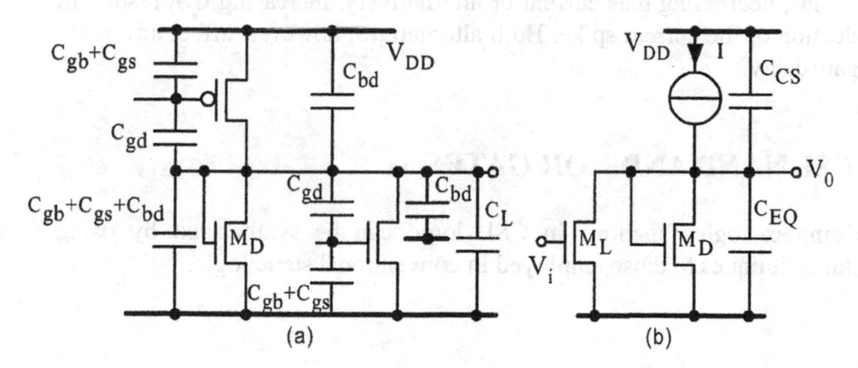

Figure 10-4. Equivalent circuit of the CSL inverter used to analyze the current spikes on the power supply line. (a) Schematic showing transistor and load capacitances. (b) Equivalent circuit used to find the expressions for the current spikes.

The current spikes were calculated for a step signal applied to the input of the CSL inverter. These spikes occur for both the rising and falling edges of the output. Therefore, two expressions are necessary to entirely characterize the current spikes. The following notation is adopted for these two expressions:

- I'MAX: peak value of the current spike associated with the rising edge of the input.
- I"MAX: peak value of the current spike associated with the falling edge of the input.

Consider the equivalent circuit of *Figure 10-4(b)*. Let us assume the current through C_{CS} is the main component of the current spikes. Solving the differential equation for the transient behavior and, using the EKV MOSFET model in strong inversion, the resultant expressions for I'_{MAX} and I''_{MAX} are given by:

$$\left|\frac{I'_{MAX}}{I}\right| = \frac{C_{CS}}{C_{EQ}+C_{CS}} \cdot [Gv-1] \tag{13}$$

$$\left|\frac{I''_{MAX}}{I}\right| = \frac{C_{CS}}{C_{EQ}+C_{CS}} \tag{14}$$

From (13) and (14), the reduction of C_{CS} is the most efficient way to reduce the amplitude of the current spikes. In both expressions, the perturbation is proportional to the bias current, and inversely proportional to C_{EQ}. Thus, decreasing bias current or alternatively, increasing C_{EQ} results in a reduction of the current spike. Both alternatives, however, affect adversely the gate delay.

4. CSL NAND AND NOR GATES

Complex logic functions in CSL logic can be synthesized by using similar techniques to those employed in conventional static logic.

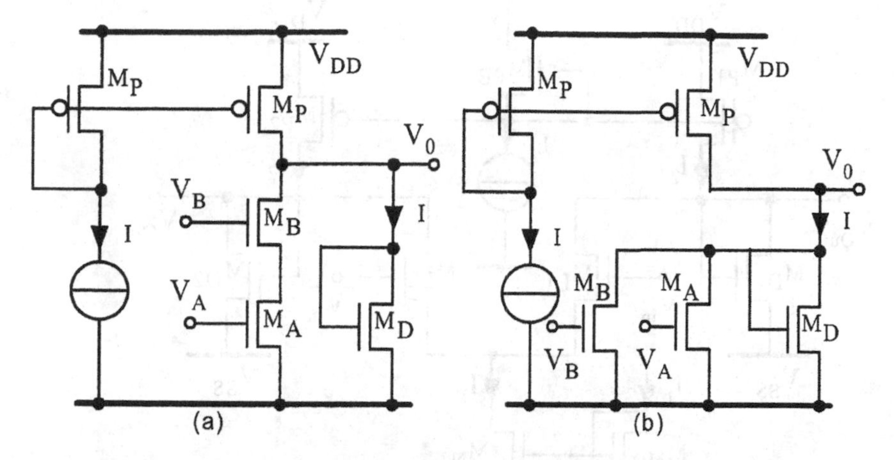

Figure 10-5. (a) Two-input CSL NAND gate. (b) Two-input CSL NOR gate.

5. FSCL INVERTER

A complete scheme of the CMOS Folded Source Coupled Logic (FSCL) inverter (*Figure 10-6*), including bias circuits, is shown in [6]. An external current source, biases the diode connected transistor (M_{NB}) with a current I_1. The same strategy is used to bias the p-channel current sources.

The outputs of the current mirrors are connected to the differential pair, where I_1 is divided between transistors M_{L1} and M_{L2}. At any time, $2 \cdot I_2$ flows through the differential structure. This method reduces dramatically current variations on power supply lines during state transitions. This reduction is one of the greatest advantages of the FSCL technique. The differential structure of the FSCL inverter allows reliable operation at a small logic swing. This characteristic becomes thus a new degree of freedom to be exploited in static as well as in dynamic characteristics of this technique. Alternative versions of the FSCL inverter are possible and can be found in [6]. The version analyzed in this chapter presents the best performance in terms of power delay product, an essential characteristic for low power applications.

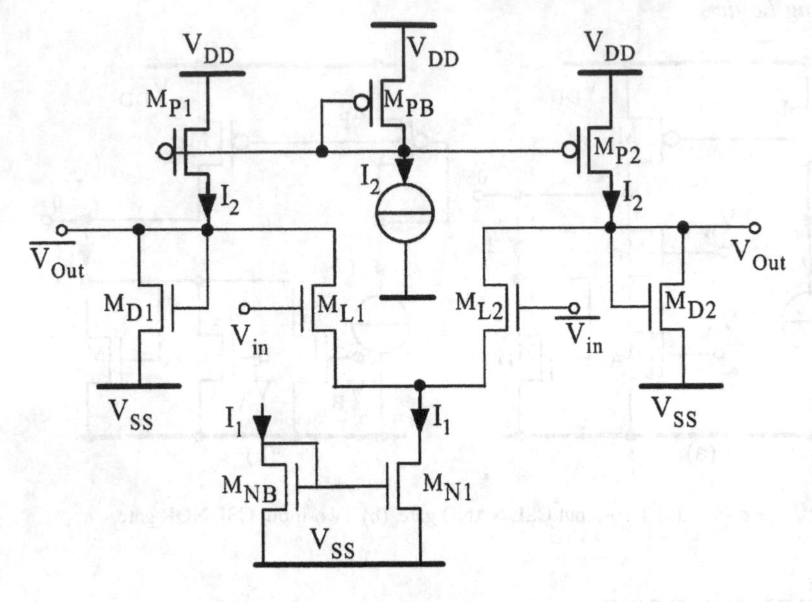

Figure 10-6. CMOS FSCL Inverter

5.1 Static Characteristics

The logic levels in the FSCL inverter are generated by the currents that flow through M_{D1} and M_{D2}. These currents correspond to a difference between I_2 and a component that is sunk by the differential pair. For a proper operation of the inverter, I_2 must be larger than I_1. So, a design parameter ξ is defined by:

$$\xi = \frac{I_1}{I_2} \tag{15}$$

Therefore: $$V_{OH} = V_{T0N}^{D} + \frac{\Delta V_L}{1-\sqrt{1-\xi}} \tag{16}$$

$$V_{OL} = V_{T0N}^{D} + \Delta V_L \cdot \frac{\sqrt{1-\xi}}{1-\sqrt{1-\xi}} \tag{17}$$

Logic swing is defined as the difference between logic levels. It will be referred to as ΔV_L in the sequel; its expression is given by:

$$\Delta V_L = V_{OH} - V_{OL} = \sqrt{\frac{2 \cdot n \cdot I_2}{\beta_d}} \cdot \left(1 - \sqrt{1 - \xi}\right) \quad (18)$$

$$V_{TH} = V_{TON}^{D} + \frac{\sqrt{1 - \frac{\xi}{2}}}{1 - \sqrt{1 - \xi}} \cdot \Delta V_L \quad (19)$$

5.2 Noise Margin

Considering that, I_{ML1} and I_{ML2} are the currents of transistors M_{L1} and M_{L2} (*Figure 10-6*) normalized to the tail current I_1.

$$I_{ML1(2)} = \frac{I_1}{2} \cdot \left(1 + (-) \sqrt{\frac{\beta}{2 \cdot n \cdot I_1}} \cdot Vd \cdot \sqrt{2 - \frac{\beta}{2 \cdot n \cdot I_1} \cdot Vd^2} \right) \quad (20)$$

With $V_d = V_{in} - \overline{V_{in}}$

If $V_d/(2 \cdot n \cdot I_1/\beta)^{1/2}$ equals to ± 1, I_1 is completely steered to one transistor of the differential pair. The factor that normalizes V_d is frequently referred to as saturation voltage of the differential pair. Its expression is given by:

$$V_{SAT} = \sqrt{\frac{2 \cdot n \cdot I_1}{\beta}} \quad (21)$$

According to *Figure 10-7*, V_{OH} and V_{OL} denote the output voltages when one of the input transistors is cut off and the other conducts the whole tail current I_1. V_{IH} and V_{IL} are here defined as $\pm V_{SAT}$, the saturation voltage of the differential pair, alternatively to the conventional definition of slope -1. Rigorously, V_{SAT} is not equal to the input voltage for which $dV_O/dV_I = -1$. However for practical purposes the two voltages are about the same.

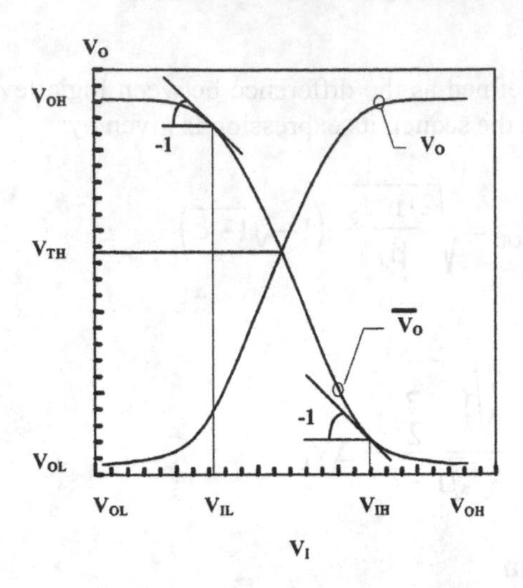

Figure 10-7. Voltage transfer characteristic of a differential structure.

Therefore, the definition of noise margins for the FSCL inverter is given by the saturation voltage criterion that leads to satisfactory results when compared to the slope criterion and greatly simplifies analytical expressions.

Furthermore, to ensure a safe operation of a FSCL logic gate, ΔV_L should be considered as:

$$NM = \Delta V_L - V_{SAT} \tag{22}$$

5.3 Dynamic Characteristics

To find the output rise (tp_{LH}) and fall (tp_{HL}) propagation times, current sources are assumed to be ideal and each output of the FSCL inverter is loaded by a capacitance C_L. The inverter is excited by complementary step signals, and propagation delays are computed from a given logic level (V_{OH} or V_{OL}) to the logic threshold voltage (V_{TH}).

During the transient, the differential pair is assumed instantaneously saturated. Using the EKV MOSFET model in strong inversion gives the rise and fall propagation time respectively.

$$tp_{LH} = \frac{\Delta V_L}{2} \cdot \frac{1}{I_2} \cdot \frac{CL}{1-\sqrt{1-\xi}} \cdot \ln\left\{ \frac{\sqrt{1-\xi}-1}{\sqrt{1-\xi}+1} \cdot \frac{\sqrt{1-\frac{\xi}{2}}+1}{\sqrt{1-\frac{\xi}{2}}-1} \right\} \tag{23}$$

$$tp_{HL} = \frac{\Delta V_L}{2} \cdot \frac{1}{I_2} \cdot \frac{CL}{\sqrt{1-\xi}-1+\xi} \cdot \ln\left\{ \frac{\sqrt{1-\frac{\xi}{2}}+\sqrt{1-\xi}}{\sqrt{1-\frac{\xi}{2}}-\sqrt{1-\xi}} \cdot \frac{1-\sqrt{1-\xi}}{1+\sqrt{1-\xi}} \right\} \quad (24)$$

5.3.1. Current Spikes

FSCL circuits are biased by constant currents, which allow for a dramatic reduction of current spikes on the power supply line. The residual current spikes result from a complex net of parasitic or functional capacitances, represented for a section of the entire inverter in *Figure 10-8(a)*. In *Figure 10-8(b)*, a simplified model is displayed. It will be used to derive some expressions concerning the current spikes.

Each FSCL inverter output presents a connection with the V_{DD} power supply line at least through the PMOS drain to bulk capacitance. When the logic gate switches between states, the voltage variation across C_{CS} produces a current spike on the power supply line. The analysis is split into two cases: one for output rising transitions (IN_R) and one for output falling (IN_F) transitions.

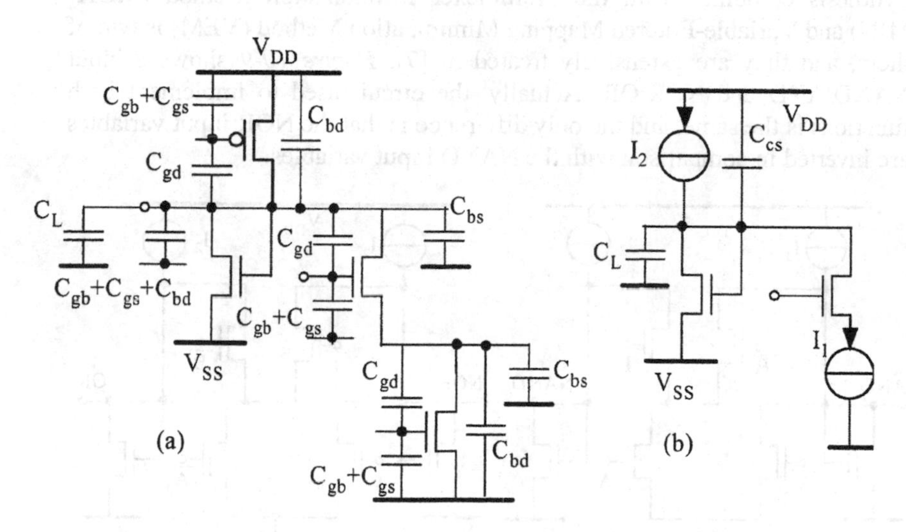

Figure 10-8. (a) Schematic showing the parasitic capacitances of each transistor; (b) simplified model used to derive the current spikes.

$$IN_R = -\frac{C_{CS}(\xi)}{C_L(\xi)+C_{CS}(\xi)} \cdot I_2 \cdot \xi \tag{25}$$

$$IN_F = +\frac{C_{CS}(\xi)}{C_L(\xi)+C_{CS}(\xi)} \cdot I_2 \cdot \xi \tag{26}$$

When the FSCL inverter changes logic state, the outputs present complementary transitions. As far as ideal symmetry is considered, IN_R and IN_F cancel each other. In practical implementation, however, this symmetry does not exist and the total cancellation is not achieved. Causes for the asymmetry of rise and falling spikes are mismatch and nonlinear capacitances.

5.4 COMPLEX GATES IN FSCL

There are various methods to generate complex logic gates in FSCL and reduce the number of transistors per logic function. The Series-Gating Synthesis combined with the Multiplexer Minimization Method (MUX-MIN) and Variable-Entered Mapping Minimization Method (VEM) is two of them, and they are extensively treated in [7]. *Figure 10-9* shows 2-input NAND/AND and NOR/OR. Actually, the circuit used to implement both functions is the same, and the only difference is that the NOR input variables are inverted in comparison with the NAND input variables.

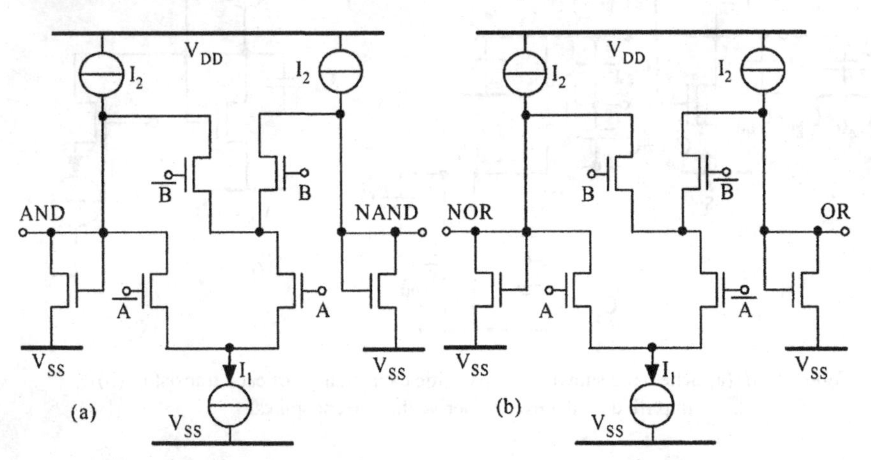

Figure 10-9. (a) FSCL NAND/AND function. (b) FSCL NOR/OR function.

6. EXPERIMENTAL COMPARISON BETWEEN STATIC LOGIC AND CSL

Two experimental circuits have been designed to compare current steering approach to standard static logic.

6.1 Switching noise sensing

The comparison of noise generation depicted below is based on measurements. An integrated circuit featuring noise sensors (NMOS transistors) and noise sources (logic inverters) has been integrated using N-Well epi-2µm technology on a Sea of gate structure. *Figure 10-10* shows the schematic of the noise measurement circuit.

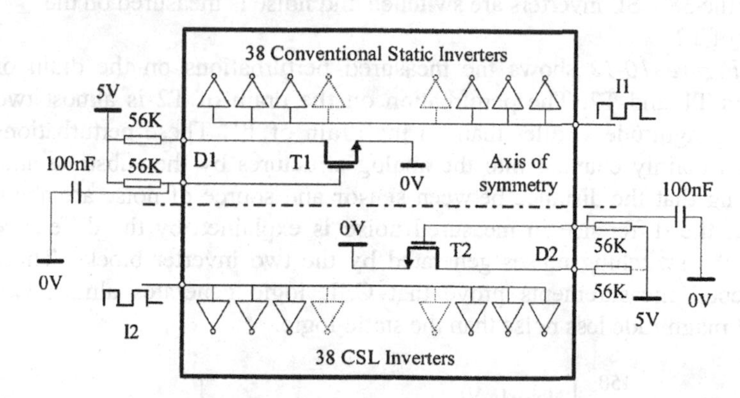

Figure 10-10. Schematic diagram and relative position of elements in test chip.

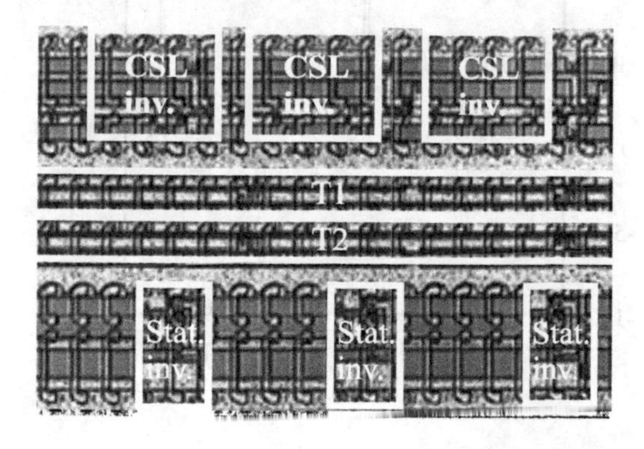

Figure 10-11. Microphotograph of a part of Sea of gate 2µm test chip.

The noise sensor is implemented with NMOS transistors (T1 and T2) with W/L = 140/20 biased as common source amplifiers. Sensors are placed equidistant from a symmetry axis. Two separate logic blocks implemented using CMOS Static and CSL logic families, with 38 inverters each were also placed symmetrically to the same axis. To minimize capacitive coupling between wires, the chip has been carefully laid out. For instance, the coupling between digital input signals and power supply lines has been reduced. In order to avoid power supply fluctuations, input and output buffers have not been used. The power supply line and the transistor gates have been filtered in order to reduce off-chip perturbations.

The input signals (I1, I2), used to excite the inverter chain have been alternatively switched. The measurements were taken in two steps:

- The 38 static inverters are switched and the coupled noise is measured on the drain of T1.
- Then the 38 CSL inverters are switched and noise is measured on the drain of T2.

The *Figure 10-12* shows the measured perturbations on the drain of transistors T1 and T2. The perturbation on the drain of T2 is almost two orders of magnitude smaller than on the drain of T1. These perturbations have been mainly coupled into the analog structures by the substrate and, considering that the distance between sensor and source of noise are about the same, the difference in measured noise is explained by the difference between the switching noises generated by the two inverter blocks. These experimental measurements prove that CSL logic generates almost two orders of magnitude less noise than the static logic.

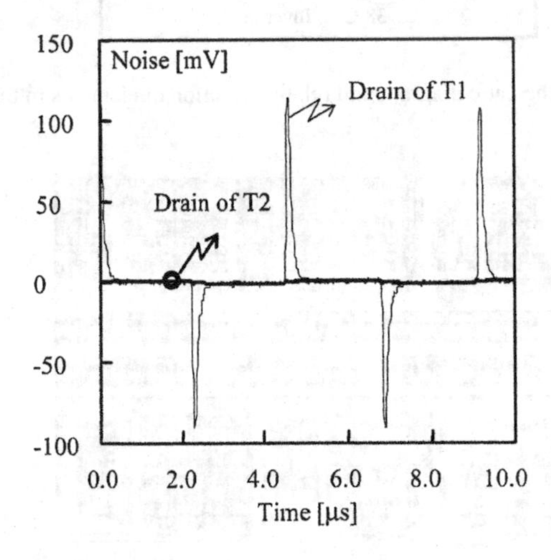

Figure 10-12. Measured noise perturbations on drains of T1 and T2.

6.2 Comparison between CSL and Standard static logic in a Mixed-mode application.

A CSL library has been used to implement an algorithm to compensate the offset of an operational amplifier. The main objectives of this system are: to provide 100% of duty cycle for signal processing, to process low frequency low voltage input signals and to have high robustness against temperature variation and mismatch.

The scheme consists of two matched operational amplifiers (OA1 and OA2) working in ping-pong architecture [8]. In this architecture, one OA is processing a signal while the other one is under offset calibration. The auto-zero technique uses a successive approximation algorithm.

The amplifier is very sensitive to any kind of internal and external noise owing to its high gain and low offset voltage. Note that in this integrated circuit the power supply of the analog and digital parts are the same. No shielding technique has been used to protect the analog part from the digital one.

The offset compensation scheme has been selected in order to verify the efficiency of the developed CSL library and to compare CSL to CMOS standard static logic. This comparison is done in terms of area and switching noise.

Test devices were fabricated using ATMEL-ES2 0.7μm CMOS technology. The total circuit area is 0.5x1 mm^2 for CSL implementation as well as for CMOS standard static implementation. Both digital circuitries have been placed and routed automatically.

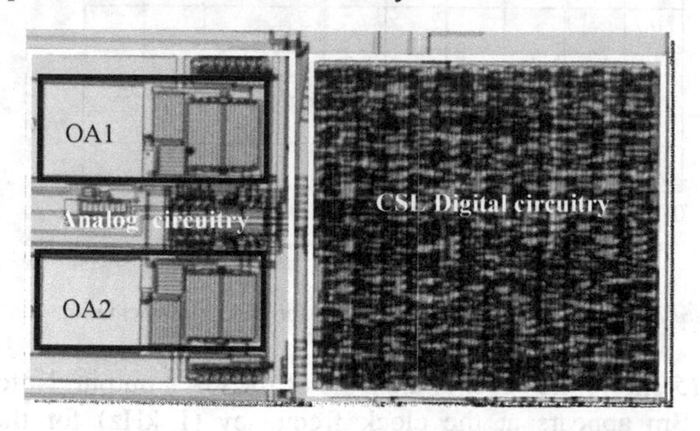

Figure 10-13. Photomicrograph of the CSL implementation.

The only special precautions applied at the layout level of this circuit are:

- separation of the digital from the analog parts (*Figure 10-13*),

– careful routing of the clock signal to avoid cross talk between the analog circuits and clock.

In order to measure the switching noise generated by the digital CSL circuitry, one of the integrated operational amplifier was connected in open loop configuration and used as a noise sensor, when the digital circuitry is active.

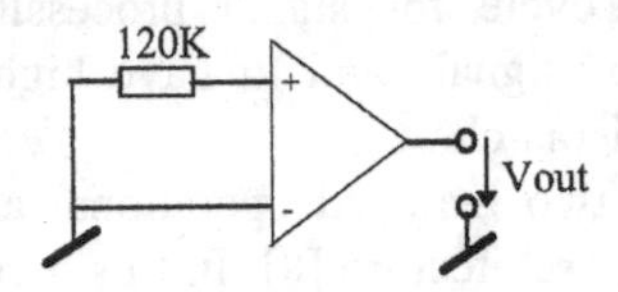

Figure 10-14. Open loop OA configuration used as noise sensor.

V_{out} has been measured under the following conditions:

- OA Offset voltage = 4μV.
- DC open loop gain = 75 dB.
- Clock frequency = 1kHz.

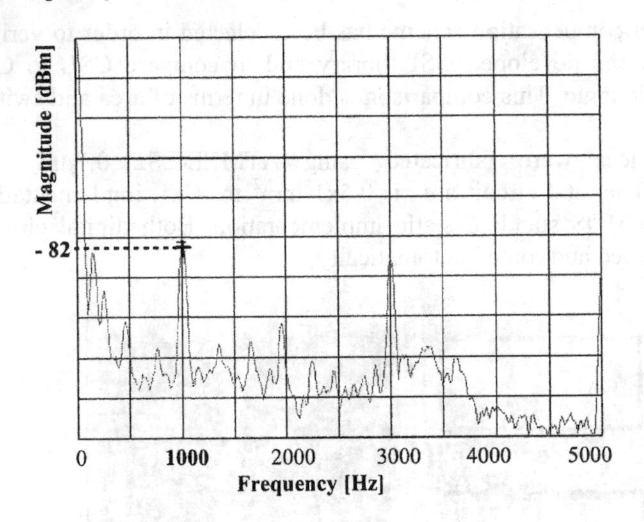

Figure 10-15. Measured OA output signal for CSL implementation

Figure 10-15 shows the spectral analysis of the OA's output. Note that a peak of -82 dBm appears at the clock frequency (1 kHz) for the CSL implementation. The same measurement for standard static logic implementation displays -32 dBm. Considering the proximity between analog and digital parts, and the high gain of the amplifier at the clock frequency, the measured amplitude is very small. This experiment has demonstrated that the switching noise generated by the CSL circuitry is very low compared to standard static logic when the area is the same.

7. COMPARATIVE EVALUATION OF CSL, FSCL AND CONVENTIONAL STATIC LOGIC

The power consumption crisis has changed digital designers' mind in the last seven years in a revolutionary way. The inclusion of the power consumption criterion in the conventional tradeoff between speed and area has become a fertile field where designers' opinions are passionately discussed. The necessary reduction in power supply voltage, and the imperative reduction in power consumption, fundamentally due to technological limits, has created a new discipline that has been spread through all abstraction levels of the integrated circuit design. The most effective way to deal with the power consumption issue is at the architecture and circuit level. It requires low investment and gives returns in the short term. Analog circuit designers, on the other hand, have always managed power consumption. New in this discipline is the reduction of power supply voltage. In mixed-mode circuits there is still another dimension to manage, namely, the amount of noise generated by digital circuits and coupled into analog circuits. In some cases the deleterious action of this perturbation degrades the performance of analog circuits. This section presents a comparative evaluation of CMOS CSL, FSCL and conventional static logic, considering the trends in technology and circuit constraints such as low voltage, low power, and low digital switching noise.

7.1 Power Consumption in CMOS Conventional Static Logic

The power consumption in conventional static logic is composed of:

- Dynamic power consumption - P_D
- Short circuit power consumption - P_S
- Leakage power consumption - P_L

P_D is related to the charge necessary to change the voltage across capacitance CL of ΔV in a time interval t_{ck}. The expression for P_D is

$$P_D = p_t \cdot CL \cdot \Delta V \cdot Vdd \cdot f_{CK} \quad (27)$$

- pt - probability that a transition occurs (activity factor).

- CL - Load capacitance.
- ΔV - Voltage swing.
- Vdd - Power supply voltage.
- f_{CK} - Clock frequency.

It is widely accepted that the predominant part of the power consumption is P_D.

7.2 Power Consumption in CMOS CSL and FSCL

The CSL power consumption presents a dominant component which is the static power dissipation. In a first order approximation, power consumption in CSL is given by the bias current times the supply voltage.

$$P_{CSL} = I.V_{DD} \tag{28}$$

Positive and negative current spikes that could be considered as a dynamic consumption present effects that cancel each other. Therefore, power optimization for this family is generally achieved by a reduction in supply voltage. In order to keep a constant gate delay, the bias current has to be kept constant. The reduction in supply voltage is strongly related to the current source saturation voltage, and high output logic level. The smaller the saturation voltage, the wider the PMOS transistor needs to be, and consequently, the higher is the switching noise injected into the power bus.

Similarly to the CSL family, the static dissipation is the dominant component of the power consumption in a FSCL logic gate. It is calculated by the sum of the bias currents of the logic gate, multiplied by the supply voltage.

Therefore, a solution to reduce the power consumption of CSL logic gates is convenient for FSCL logic gates as well. One of the methods to compare the performance of CSL and FSCL is to characterize their gate delays at the same power consumption, under equivalent loading conditions.

Five different designs for both the CSL and FSCL inverters, corresponding to different values of G_v and ξ, have been made. Propagation delay has been investigated for each of the designed inverters. Theoretical results have been determined based on expressions derived in paragraph 3 and 5 for the CSL and FSCL inverter, and measurements have been extracted from ring oscillators implemented in ATMEL-ES2 0.7 μm technology. The results, displayed in Figure *10*-16, indicate that the CSL inverter is faster than the FSCL inverter under the specified conditions. Calculated and measured values are close to each other.

Table 10-1. Parameters used to compare CSL and FSCL

Parameters	CSL	FSCL
Power supply voltage	2.5V	2.5V
Power consumption/gate	50μW	50μW
Noise Margin	MIN{NMH;NML}≥ $\Delta V_L/2$	NM=$\Delta V_L/2$
Logic swing	Depends on G_v	ΔV_L=500mV

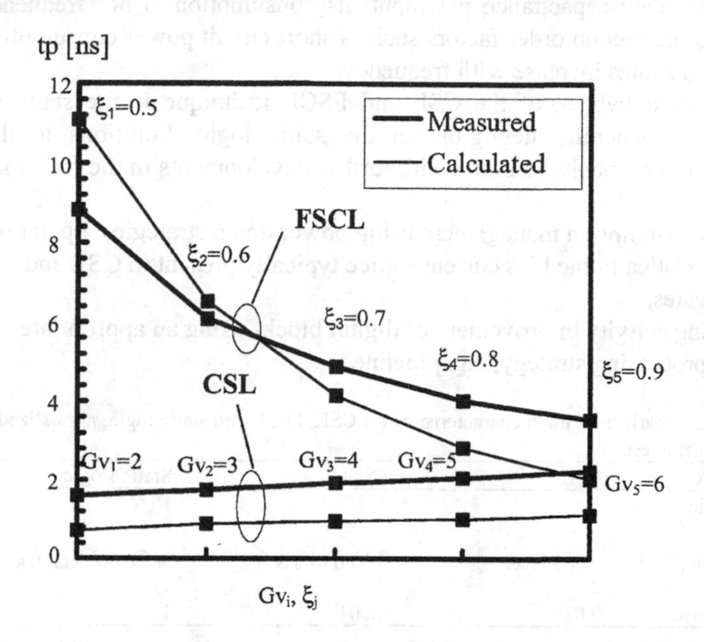

Figure 10-16. Measured and calculated propagation time for CSL and FSCL inverters. The values of Gv_i and ξ_i are displayed directly on the figure.

7.3 Summary

The current spikes produced by CSL and FSCL have been theoretically derived. This result has shown that current spikes generated by FSCL are about one order of magnitude smaller than for the CSL logic family.

The area used by the CSL logic is larger than the static logic for the inverter; however, when gate complexity increases, this factor is slightly inverted. Nevertheless, the FSCL gate area is larger than for the other two logic families, and this factor is not inverted.

For the same power consumption per gate, under the same power supply voltage, the CSL logic presents a propagation delay two times smaller than the FSCL.

Comparison of consumption for CSL and static logic is not straightforward since the consumption of static logic depends on many factors, such as frequency and activity factor. Considering the same power supply voltage and a high activity factor, over certain frequency and for certain functions such as NOR gates, the CSL gate has a lower power consumption that the static logic gate. This comes from the fact that CSL presents less gate capacitance per input, its consumption is not frequency dependent, and second order factors such as short circuit power consumption of static logic gates increase with frequency.

The main drawback of the CSL and FSCL technique is the stand by consumption, generally negligible in the static logic. Solutions to this problem could probably be found with further developments in the following topics:

- power consumption management using power down strategies, operating opportunistically the bias current source typically present in CSL and FSCL gates,
- switching activity improvement of digital blocks using an appropriate signal processing strategy, e.g. pipeline.

Table 10-2. Comparison of main characteristics for CSL, FSCL and static logic, normalized to static logic characteristics.

Characteristic	CSL	FSCL	Static Logic
Area per logic function	1	2	1
Power consumption	$V_{DD} \cdot I_{BIAS}$	$2\ V_{DD} \cdot I_{BIAS}$	$p_t \cdot C_L \cdot \Delta V \cdot V_{DD} \cdot f_{CK}$
Gate delay	2	4	1
Switching noise	0.01	0.001	1

8. CSL DESIGN AND LAYOUT CAD TOOLS

Two CAD tools were developed to assist designers throughout the creation and retargeting of circuits using the CSL approach: CSL_ASLIB (CSL Application Specific LIBraries design) is an automatic library designer and CellEdit is a layout generator. Both can be downloaded from http://legwww.epfl.ch/CSL.

8.1 CSL Libraries design CAD tool

The aim of this tool is to design optimal CSL libraries in terms of power consumption and transistor area for a given frequency and load capacitance.

This program is composed of a set of design procedures based on the analytical approach presented in this chapter that provide a fast initial transistor sizing. Then an embedded simulator engine automatically helps to verify and refine the initial results in a successive approximation scheme. Finally a Spice-like output file is generated. The designer is free to mix different cell performances in terms of speed in accordance with the requirements of the application. In this way, the cells on the critical path of the circuit are designed with the fastest gates whereas the other parts are based on low-power gates.

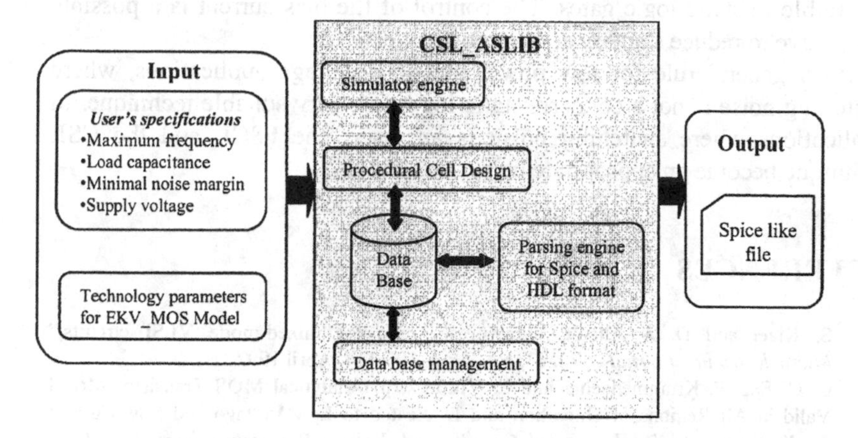

Figure 10-17. Structure and environment of CSL_ASLIB.

8.2 CSL Layout generator CAD tool

CellEdit is a CAD tool for automatic layout generation of CMOS digital libraries. When linked to the CSL_ASLIB tool, CellEdit automates the layout of the CSL cells once the dimensioning phase is done by CSL_ASLIB. For a given technology and starting from a symbolic representation of a cell, CellEdit computes an optimised layout of this cell. This makes the layout phase very easy and fast. Moreover, retargeting isn't a problem any more, because CellEdit can automatically make several layouts of a same cell using different technologies.

9. CONCLUSION

This chapter has presented the static and dynamic analysis for the CSL and FSCL inverters.

Simplification strategies have been adopted to derive analytical expressions for the static and dynamic characteristics. Simulated as well as measured results confirm the accuracy of the derived expressions. The EKV MOSFET model has been employed to derive theoretical expressions and for electrical simulation as well.

The main drawback is the standby current of FSCL and CSL, which is negligible in static logic gates. The control of the bias current is a possible alternative to reduce standby consumption.

As a general rule for low power and low voltage applications, where switching noise is not a concern, static logic is a very suitable technique. In applications where switching noise is an issue, the FSCL and the CSL technique become an appropriate help.

REFERENCES

[1] S. Kiaei and D. J. Allstot, "Low-noise logic for mixed-mode VLSI circuits," *Microelectronics Journal*, Vol. 23, No. 2, pp. 103-114, April 1992.

[2] C. C. Enz, F. Krummenacher and E. Vittoz, "An Analytical MOS Transistor Model Valid in All Regions of Operation and Dedicated to Low-Voltage and Low-Current Applications," *Analog Integrated Circuits and Systems Processing Journal on Low-Voltage and Low-Power Circuits*, Vol. 8, pp. 83-114, July 1995.

[3] E. A. Vittoz, "The Design of High-Performance Analog Circuits on Digital CMOS Chips," *IEEE J. Solid-State Circuits*, Vol. SC-20, No. 3 , pp. 657- 665, June 1985.

[4] J. P. Uyemura, *Fundamentals of MOS Digital Integrated Circuits*, Addison-Wesley Publishing Company, 1988.

[5] T. A. Demassa and Z. Ciccone, *Digital Integrated Circuits*, John Wiley&Sons, 1996.

[6] D. J. Allstot, S. H. Chee, and S. Kiaei, "Folded Source-Coupled Logic vs. CMOS Static Logic for Low-Noise Mixed-Signal ICs," *IEEE Trans. Circuits and Systems I*, Vol. 40, No. 9, pp. 553-563, September 1993.

[7] S. R. Maskai, S. Kiaei, and D. J. Allstot, "Synthesis Techniques for CMOS Folded Source-Coupled Logic Circuits," *IEEE J. Solid-State Circuits*, Vol. 27, No. 8, pp. 1157-1167, August 1992.

[8] M. Kayal, R. T. L. Sáez & Prof. M Declercq : An Automatic Offset Compensation Technique Applicable to Existing Operational Amplifier Core Cell, *Custom Integrated Circuits Conference (CICC),* Santa Clara, 11-15 May 1998.

Chapter 11

LOW-NOISE DIGITAL DESIGN TECHNIQUES

Mustafa Badaroglu and Stéphane Donnay
IMEC – DESICS, Kapeldreef 75, B-3001 Leuven, Belgium

Abstract: In most cases ringing of the power supply is the major source of substrate noise generation, as shown in chapter 2. Techniques targeting at changing the shape of the supply current or its transfer function to the substrate can reduce substrate noise generation significantly. This chapter describes such reduction techniques, which modify the supply current and its transfer function. A mixed-signal ASIC was fabricated in a 0.35µm CMOS-EPI process in order to evaluate these low-noise design techniques. The test chip contains one reference design and two digital low-noise designs with the same basic architecture and functionality. Measurements show more than a factor of 2 on average in RMS noise reduction with penalties of 3% in area and 4% in power for the low-noise design employing a supply-current waveform shaping technique based on a clock tree with latencies. The second low-noise design employing separate substrate bias for both n and p-wells, dual-supply, and on-chip decoupling achieves more than a factor of two reduction in RMS noise, with however a 70% increase in area but with a 5% decrease in power consumption.

1. INTRODUCTION

Substrate noise degrades the performance of analog circuits integrated on the same substrate with switching digital circuits. In these mixed-signal ICs, the low cost and lower static power consumption of CMOS logic are overshadowed by the high substrate noise generation due to the large rail-to-rail voltages and the sharp current spikes during switching. There are a number of techniques to reduce substrate noise coupling. These techniques can be grouped into the following four categories (see *Figure 11-1*):

1. Techniques reducing the noise generation from the noise generator (e.g. switching digital circuits),

2. Techniques changing the transfer function from the noise generator to its bulk node,
3. Techniques changing the propagation of the noise to the sensitive circuits through the substrate (not effective in low-ohmic EPI substrates),
4. Techniques desensitizing the analog circuit by changing the transfer function from its bulk node to its sensitive circuit parameters which cause performance degradation.

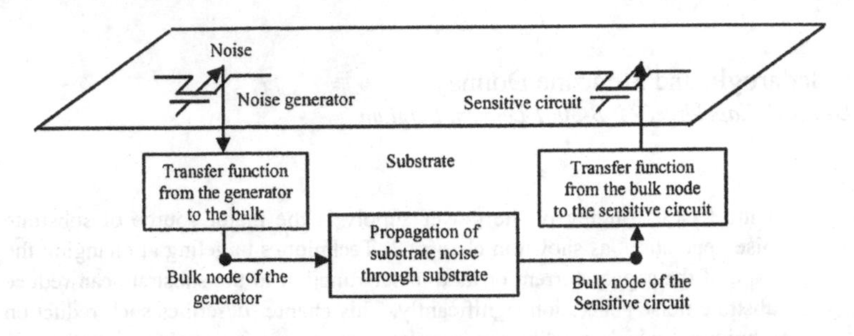

Figure 11-1. Four ways to reduce substrate noise.

Techniques often used for substrate noise reduction mostly concentrate on the last three categories listed above. Decoupling, packaging and power distribution can be optimized for reducing the noise transfer function to the substrate. Technologies such as SOI and triple-well or the use of guard rings (see also chapter 13) can be ways to change the propagation of the noise through the substrate. Circuit level techniques such as differential, high CMRR and PSRR analog circuit design and decoupling can be employed for desensitizing the sensitive analog circuits.

On the other hand, only a few publications concentrate on using the first category of techniques, which reduce the noise generation at its source. There are publications on low-noise logic cell design, such as low-voltage logic [1], current-mode logic [2] (see also chapter 10), and CMOS gates with guard wiring and decoupling [3]. Speed degradation and lower noise margins are the drawbacks for low-voltage logic. Static power consumption increase is an important drawback of current-mode logic, usually not tolerable in large digital systems. The gates with guard wiring and decoupling have a drawback of increase in area and additional supply rails. Up till now, no good methodologies exist to reduce the substrate noise at its source without drawbacks such as increase in area and power or speed degradation.

In large digital circuits, high peaks and fast slew rates on the supply current create ringing (Ldi/dt noise) in the supply network due to the damped LC-tank, formed by the on-chip capacitance between VDD and VSS and the package inductance with series resistance in the supply connection. On a typical p-type substrate, this noisy supply couples into the substrate capacitively from VDD via the n-well junction capacitance and resistively via the substrate contacts from VSS. Fast switching of the CMOS gate outputs (CdV/dt noise) couples into the substrate from the drain of the transistors via the diffusion capacitance. Decreasing the peak and the slope of the supply current will reduce the substrate noise since a large part of the noise is generated due to the noisy supply, as shown in chapter 2. This chapter specifically addresses the tuning of the supply current and its transfer function to the substrate for less substrate noise. These techniques are experimentally validated, considering the trade-offs in area, power, speed and substrate noise generation under given technology and package options.

Section 2 explains noise reduction techniques focusing on the generation part. Section 3 specifically explores supply current waveform shaping at system level by using a clock tree with latencies. Section 4 describes the measurements from a test ASIC evaluating the low-noise digital design techniques. Finally, we draw conclusions in section 5.

2. REDUCING SUBSTRATE NOISE GENERATION

2.1 Supply current transfer function to the substrate

The extraction of a chip-level substrate macro model for low-ohmic EPI-type substrates was presented in chapter 6. In this model the noise injection mechanisms are represented by two independent current sources: the bulk (Ibulk) and supply (Isupply) current sources (see *Figure 11-2a*).

For a frequency domain analysis of substrate noise these current sources can be represented by their Laplace transforms, Isupply(s) and Ibulk(s). The Laplace transform of the substrate noise, Vsub(s), can then be computed as a superposition of the outputs generated from the product of Isupply(s) and Ibulk(s) with their corresponding transfer functions G(s) and H(s) to the substrate (see *Figure 11-2b*). These transfer functions can be easily derived by solving the extracted chip-level substrate model shown in *Figure 11-2a*.

In order to reduce the generated substrate noise, one can modify the spectra of the noise sources, Isupply(s) and Ibulk(s) and/or their corresponding transfer fucntions to the substrate. Since the supply current is around two orders of magnitude larger than the bulk current, the supply current will be the dominant noise source in packaged ICs (see also chapters 2 and 6). Thus this section considers the following options to reduce the generated substrate noise (see *Figure 11-3*):

- Reducing the supply noise ($I_{supply}(s)$) (e.g. by flattening the supply current or reducing the supply voltage): this will be discussed in section 2.2.
- Changing the transfer function of the supply current source to the substrate (H(s)) (e.g. by increasing the decoupling to reduce the effect of switching capacitance, by increasing the damping of the oscillations, or by employing a dedicated substrate bias to block the noisy supply currents from going into the substrate): this will be discussed in section 2.3.

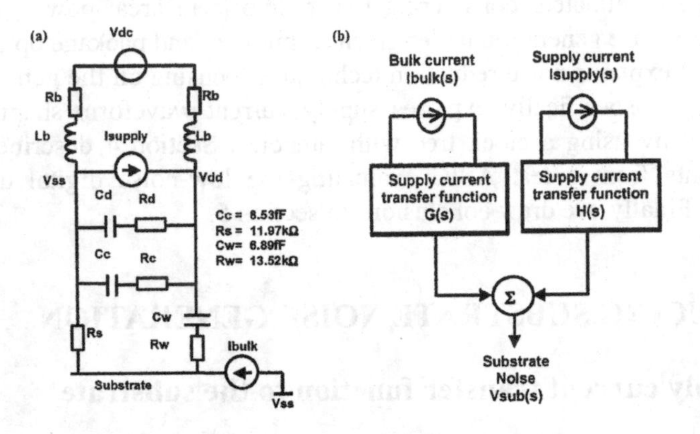

Figure 11-2. (a) An inverter substrate macro model in a 0.35µm CMOS on an EPI-type substrate. (b) Block level modeling of substrate noise generation.

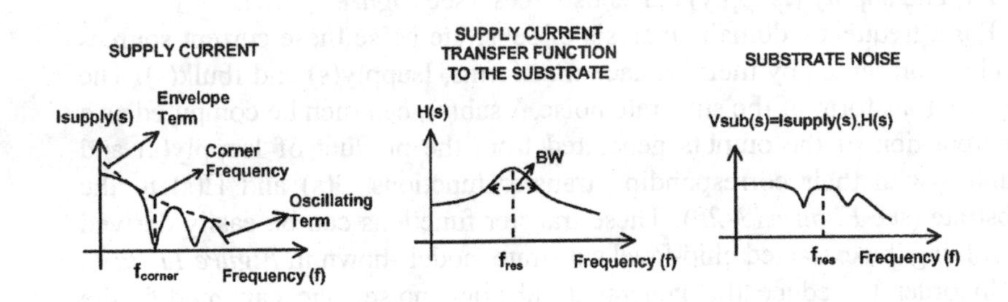

Figure 11-3. Options to reduce the substrate noise generation.

2.2 Shaping the supply current

Supply current waveform shaping is a noise reduction technique based on avoiding large current peaks on the supply lines, e.g. by spreading simultaneous switching events in time or by reducing the supply voltage. Reducing the total spectral power of the supply current will also reduce the generated substrate noise as the RMS value of the substrate noise is proportional to the integral of its power spectrum, resulting from the multiplication of the supply current spectrum and the supply current transfer function to the substrate. The total spectral power of the supply current can be reduced by:

- reducing the magnitude of the supply current spectrum, or
- moving the local minima points (corner frequency) of the supply current spectrum to lower frequencies.

The designer can achieve this by reducing two independent parameters out of the set of three dependent parameters: (1) the amplitude (2) the slope and (3) the energy of the supply current in the time domain.

For synchronous CMOS circuits, the total supply current in the time domain can be approximated by a triangular waveform. In this approximation we define I_p, t_r, t_f and E as the peak current, rise time, fall time and total charge of the supply current respectively. The Fourier transformation of this triangular supply current gives:

$$I_{supply}(f) = \frac{E}{2\pi^2.(t_r + t_f).f^2}\left[\frac{(1-\exp(j.2\pi f.t_r))}{t_r} + \frac{(1-\exp(-j.2\pi f.t_f))}{t_f}\right] \quad (1)$$

We define the corner frequency (f_{corner}) in the supply current spectrum as the minimum of $1/t_r$ and $1/t_f$. Moving the corner frequency of the supply current spectrum below the major resonance frequencies in the supply current transfer function H(s) will reduce the substrate noise generation significantly since most of the noise power is a result of this resonance behavior. The optimum value for the rise/fall time is given by:

$$\left(f_{corner} = \min\left(\frac{1}{t_r}, \frac{1}{t_f}\right)\right) < \left(f_{res} = \frac{1}{2\pi\sqrt{2L_b(C_w + C_c)}}\right) \quad (2)$$

In order to satisfy equation (2), the designer can increase the rise/fall time or alternatively increase the major resonance frequency by decreasing the inductance and/or the circuit capacitance. This procedure comes with a number of drawbacks:

- the difficulty to meet the timing constraints to achieve a minimum rise/fall time
- the need for a larger number of supply pads in large digital circuits
- reducing the on-chip capacitance, which can be used as the decoupling.

To allow a margin for errors in the estimation of the major resonance frequency, the corner frequency in the supply current is made as low as possible. For example, a 1M-gate circuit, which is implemented in a 0.35μm CMOS process on an EPI-type substrate, will require 30 VDD/VSS pairs in a BGA package with 1nH inductance for a single bondwire, when a maximum of 5ns is allowed for the rise/fall time on the supply current due to timing constraints. On the other hand only 1 supply pair is enough for a 15K-gate circuit under the same conditions. For large circuits the large on-chip capacitance would require smaller inductance values, therefore more supply pads, to achieve a certain timing constraint. The supply current shaping technique described above is therefore most practical for small to medium scale circuits (up to 1M gates).

A number of noise reduction techniques using supply current shaping will be described next:

- Reducing supply voltage (section 2.2.1)
- Decreasing input slope (section 2.2.2)
- Spreading switching activities (section 2.2.3)
- Reducing switching activities (section 2.2.4)

2.2.1. Reducing supply voltage

One of the most effective techniques to reduce the supply current noise is reducing the supply voltage [1]. Equation (1) also shows that any low-power design technique reduces the substrate noise linearly as E decreases linearly. Reducing the supply voltage has two effects: reduction in energy (E.VDD) and increase in rise/fall times (t_r, t_f) of the supply current under the same load conditions (see *Figure 11-4*).

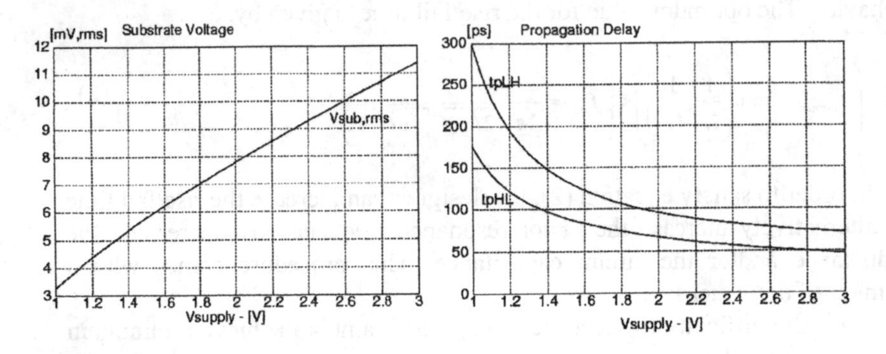

Figure 11-4. Substrate noise (left) and propagation delay (right) for a minimum size inverter in a 0.35μm CMOS-EPI process as a function of supply voltage from SPICE simulations.

The supply voltage dependence of the substrate noise is also influenced by the location of the corner frequency of the supply current relative to the major resonance frequency of the supply current transfer function. This dependence is linear only when the supply current corner frequency is far from the major resonance frequency of the circuit.

The rise/fall times are linearly proportional to 1/VDD. In order to decrease the supply voltage without trading the speed performance on a system-on-chip, the design can be partitioned into two groups: the slow cells (non-critical path cells), and the fast cells (critical path cells). The fast cells are supplied with the normal supply voltage while the slow cells are supplied with a lower supply voltage. The minimum value of the supply voltage is determined by the condition that the worst-case delay of the slow cells still stays smaller than the critical path delay of the overall circuit. However, the output levels of the slow cells driving the fast cells having higher logic levels have to be restored.

2.2.2. Decreasing input slope

Another technique to reduce the substrate noise generation is to increase the input transition time. In this way, the slope of the supply current will decrease and this will decrease Ldi/dt and CdV/dt noise, however, with a trade-off of increasing delay. This technique is mostly used in slew-rate control of the output I/O buffers. Slow slew-rate output I/O buffers are preferred for less substrate noise generation if chip-to-chip delays are not critical. One can also reduce the slope of the system clock driving simultaneously switching flip-flops. However, the delay penalty at the output transition of a gate is not acceptable for timing critical applications.

2.2.3. Spreading switching activities

These techniques target at changing the supply current waveform of the total digital circuitry rather than the supply current of a single gate. The idea is to spread the switching activities as much as possible in time so that no sharp current peaks will exist due to the large number of simultaneous switching activities. One way to achieve this is to increase the combinatorial logic depth. Another way to spread the switching activities can be done by the introduction of different latencies in the branches of a clock tree driving a synchronous digital circuit. This technique will be described more thoroughly later in section 3. Asynchronous logic is also another way to spread switching activities [4]. However the performance of self-timed logic is highly dependent on the delay, and therefore the design of complex systems is more difficult with self-timed logic than with synchronous logic.

2.2.4. Reducing switching activities

Most low-power design techniques target at reducing the number of switching activities. These techniques can also be used for substrate noise reduction. The efficiency of these techniques is highly dependent on the design.

One technique that makes use of redundancies in Boolean mapping considers "don't cares" in the Karnough mapping in order to find a circuit, which generates less switching activity. This can be done by using an operand isolation technique or a precomputation logic in order to eliminate the redundant switching activity within a combinatorial logic [5],[6]. The idea of this technique is to block the input signal just before complex combinatorial logic, whenever the input does not cause a change at the output. However, in data dependent applications the redundant logic inserted for blocking the input signals may become quite complex whenever the output of the original logic is highly dependent on the input word. In this case the noise generation due to this block will exceed the reduction gains due to the input blocking.

2.3 Changing the supply current transfer function to the substrate

Sharp supply currents generated by the digital circuit go through a transfer function determined by the package and PCB parasitics, the decoupling capacitors and the substrate model of the circuit. Some techniques to change the supply current transfer function to the substrate are:

- decoupling the noisy power supply in order to provide a direct return path for the power supply noise to the ground of the power supply before it couples into the substrate via the substrate contacts (see section 2.3.1).
- providing a dedicated n-well and p-well bias in order to prevent the coupling of the power supply noise (see section 2.3.2).

2.3.1. Decoupling the noisy power supply

When the gates switch, charge redistribution through the switching capacitance generates ground noise. In large digital designs the resistance between the digital ground and the low-ohmic bulk can be very low, so all noise present on the digital ground will also appear on the substrate. Increasing the ratio of non-switching capacitance to the switching capacitance will decrease ground noise and therefore substrate noise.

Assuming a symmetric noise on VDD and VSS, the peak value of the ground noise (ΔVSS) due to charge redistribution is given as follows [7]:

$$\Delta VSS = \frac{VDD.\alpha.C_c}{2.[C_w + C_D + (1-\alpha).C_c]} \qquad (3)$$

where α is the switching activity, which is dependent on the circuit functionality and is around 5-30% [8]. Equation 3 is a good approximation when the non-switching capacitance, Cw + Cd + (1-α)Cc, is larger than the switching capacitance, αCc. This is usually the case for the core cells in a large system. The effect of an on-chip decoupling capacitor (see *Figure 11-2*) on these oscillations is given by:

$$\omega_o = \frac{1}{\sqrt{2L(C_D + C_C)}}, \quad \zeta = \frac{R_D}{2}\sqrt{\frac{C_D + C_C}{2L}}, \quad Q_S = \frac{1}{R_D}\sqrt{\frac{2L}{C_D + C_C}} \qquad (4)$$

where L, C_C, C_D, R_D are the supply inductance, circuit capacitance, decoupling capacitance and decoupling resistance respectively. ω_o, ζ, Q_s are the resonance frequency, damping factor and the quality factor for the generated substrate noise.

It is important to note that the total decoupling capacitance is determined not only by the on-chip decoupling but also by the circuit capacitance due to the core cells, supply and I/O pads. Increasing the decoupling capacitance will reduce the peak noise and also the resonance frequency [9]. The latter result requires a special attention when the resonance frequency shifts towards the operating clock frequencies of the circuit. Also on-chip parasitics increase due to the addition of the decoupling capacitor. This shift of the resonance frequency will cause an accumulation of the noise as the oscillations may not be damped completely before the next supply current spike occurs on the supply rail. To avoid this, the damping ratio has to be increased via increasing the series resistance of the decoupling capacitor. The quality factor of the oscillations is reduced whenever there is a series damping resistor with the decoupling capacitor. The optimum value of this resistance is found by comparing the time constant of the circuit impedance between VDD and VSS [7]. For larger values of this resistance, the voltage drop can be a problem when the sharp current of switching circuits are injected into the resistance inside the decoupling capacitor rather than charging the switching gates.

2.3.2. Dedicated substrate biasing

Important for the substrate noise coupling is also the way in which the substrate is connected to the ground. Substrate biasing is important in order to have a non-floating bulk/well region and latch-up prevention. In traditional CMOS cells digital (core) ground and substrate are short-circuited by substrate contacts. On a p-type substrate, a dedicated bias for the p-well is more effective than a dedicated bias for the n-well. For the p-well, the resistive contact is disconnected from VSS, resulting in capacitive coupling from VSS to the substrate rather than resistive coupling. On the other hand, the removal of the n-well resistive contact will not give a similar reduction as the p-well biasing, since the noisy VDD line is already blocked from the substrate by the well capacitance. The removal of the supply contacts will reduce the effective decoupling capacitance between VDD and VSS. So the dedicated substrate bias is effective for avoiding coupling of the supply noise into the substrate at the expense of a reduction in free decoupling.

3. CLOCK TREE WITH DIFFERENT LATENCIES

3.1 Introduction

Existing design methodologies generally target a zero clock latency between clock regions [10],[11]. However, one could introduce a different latency to each clock region with respect to the master clock to make the total supply current waveform flatter. The clock regions are generated by an equal split of the cells in such a way that each region has equalized supply current profiles. The latency defined for each clock region causes a shift of the switching activities in time. During the optimization of the latencies, the use of the total transient of the supply current is not feasible due to the complexity of such an exhaustive search. It is therefore necessary to find a representative current waveform of all clock cycles for each clock region to reduce the number of points used in the optimization.

There are techniques using clock latencies in order to reduce the peak current [12] and the ground bounce [13]. However, they suffer from a large number of constraints, equal to the total number of flip-flops. Moreover, they do not give a value for the number of clock regions that should be used, which is dependent on the relation between the major resonance frequency of the circuit and the rise/fall time of the supply current.

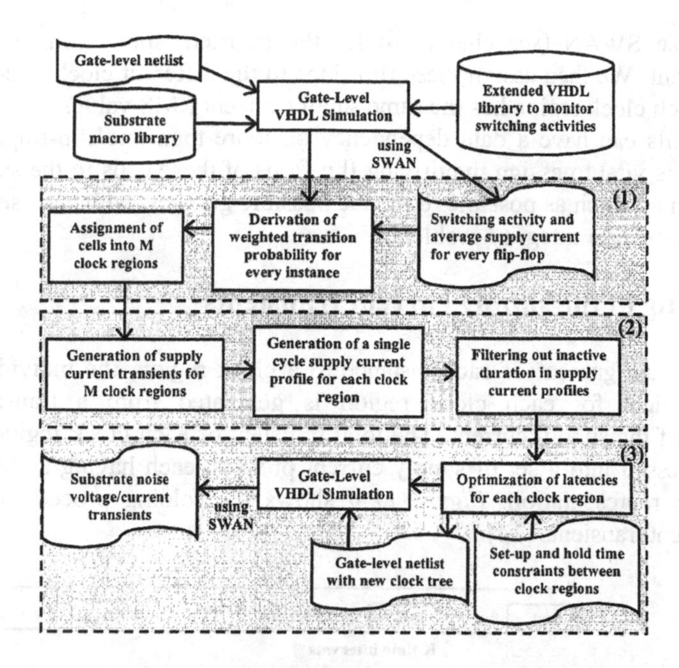

Figure 11-5. Clock tree latency optimization methodology.

Our methodology allows to take into account these effects. It uses a lower number of constraints due to the folding of the supply current transients. The optimum number of clock regions can be determined by performing a frequency spectrum analysis on the supply current. The methodology flow is shown in *Figure 11-5*. The latency optimization procedure consists of three main parts [14]:

1. assignment of cells into M clock regions,
2. folding of supply current transients in each clock region, and
3. optimization of latencies in each clock region.

The next three sections will describe these steps in detail. The experimental results will be presented in section 3.5.

3.2 Clock region assignment

It is important to balance the gate instances over the clock regions for a significant reduction in the substrate noise generation. Each clock region must have the same supply current energy.

We first simulate the RMS value of the supply current for each flip-flop and the set of all instances that have a data dependency on this driving flip-flop. We use SWAN (see chapter 6) for the transient simulation of the supply current. We then assign these flip-flops to the different clock regions such that each clock region has the same supply current RMS value.

Some cells can have a data dependency on more than one flip-flop. In this case, it is vital to assign the driving flip-flops of these cells to the same clock region as much as possible to reduce possible glitches, which cause an increase in power or integrity problems.

3.3 Folding of the supply current transients

After the assignment of each instance to a clock region, the individual supply transient for each clock region is generated from a transient simulation of the supply current. The total transient of each clock region is then compressed into a set of supply current profiles, each having a single clock cycle representation. *Figure 11-6* shows the folding procedure of supply current transients [14].

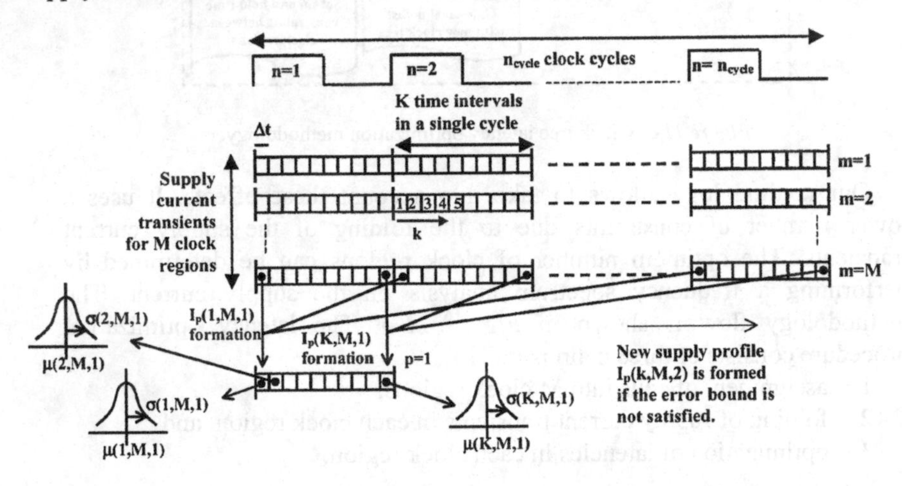

Figure 11-6. Folding algorithm for forming the current profiles.

Every clock region contains at least one current profile defined over every time interval. If a user error bound is given, the compression can create more than one profile in a clock region. Each current profile has statistical properties such as mean, standard deviation and probability density function at each profile point $I_p(k,m,p)$ using all the points in the actual waveform. $I_p(k,m,p)$ contains a set of statistical functions given as:

$$I_p(k,m,p) = \{\mu(k,m,p), \sigma(k,m,p), h(k,m,p)\} \quad (5)$$

where μ(k,m,p): mean of I(k,m,n),
σ(k,m,p): standard deviation of I(k,m,n),
h(k,m,p): histogram of I(k,m,n) distribution,
over all clock cycles (n=1..n_{cycle}).

3.4 Clock latency optimization

The optimization of the clock tree latencies using the supply current profile of each clock region takes timing constraints into account. The latencies have to be constrained by setup and hold timing constraints between the different clock regions and also the clock uncertainty due to the unexpected skew coming from the clock interconnect. The optimization is then performed on the constraint space formed by the latencies ($l_1, l_2, \ldots, l_M$) as follows [14]:

$$\begin{aligned}
&\min_{l} f_{cost}(l_1, l_2, \ldots, l_M) = Amplitude.Slope \\
&Amplitude = \max\left(\sum_{m=1}^{M} I_p([1,K]-l_m, m)\right) \\
&Slope = RMS\left(\sum_{m=1}^{M} \left(I_p([1,K]-l_i, m) - I_p([1,K]-l_i-1, m)\right)\right) \\
&s.t.\ l_1 = 0,\ t_{max}(i,j) < l_i - l_j < t_{min}(i,j) \quad \forall i,j \in [1,M]x[1,M]
\end{aligned} \quad (6)$$

where l_m is defined as the latency value of clock region-m. One can freely set one of the latencies to zero, such that one of the clock regions is aligned to the edge of the clock. At each latency value, the cost function is evaluated as the product of the peak value and the slope of the total supply current. This is a direct result from equation 1, which states the direct proportionality of the supply current spectrum to the product of its peak and slope. The optimization tries to minimize this factor in order to reduce the spectral energy of the supply current, and therefore the RMS value of the substrate noise as described in section 2.2.

3.5 Experimental results

The methodology is first illustrated for a 4-bit Pseudo-Random-Noise-Sequencer (PRBS) implemented in a 0.35-μm CMOS process on an EPI-type substrate at 3.3V supply. The supply current transfer function to the substrate has a resonance frequency of 2.3GHz. The 3dB bandwidth of the resonance stretches from 1.3GHz to 3.2GHz. The design has a clock period of 4ns and a supply line parasitics of 5nH+0.5Ω. The supply current spectrum has a periodicity at 250MHz determined by the clock frequency. The corner frequency corresponding to the rise/fall time of the supply current is 3.8GHz as computed using equation (2). Thus we can safely choose to use 4 clock regions in order to shift this corner frequency well below the resonance frequency at 2.3GHz.

For each of the 4 clock regions, supply current profiles have been constructed using the actual supply current data from a transient simulation of 105 clock cycles using SPICE. *Figure 11-8* shows the transients (a) and the profiles (b) of the total supply current with/without latencies. This total supply current has been constructed by summing the individual profiles of each clock region. Using the profiles of the different clock regions, the optimum latencies are computed by taking into account timing constraints. The latencies are then implemented using a clock delay line (see *Figure 11-7*). For a larger reduction of the substrate noise, sometimes the timing constraints have to be relaxed. Afterwards a timing correction module is used in order to correct timing between the clock regions.

The quality of the optimization results towards clock skew are evaluated within a skew radius around the optimum point of the latencies. We then exhaustively compute the cost function in equation 6 within the skew radius. These values are then compared with the cost function value evaluated at zero latencies. The latencies found for the PRBS design in this example is immune to the clock skews below 150ps.

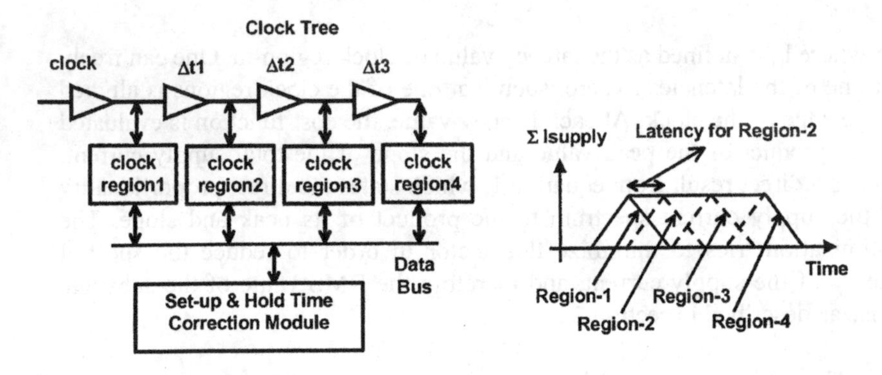

Figure 11-7. Implementation of clock tree with different latencies (left) and the construction of the supply current waveform under different latencies (right).

The design with optimized latencies achieves factors of 2.10 and 1.75 reduction in the peak-to-peak and the RMS value of the substrate noise respectively. The design with optimized latencies has a reduction of 8.9dB, 20dB, 18dB, 14dB, and 13dB at the fundamental, 2nd, 3rd, 4th and 5th harmonics of the clock respectively. This is due to the reduction of the spectral power of the supply current (*Figure 11-9*).

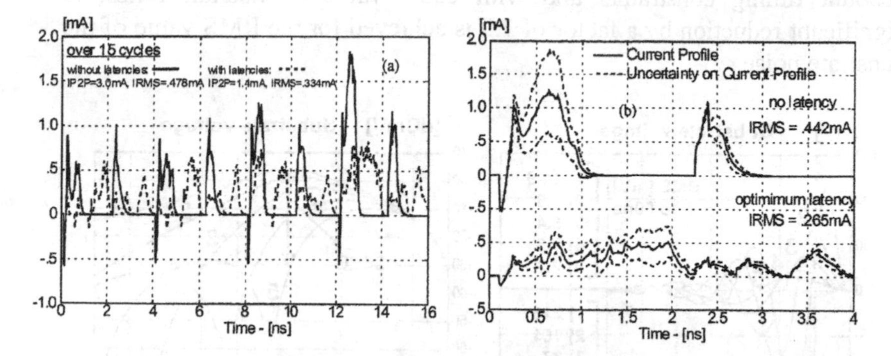

Figure 11-8. Simulated (SPICE) supply current (a) transients, (b) their profiles with/without latencies in the PRBS circuit.

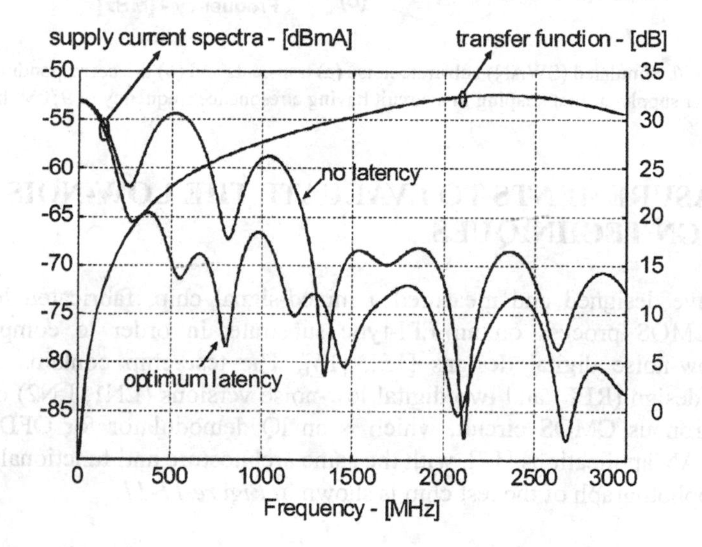

Figure 11-9. Supply current spectra with/without latencies and their transfer function to the substrate in the PRBS circuit.

The reduction becomes more significant when the initial corner frequency is far above the major resonance frequency and is then shifted below the major resonance frequency by applying the above mentioned techniques. *Figure 11-10* shows the transients and the corresponding spectra of a circuit, having a resonance at 950MHz, as a result of changing the amplitude and slope of its supply current. The supply current has been changed from (50mA, 200ps) to (10mA, 1000ps), without taking into account timing constraints and with equal value of rise/fall times. A significant reduction by a factor of 7.9 is achieved for the RMS value of the substrate noise.

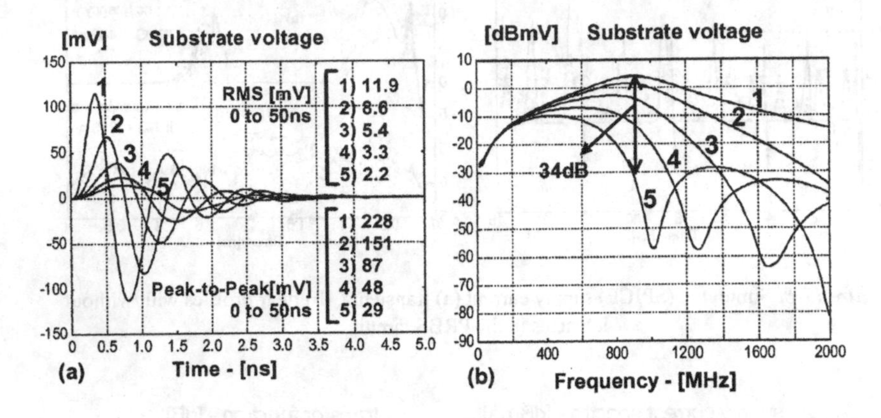

Figure 11-10. Simulated (SWAN) substrate noise (a) transients and (b) the corresponding spectra after supply current shaping in a circuit having a resonance frequency of 950MHz.

4. MEASUREMENTS TO EVALUATE THE LOW-NOISE DESIGN TECHNIQUES

We have designed and measured a mixed-signal chip, fabricated in a 0.35μm CMOS process on an EPI-type substrate, in order to compare several low-noise digital designs [15], [16]. The test chip contains one reference design (REF) and two digital low-noise versions (LN1, LN2) of a 5K synchronous CMOS circuit, which is an IQ demodulator for OFDM-based WLAN applications [17], with the same architecture and functionality. The microphotograph of the test chip is shown in *Figure 11-11*.

4.1 Overview of the simulated reduction factors for the generated substrate noise

In LN1 the substrate noise was reduced by decreasing the peak and the slope of the supply current as described in the previous sections. To make the supply current flatter we employed an optimized clock tree, where each clock region latency is optimized using the supply current profile statistics computed by SWAN. Without looking at the timing implications and for the 12nH bondwire inductance that is used, the simulated reduction becomes 6.2dB, 15.3dB, 23.7dB and 24.5dB for 2 to 5 clock regions respectively (see *Figure 11-12*). Note that timing constraints on the design may prevent the designer from reaching these optimal figures. In our final design [15] we selected 4 clock regions. The optimum latencies are then implemented using a clock delay line. Because of the strict timing constraints in this design, we obtained a simulated substrate noise reduction of only 6dB instead of the maximum reduction of 23.7dB, which would be possible in theory without timing constraints. Strict timing constraints of our design are due to the large number (79% of the gate equivalent area) of the cells being on, which are in the critical path of the design.

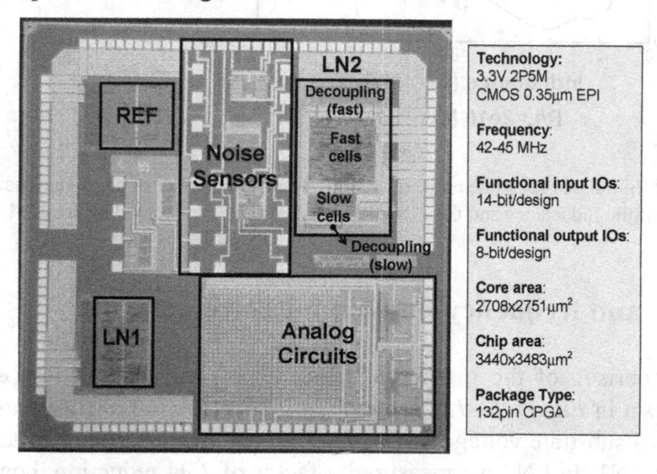

Figure 11-11. Microphotograph of the test chip and its specifications. The chip consists of a reference design and 2 low-noise designs with the same functionality. It also contains 2 substrate noise sensors. The analog circuits are comparator arrays to evaluate the impact (see chapter 7).

LN2 employs a separate substrate bias in order to prevent direct coupling of the noisy supply currents into the substrate. Further it uses a separate lower supply voltage for the cells that are not critical for the speed performance of the circuit. Finally, it also uses on-chip decoupling. To determine an optimum value for the on-chip decoupling capacitors employed by LN2, we have used the chip-level substrate model extracted by SWAN, taking into account the multiple supply domains. The simulated reduction is 6-7dB for the separate substrate bias. It is 3dB for the on-chip decoupling, while it is just 0.5dB for the technique employing a separate lower supply voltage for non-critical cells.

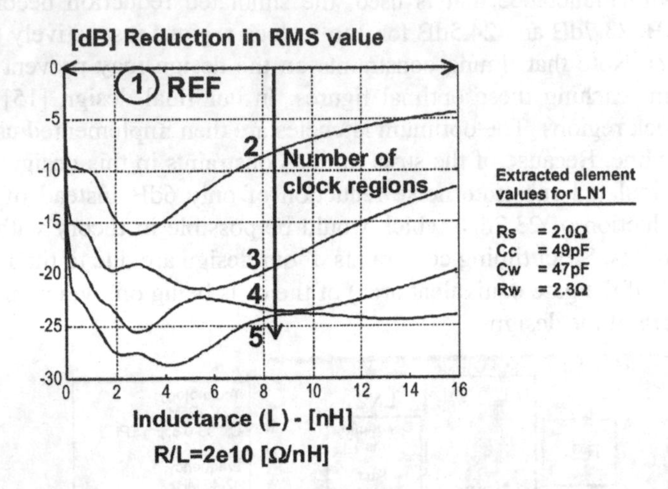

Figure 11-12. Simulated substrate noise reduction versus the number of clock regions as a function of the inductance and the extracted macro model element values for LN1.

4.2 Time- and frequency-domain measurements

The comparison of the measured transients of the generated substrate noise is shown in *Figure 11-13*. A factor of 2.14 reduction in the measured RMS value of substrate voltage with only 3% area and 4% power penalty is obtained for LN1. In LN2 we measured a factor of 2.94 reduction, however with a factor of 1.72 increase in area. The oscillations in LN2 are damped more compared to REF and LN1 due to the damping resistance in series with the decoupling capacitor.

The measured frequency spectrum comparison is shown in *Figure 11-14*. The largest substrate noise peaks in the spectrum are 19 dBmV, 13 dBmV and 11 dBmV at the 3rd, 2nd and fundamental clock harmonics for REF, LN1 and LN2 respectively. For LN1 the peaking measured at 125MHz is attenuated as a result of the supply current shaping. For LN2 the measured resonance frequency shifts to 105MHz due to the on-chip decoupling capacitors and the supply distribution for fast and slow cells. The spectral peaks are 35-40 dB above the substrate noise floor at the first four clock harmonics.

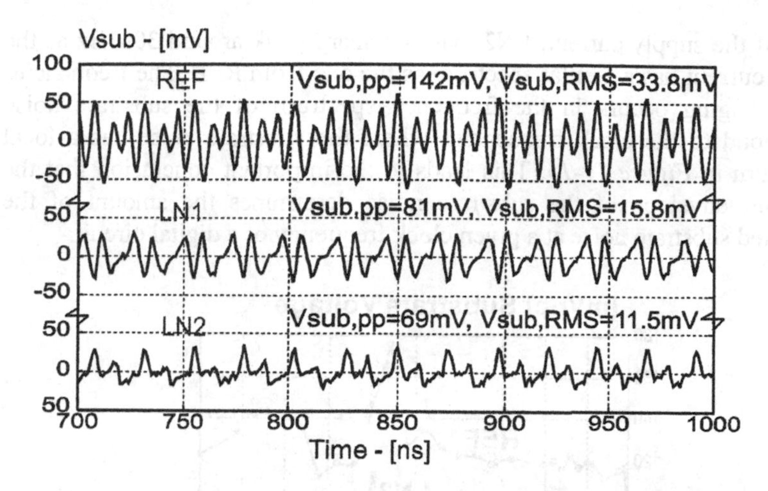

Figure 11-13. Comparison of measured substrate noise transients for the reference design (REF) and its two low-noise versions (LN1, LN2).

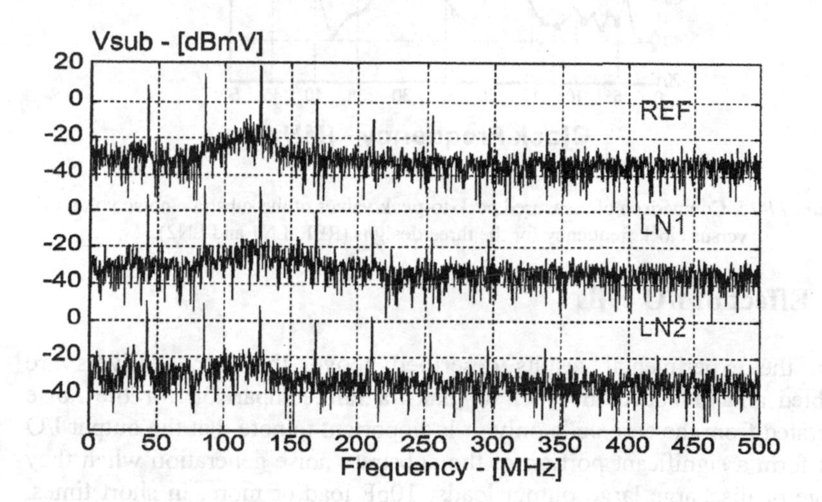

Figure 11-14. Comparison of measured substrate noise spectra for the reference design (REF) and its two low-noise versions (LN1, LN2).

The clock dependency of the substrate noise is shown in *Figure 11-15*. At all clock frequencies, the substrate noise generated by LN1 and LN2 is lower than REF, by a factor of 2 on average in both RMS and peak-to-peak values. The figure also shows the resonance behavior at some clock frequencies. Two core supply pairs have been used for each design to have the same inductance. Note that LN1 shows less resonance due to the reduced slope of the supply current. LN2 makes a sharp peak around 30MHz as the supply current has a similar spectrum as the one from REF. The frequencies where ringing occurs in the frequency spectrum of the substrate noise correspond to the clock frequencies where the substrate noise has a local maximum in *Figure 11-15*. This leads to the important conclusion that the spectrum envelope of the substrate noise determines the amount of the generated substrate noise at a given clock frequency of a digital circuit.

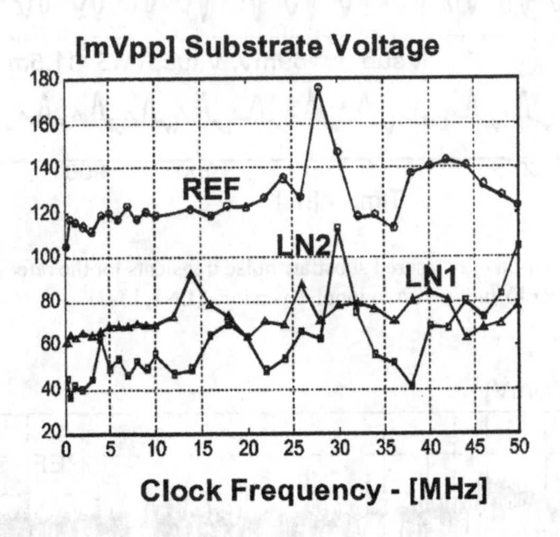

Figure 11-15. Comparison of measured peak-to-peak values of the substrate noise voltages versus clock frequency for the three designs (REF, LN1 and LN2).

4.3 Effect of I/O cells

In the experimental results described above, the output drivers are disabled and unloaded in order to make a fair comparison for the noise generated from the core cells only. It is important to note that the output I/O cells form a significant portion of the substrate noise generation when they charge or discharge large output loads, 10pF load or more, in short times, e.g. 1.5ns. On-chip decoupling does not help as the current loop flows externally off the package. There are techniques to reduce the ringing on output I/O cells [18], [19]. The substrate noise measurements showing the impact of the I/O cells are shown in *Figure 11-16*. The measured noise contribution of the substrate noise from the core logic is only 9.1mVrms and 85mVpp, while it is 18.1mVrms and 150mVpp for the core and a single additional output I/O pad. The output I/O has a 12nH+1.5Ω bondwire and a load of 12pF in parallel with 100kΩ. In large digital circuits, the noise contribution of the I/O cells however relatively becomes less important compared to the one generated from the simultaneous switching of the core cells [11].

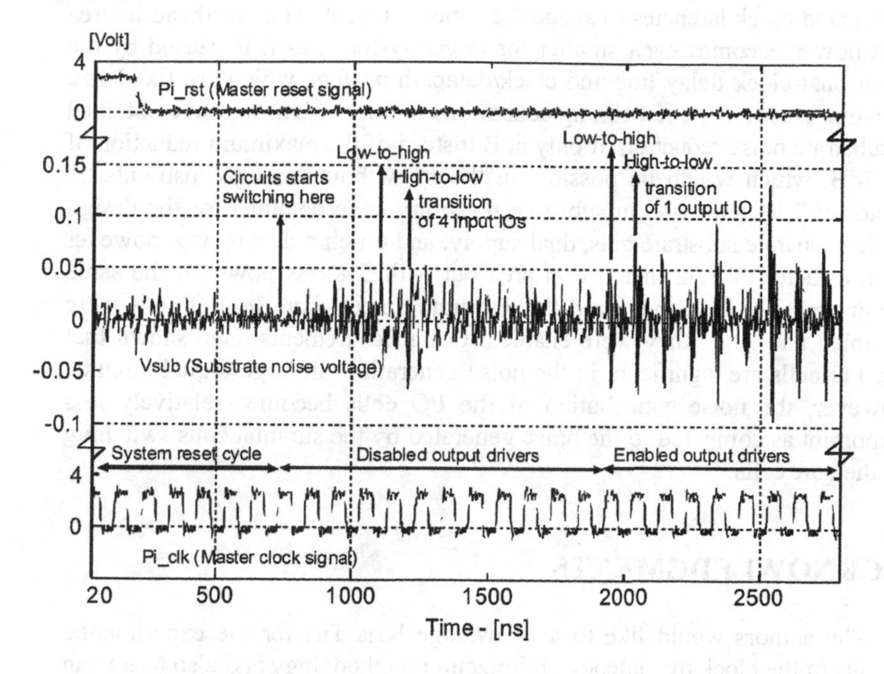

Figure 11-16. Substrate noise measurements showing the impact of the I/O cells.

5. CONCLUSIONS

Digital substrate noise reduction techniques targeting at the supply current and its transfer function to the substrate are very appealing as coupling from the ringing supply into the substrate is dominant in packaged ICs. Techniques using supply current shaping have been shown to be highly effective in small to medium scale circuits (1 gate up to 1M gates) while having no significant increase in area and power. But timing conditions should allow such shaping. The use of on-chip decoupling capacitance can always be used when there is no strict area limitation. Dedicated substrate biasing is an interesting option to filter the supply noise but with the drawback of reducing the free on-chip decoupling capacitance of the non-switching circuits. For larger reduction these techniques can be used together if these drawbacks are tolerable.

We have evaluated these low-noise digital design techniques in a test ASIC fabricated in a 0.35μm CMOS process on an EPI-type substrate. Our measurements have shown that we have more than a factor of 2 reduction in substrate noise with only 3% area and 4% power increase in the design with optimized clock latencies to shape the supply current. This overhead in area and power becomes even smaller for larger systems, as it is caused by the additional clock delay line and clock/datapath buffers, which are fixed-size circuits. Due to the strict timing constraints in this design, we have obtained a substrate noise reduction of only 6dB instead of the maximum reduction of 23.7dB, which would be possible in theory without timing constraints. A factor of 2.94 reduction in substrate noise has been obtained for the design with a separate substrate bias, dual supply, and on-chip decoupling, however with a factor of 1.72 increase in area but with 5% less power at the same frequency. We have measured the substrate noise when the I/O cells were disabled and when they were enabled. Our measurements have shown that the I/O cells are significant in the noise generation. In large digital circuits, however, the noise contribution of the I/O cells becomes relatively less important as compared to the noise generated by the simultaneous switching of the core cells.

ACKNOWLEDGMENTS

The authors would like to acknowledge Kris Tiri for the experimental results of the clock tree latency optimization methodology and also Marc van Heijningen and Vincent Gravot for their invaluable support in this work. This work was supported in part in the frame of the ESPRIT Project-BANDIT, funded by the European Commission.

REFERENCES

[1] A. Matsuzawa, "Low-voltage and low-power circuit design for mixed analog/digital systems in portable equipment," *IEEE J. Solid State Circuits*, vol. 29, no. 4, pp.470-480, April 1994.

[2] D. J. Allstot, H. C. San, S. Kiaei, and M. Shriyastawa, "Folded source-coupled logic vs. CMOS static logic for low-noise mixed-signal ICs," *IEEE Trans. on Circuits and Systems I*, vol. 40, no.9, pp.553-563, Sep. 1993.

[3] M. Nagata, K. Hijikata, J. Nagai, T. Morie, and A. Iwata, "Reduced substrate noise digital design for improving embedded analog performance," *ISSCC Digest of Technical Papers*, pp.224-225, Feb. 2000.

[4] J. Butas, C. Chiu-Sing, J. Povazanec, and C. Cheong-Fat, "Asynchronous cross-pipelined multiplier," *IEEE J. Solid State Circuits*, vol. 36, no. 8, pp. 1272-1275, Aug. 2001.

[5] J. Monteiro, S. Devadas, and A. Ghosh, "Sequential logic optimization for low power using input-disabling precomputation architectures," *IEEE Trans. on Computer-Aided Design of Integrated Circuits and Systems*, Vol. 17 Issue 3, March 1998, pp. 279-284.

[6] M. Munch, B. Wurth, R. Mehra, J. Sproch, and N. When, "Automating RT-level operand isolation to minimize power consumption in datapaths," *Proc. of Design, Automation and Test in Europe Conference*, pp. 624-631, March 2000.

[7] P. Larsson, "Resonance and damping in CMOS circuits with on-chip decoupling capacitance," *IEEE Trans. Circuits and Systems I*, vol. 45, no. 8, pp. 849-858, August 1998.

[8] D. Liu, C. Svensson, "Power consumption Estimation in CMOS VLSI Chips," *IEEE J. Solid-State Circuits*, Vol. 29, No. 6, pp. 663-670, June 1994.

[9] M. Ingels and M. Steyaert, "Design strategies and decoupling techniques for reducing the effects of electrical interference in mixed-mode IC's," *IEEE J. Solid-State Circuits*, vol. 32, pp. 574-576, Apr. 1997.

[10] F. Minami and M. Takano, "Clock tree synthesis based on RC delay balancing," *IEEE Proc. of Custom Integrated Circuits Conference*, pp.28.3.1-28.3.4, May 1992.

[11] S. Tam, S. Rusu, U. N. Desai, R. Kim, J. Zhang and I. Young, "Clock Generation and Distribution for the First IA-64 Microprocessor," *IEEE J. Solid-State Circuits*, Vol. 35, No. 11, pp. 1545-1552, November 2000.

[12] P. Vuillod, L. Benini, A. Bogliolo, and G. De Micheli, "Clock-skew optimization for peak current reduction," *IEEE Int. Symp. On Low Power Electronics and Design*, pp. 265-270, 1996.

[13] A. Vittal, H. Lia, F. Brewer, and M. Marek-Sadowska, "Clock skew optimization for ground bounce control," *ICCAD Digest of Technical Papers*, pp. 395-399, 1996.

[14] M. Badaroglu, K. Tiri, S. Donnay, P. Wambacq, I. Verbauwhede, G. Gielen, and H. De Man, "Clock tree optimization in synchronous CMOS circuits for substrate noise reduction using folding of supply currents," *Proc. of Design Automation Conference*, June 2002.

[15] M. Badaroglu, M. van Heijningen, V. Gravot, J. Compiet, S. Donnay, M. Engels, G. Gielen, and H. De Man, "Methodology and Experimental Verification for Substrate Noise Reduction in CMOS Mixed-Signal ICs with Synchronous Digital Circuits," *ISSCC Digest of Technical Papers*, pp. 274-275, February 2002.

[16] M. Badaroglu, M. van Heijningen, V. Gravot, J. Compiet, S. Donnay, M. Engels, G. Gielen, and H. De Man, "Methodology and Experimental Verification for Substrate Noise Reduction in CMOS Mixed-Signal ICs with Synchronous Digital Circuits," *IEEE J. Solid-State Circuits*, to be published in November 2002.

[17] S. Signell, T. Fonden, M. Badaroglu, and S. Donnay, "Implementation of an efficient lattice digital ladder filter for up/down conversion in an OFDM-WLAN system," *Proc. of the European Solid-State Circuits Conf.*, pp. 480-483, September 2001.

[18] T. Gabara, W. Fischer, and J. Harrington, "Forming damped LRC parasitic circuits in simultaneously switched CMOS output buffers," *Proc. of the IEEE Custom Integrated Circuits Conference*, pp.277-280, 1996.

[19] S. Choy, C. F. Chan, M. H. Ku, and J. Povazanec, "Design procedure of low-noise high-speed adaptive output drivers," *Proc. of IEEE International Symposium on Circuits and Systems*, vol.3, pp.1796-1799, 1997.

[20] M. van Heijningen, M. Badaroglu, S. Donnay, G. Gielen, and H. De Man, "Substrate noise generation in complex digital systems: efficient modeling and simulation methodology and experimental verification," *IEEE J. of Solid-State Circuits*, vol. 37, pp. 1065-1072, August 2002.

Chapter 12

HOW TO DEAL WITH SUBSTRATE BOUNCE IN ANALOG CIRCUITS IN EPI-TYPE CMOS TECHNOLOGY

Bram Nauta and Gian Hoogzaad*
*IC Design group, MESA+ Research Institute, University of Twente, Enschede , The Netherlands . *with Philips Semiconductors, Delft, The Netherlands*

Abstract: Substrate noise is one of the key problems in mixed analog/digital ICs. Although measures are known to reduce substrate noise, the noise will never be completely eliminated since this requires larger chip area or exotic packages and thus higher cost. Analog circuits on digital ICs simply have to be resistant to substrate noise. A general strategy is given which can be summarized as: the supply of the analog circuits must be referred to the substrate and the analog signals must be referred to a clean analog ground. Furthermore several design constraints are given to minimize the effect of substrate noise on analog. Two bandgap circuits are discussed and it is shown that apparently minor design issues, such as the connection of an n-well of a PMOS differential pair, can have large impact on the substrate sensitivity of this circuit. This has been verified by measurements.

1. INTRODUCTION

One of the major challenges in mixed signal IC design is to deal with substrate noise and crosstalk problems. Especially in epitaxial CMOS technology the problems are serious, since both analog and digital share the same - low ohmic - substrate. As the switching speed and packing density of digital CMOS increases, and the supply voltage drops, it's more and more difficult to design analog modules with good performance. For example it is not trivial to design a 10 bit video ADC (1LSB = 1mV) or to design a low-jitter PLL while there is 300mV substrate noise.

This chapter describes techniques to make analog circuits less sensitive to substrate bounce. We restrict ourselves to epi-CMOS technology, with a low-ohmic substrate and to analog circuits which operate up to intermediate frequencies. The techniques to reduce the effect of substrate bounce in non-epi technology [1] or in RF circuits [2] are different.

This chapter describes briefly the origin of substrate noise (section 2), followed by an example of a simple current source to illustrate the problems in analog (section 3). In section 4 a general substrate strategy for analog is given and in section 5 a practical example is given on what can go wrong with the design of a simple CMOS bandgap reference. Finally conclusions are drawn.

2. SUBSTRATE NOISE

Figure 12-1 shows a well-known cross section of a standard epitaxial CMOS IC. The substrate is very low-ohmic (say 0.01 Ohm cm) on top of which an epi-layer is grown. This epi-layer is several micrometers thick and has a higher resistance (say 10 Ohm cm). On this low-ohmic substrate both analog and digital circuits are located, and due to the low substrate resistance this causes crosstalk problems. Now a brief explanation of the origin of the substrate noise in epitaxial CMOS technology is given.

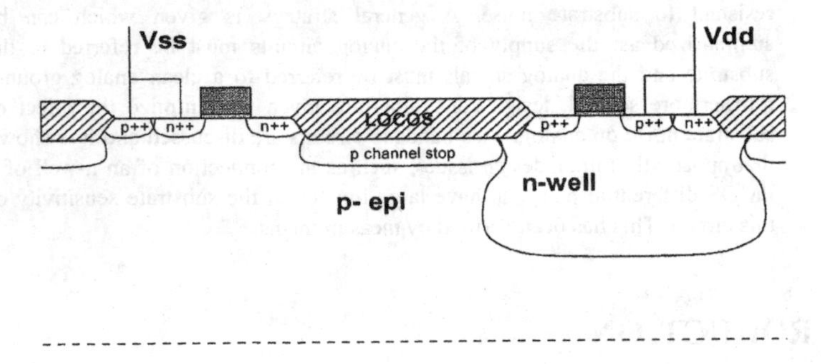

p+ SUBSTRATE

Figure 12-1. Cross section of a p-substrate p-epi CMOS IC.

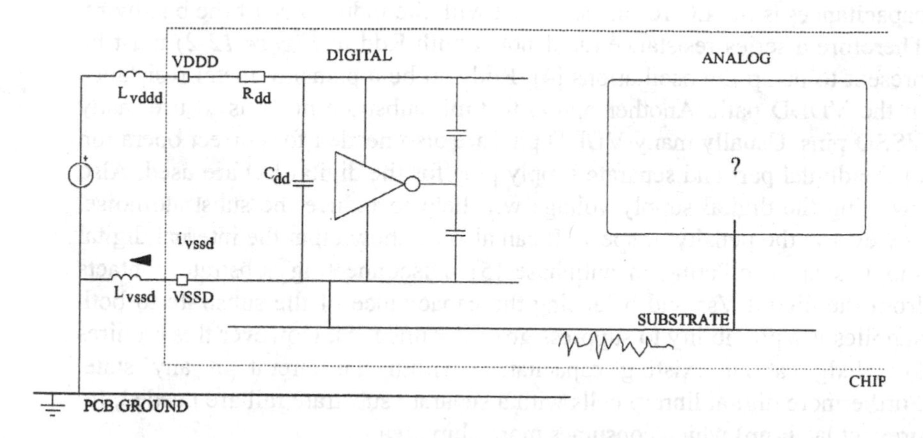

Figure 12-2. The origin of substrate noise: Switching of digital circuits results in dI_VSSD/dt, and thus in a voltage drop over the VSSD bondwire.

In *Figure 12-2* schematically the typical power routing of a digital CMOS IC is shown. The digital part of the IC is here simplified to an (huge) inverter. The digital part has its own VDDD and VSSD pins. The CMOS logic is switching and this means current spikes through the VDDD and VSSD pins. The current spikes are due to (dis)charging of capacitances and short circuit current as present in CMOS gates [3]. In the digital standard cells normally substrate contacts are present for latch-up reasons. A single substrate contact has a resistance of several kOhms, however on a large IC with - say 500k - gates, the substrate is very well connected to VSSD. The problem is now the self inductance of the VDDD and - more serious for analog - the self inductance of the VSSD pins. The self inductance of a single bondwire is typically 3nH and the resulting voltage drop over the VSSD bondwire is $V_{bondwire} = Lvssd * dIss/dt$. In a system the PCB (Printed Circuit Board) ground is usually the reference for analog and digital circuits and thus the substrate noise is equal to $V_{bondwire}$ [4].

If no precautions are taken then substrate noise can be in the order of one Volt. However in this case even pure digital circuits will not operate correctly. A practical method to limit substrate noise is to decouple the digital supplies with on-chip capacitors (as denoted with Cdd in figure 2). If these capacitors are large enough then the peak currents needed are drawn from these capacitors. A problem is that these capacitors need area, and are thus expensive. Furthermore the series resistance of these capacitors must be low (the RC time constants must be far in the GHz range). Consequently the capacitors must be merged with the logic. Another problem with these capacitances is the LC-resonance effect with the inductance of the bondwire. Therefore a series resistance (as denoted with Rdd in *Figure 12-2*) must be present to damp the oscillations [4]. Rdd can be a parasitic metal resistance in the VDDD path. Another option to limit substrate noise is to use many VSSD pins. Usually many VDDD pins are also needed for correct operation of the digital part and separate supply pins for the digital I/O are used. Also lowering the digital supply voltage will help to reduce the substrate noise, however at the penalty of speed. It can also be shown that the internal digital supplies are oscillating in antiphase [5]. Disconnecting substrate contacts from the digital Vss and balancing the capacitance of the substrate to both supplies is a possibility to suppress groundbounce [5]. However this requires knowledge about existing capacitance within the circuit in any state. Furthermore digital library cells with a separate substrate rail are needed, (to prevent latch-up) which consumes more chip area.

Substrate noise can thus be limited, be it at the cost of money; either area, pins or dedicated libraries. Therefore substrate noise will never be reduced to the millivolt level. Always some $100mV_{pp}$ will be present, dictated by proper functioning of the digital part. If the IC has a single clock domain, the frequency content of the substrate can be predicted [6] but, in general the frequency content of the substrate noise is not known, because of multiple clock domains on the same IC.

Guard rings for shielding the analog part from the digital substrate have no effect in CMOS with a low-ohmic substrate [7]. It's the same as building a fence around your house to keep it safe from earthquakes: the substrate noise comes from the bottom of the analog circuit. Substrate noise will simply be there, and the analog circuitry must therefore simply be able to deal with this interference.

3. PROBLEMS IN ANALOG

In this section the problems in analog will be discussed. Since analog circuits share the same substrate these circuits will always be affected by substrate noise. three mechanisms can be distinguished;

1) *direct uncoupling*. The - normally - high frequency substrate signal couples directly into the analog circuit. If the frequency content of the substrate signal is outside the signal band in the analog part, this needs not to be harmful as long as the analog circuit behaves like a linear circuit. This behaviour can be investigated in an AC simulation. Unfortunately true linear circuits are rare.

2) *demodulation*. If a substrate signal couples into an analog circuit which contains non-linear elements (which is normally the case) then

demodulation can occur. Even if the frequency content of the substrate noise is outside the signal band, demodulation or AM detection can result in noise in the signal band of the analog circuit. An example of this is described in section 6, where substrate noise in a bandgap reference circuit results in a DC shift of the output voltage. Transient analyses are needed to tackle these problems.

3) *sampling.* In any mixed-signal IC normally an AD converter is present, which means a sampling operation on the analog signal. If substrate noise has coupled into the analog circuits preceding the AD converter (Clamp, AGC, filter, buffer, etc.) at a frequency outside the signal frequency band, this sampling operation will fold the substrate noise back into the signal band.

In conclusion it can be mentioned that analog modules and circuits must have a good rejection for substrate signals. The rest of this paper discusses how this can be achieved.

Consider a very simple analog circuit like the current source in *Figure 12-3a*. The current source is supplied with a clean analog VSS (analog ground) and the gate has a parasitic capacitance Cgate. Due to the nature of the CMOS process, the backgate of M1 is coupled to the substrate. For high frequencies the Vgs of the MOST is fixed due to Cgate, and the substrate noise on the backgate directly modulates the drain current of M1. It is therefore better not to use a clean analog ground, but to use a VSS, equal to local substrate voltage. This is shown in *Figure 12-3b*. The AC content of the substrate noise is now present on the gate, source and backgate of M1. The drain is not coupled to the substrate and the substrate noise couples only via the drain-bulk and drain-source admittances. This is better than coupling via the backgate. For noise and matching reasons the dimensions of M1 and M2 are normally far from the minimum as dictated by the technology, resulting in still a large drain-bulk capacitance of M1 and M2. For this reason the current source is often provided with cascode transistors. *Figure 12-3c* illustrates how these cascode transistors should be biased in order to have small substrate sensitivity. Now the drain, gate, source and backgate of M1 and the gate, source and backgate of M3 have the same AC (substrate) signals. The output impedance of the current source is high including the capacitance since M3 can have small dimensions. Note that the capacitance C1 should be kept small for good substrate rejection.

Thus by choosing the right references, in this case the substrate for NMOS transistors, performance can be improved. In the next section a more general approach will be given.

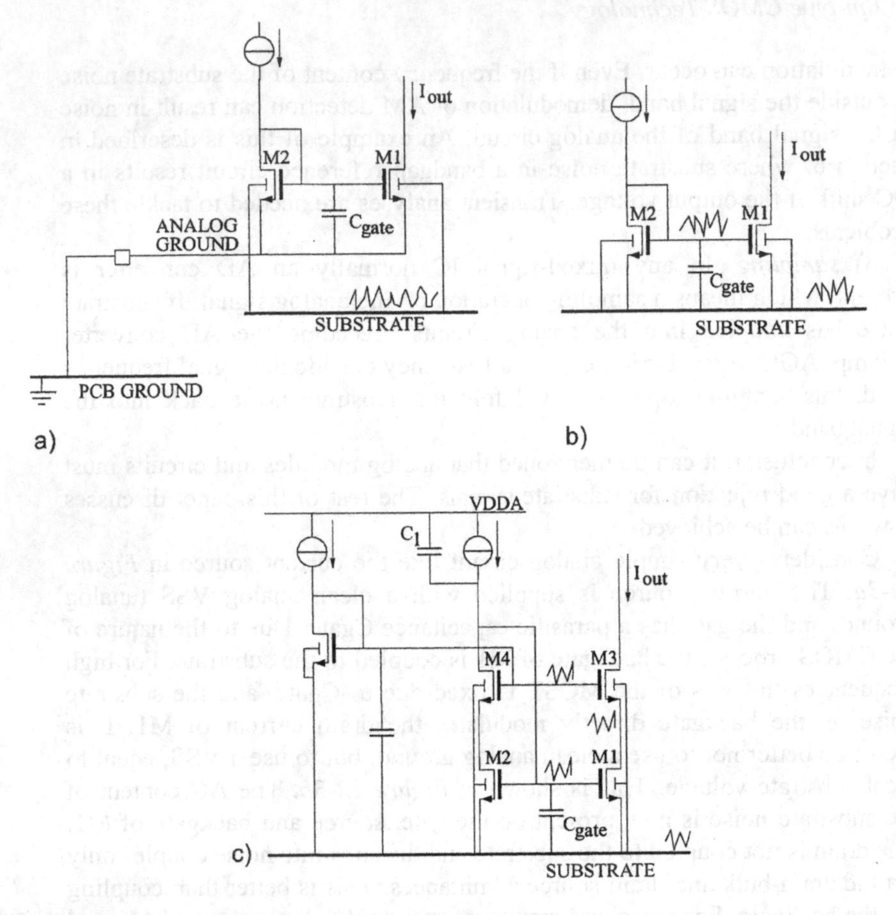

Figure 12-3. Example to illustrate the effect of substrate noise on analog circuits a)current source, realized by a current mirror, with a clean VSSA. b) improvement of the current source of 3a) with VSSA connected to the substrate c) improvement of the current source of 3b) by means of small cascode transistors.

4. STRATEGY FOR ANALOG

In this section several design rules are given to make circuits for analog signal processing less sensitive for substrate noise.[8]

. Use NMOS transistors only as DC current sources. These NMOS transistors should be referred to the substrate and not to a clean VSS.

. Use PMOS transistors for signal handling: i.e. PMOS as differential pairs and signal handling current mirrors. PMOS transistors have an n-well, which can be used to shield the transistor from the substrate. Be sure to put enough well contacts to the VDDA, and be aware of series resistances in the well, which are normally not properly modeled.

- Make analog circuits fully differential, with a possibly clean common mode level. A common mode control circuit must suppress interference. Matching and large signal behaviour is still limiting the effect of balancing.
- Use shielding of resistors and wires only for signals not referred to the substrate. Shielding can be done with n-well, poly or one of the lower metal layers, connected to VDDA. Be careful with series resistances in n-well and high-ohmic poly, since the RC time constant will limit the effect of shielding. Bondpads carrying analog signals can be shielded with n-well.

The remaining problem is now the analog interface. Analog signals are usually single-ended and defined w.r.t. the "clean" PCB ground. Inside the IC large parts of the analog circuits are referred to the substrate (all NMOS) and the signals are differential. This makes interfacing a serious matter.

Figure 12-4 shows the recommended supply and reference routing for a mixed-signal IC. The digital part has several VDDD and VSSD pins, and the substrate is contacted to VSSD in the digital part as mentioned before. The analog part has a separate VDDA, since VDDD will be polluted by the digital. The VSS of analog (VSSA) should be connected to the substrate with enough substrate contacts, and should contain the same signal as the substrate. VSSA is thus NOT clean w.r.t. PCB ground, and there is actually no difference between VSSA and VSSD. In order to interface with the outside world of the IC, a clean reference signal on chip is needed. This signal is denoted as "analog ground". Analog ground is connected to PCB ground via a bondwire and thus di/dt of this bondwire must be (almost) zero. The pin of analog ground therefore may only carry DC signals or signals with relatively low frequencies, depending on the demands. The analog ground wires on chip, must be shielded from the substrate (with n-well or lower interconnect layers, connected to VDDA) and no unwanted signals ought to couple into analog ground.

The input signal is referred to PCB ground and is fed into a first stage on the IC, which is, in the example of *Figure 12-4*, a transconductance amplifier. Important is that this first stage has analog ground as a reference for the signal. The output of the first stage is preferably differential, in order to be less sensitive to substrate or supply noise. The common mode level of the differential signals should be clean. After optional analog preprocessing the analog signal can be converted to the digital domain for further signal processing.

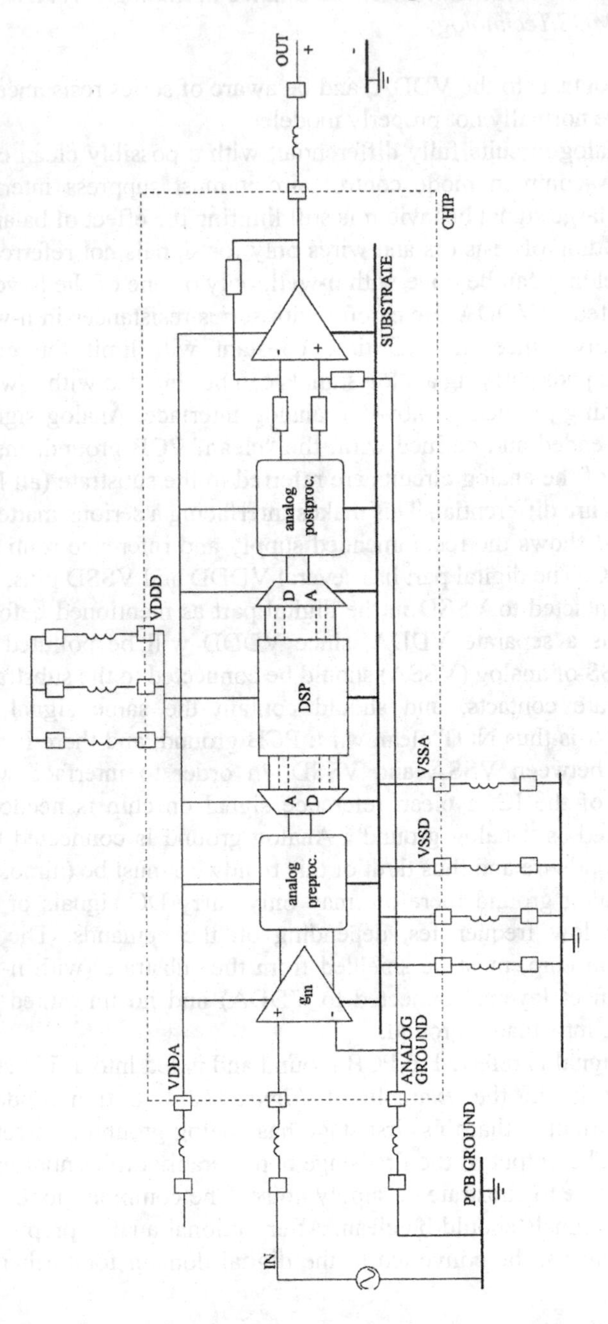

Figure 12-4. Recommended supply and reference routing for a mixed signal IC.

The signal can be converted into the analog domain via a DA converter and may be followed by postprocessing functions, such as smooth filtering. The differential signal must be converted to a single-ended signal referred to analog ground. This is also illustrated in *Figure 12-4*. It would even be better if the "outside world" signals of the IC are differential, however this is often not feasible for cost reasons.
Summarizing it can be said that: the supply of the circuits is referred to the substrate and the signals are referred to a clean analog ground.

5. EXAMPLES

In this section a practical example of the effect of substrate noise on a bandgap circuit is illustrated. *Figure 12-5a* shows a well known bandgap voltage reference [9], generating a reference voltage V_{bg} with respect to the substrate. This means that the output voltage of the bandgap circuit V_{bg} is constant w.r.t. the substrate. (and thus not constant w.r.t. analog ground, if there is substrate noise). The core of the circuit is R1, R2, R3, Q1 and Q2. The folded cascode amplifier (all MOSFETS) is responsible for the proper feedback, needed for correct bandgap operation. This bandgap is denoted here as bandgap "A".

For correct operation of bandgap A, all NMOS transistors are referred to the substrate, as illustrated in the example of a current mirror of *Figure 12-3b*. The bases of the parasitic vertical pnp transistors are also connected to the substrate. In order to keep the feedback loop stable a capacitor Cc has been added. A further advantage of Cc is that the gate-source voltage of M9 is low-pass filtered, and thus filters more or less uncoupling substrate noise.

It appeared to be very important during the evaluations of these types of bandgap circuits how the n-well of the input differential pair is connected. One can connect the n-well to VDDA or to the common source node of the differential pair. *Figure 12-6a* shows the simulation results of the two possibilities. The figure shows the bandgap voltage versus time while a 10MHz clock signal of 400mVpp is present on the substrate. If the n-well is connected to the VDDA the bandgap voltage starts drifting away from the nominal value (*Figure 12-6a*-curve 1). Note that even if the ripple is low-pass filtered, the DC value is not correct and will depend on the substrate noise, which is highly unwanted. The explanation is that the base and emitter of Q1 and Q2 follow the substrate noise. So do the gates, sources and drains of M1 and M2. If the n-well is connected to the VDDA, the source-well capacitances of M1 and M2 conduct a current related to the substrate noise and thus modulate the currents of M1 and M2. Due to non-linear effects (see section 4) this results in DC shift of the bandgap voltage. If the n-well is connected to the common source, as shown in *Figure 12-5a*, then all terminals (drain, gate, source and well) of M1 and M2 have the same substrate-related signal and the modulation does not occur. Important is that the capacitance from common source to VDDA is held small. The results of the simulations with n-well connected to the common source node is shown in *Figure 12-6a*, curve 2.

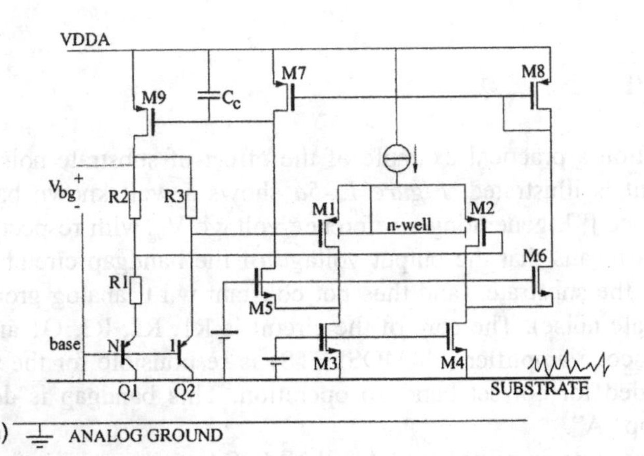

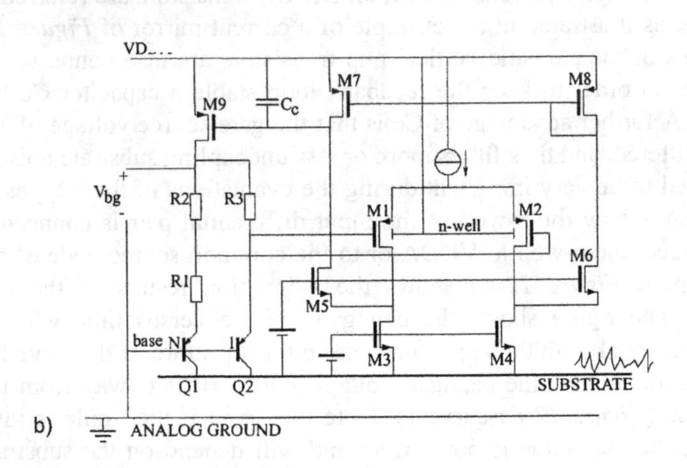

Figure 12-5. a) bandgap type A, generates a reference voltage w.r.t. substrate b) bandgap type B, generates a reference voltage w.r.t. analog ground.

The other bandgap - type B - is shown in *Figure 12-5b*. The reference voltage is now wanted w.r.t. the "clean" analog ground, and the bases of Q1 and Q2 are now connected to this analog ground. If the n-well of the differential pair M1, M2 would be connected to the common source node, then the bandgap

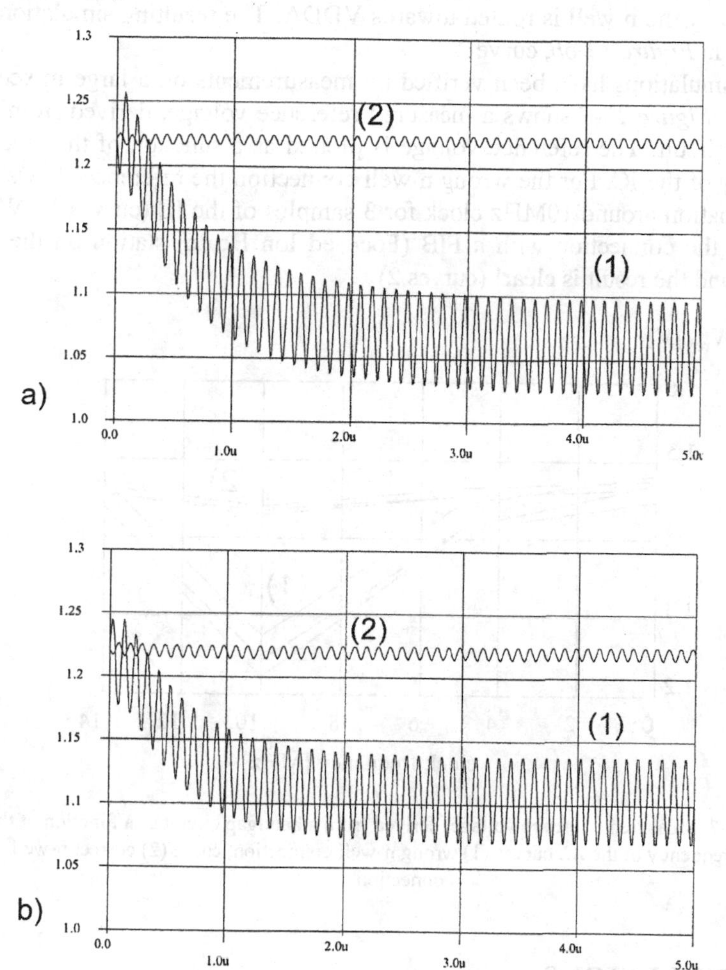

Figure 12-6. Simulation results of the bandgap reference. a) Bandgap circuit A, with output referred to the substrate. Curve 1 with n-well to VDD (wrong), curve 2 with n-well to common source node of the differential pair (correct). b) Bandgap circuit B, with output referred to the ground Curve 1 with n-well to common source node of differential pair (wrong), curve 2 with n-well to VDDA (correct).

voltage would drift away as shown in *Figure 12-6b*, curve 1. This is because the n-well to substrate capacitance picks up the substrate noise and pollutes the common source node, which should not follow the substrate noise in this case. The result is again modulation of the currents in M1 and M2. The correct connection of the n-well in bandgap B is to the VDDA. Now the main terminals of M1 and M2 (gate, source, well) are "clean". The noise picked up by the n-well is routed towards VDDA. The resulting simulations are given in *Figure 12-6b*, curve 2.

The simulations have been verified by measurements on a large mixed-signal IC. *Figure 12-7* shows a measured reference voltage, derived from a bandgap circuit. The reference voltage is plotted as a function of the clock frequency of the IC. For the wrong n-well connection the reference shows a large deviation around 10MHz clock for 3 samples of the IC (curves 1). We modified the connection with a FIB (Focused Ion Beam) station on the 3 samples and the result is clear! (curves 2)

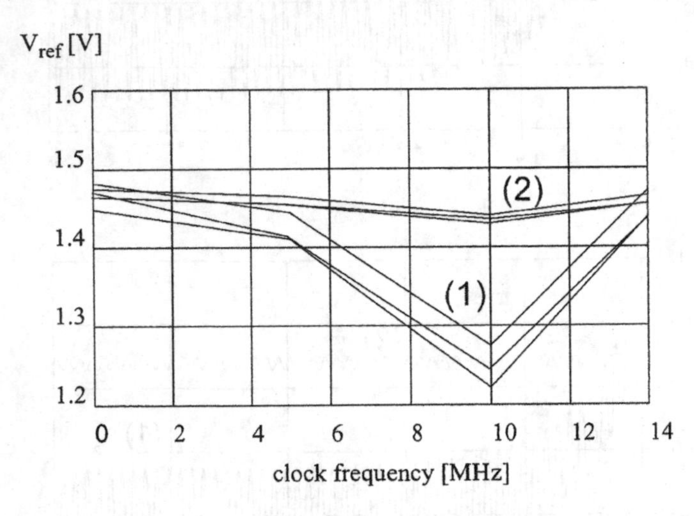

Figure 12-7. Measured reference voltage, derived from a bandgap circuit as a function of the clock frequency of the IC. curves (1) wrong n-well connection, curve (2) correct n-well connection.

6. CONCLUSIONS

Substrate noise is one of the key problems in mixed analog/digital ICs. Although measures are known to reduce substrate noise, the noise will never be completely eliminated for cost reasons. Analog circuits on digital ICs simply have to be resistant to substrate noise. A general strategy has been given which can be summarized as: the analog circuits must be referred to the substrate and the analog signals are referred to a clean analog ground. Furthermore several design constraints are given to minimize the effect of substrate noise on analog. Two bandgap circuits have been discussed and it has been shown that apparently minor design issues, such as the connection of an n-well of a PMOS differential pair, can have large impact on the substrate sensitivity of this circuit. This has been verified with measurements.

ACKNOWLEDGMENT

The techniques described in this chapter are a result of the discussions with colleagues within Philips Research and Philips Semiconductors. Special thanks are given to W. Relyveld for doing the bandgap simulations.

REFERENCES

[1] X. Aragones, A. Rubio, "Experimental comparison of substrate noise coupling using different wafer types" *IEEE Journal of Solid-State Circuits*, vol 34, pp 1405-1409, October 1999

[2] D. Leenaerts and P. de Vreede, Influence of Substrate Noise on RF performance, *Proceedings ESSCIR 2000*, 19-21 September, Stockholm, Sweden.

[3] H.J.M. Veendrick, "Short-Circuit Dissipation of Static CMOS Circuitry and its Impact on the Design of Buffer Circuits", *IEEE Journal of Solid-State Circuits*, Vol. SC-19, August 1984, pp 468-473.

[4] D.W.J. Groeneveld, "Ground bounce in CMOS", *Digest of technical papers, workshop on Advances in Analog Circuit design*, Leuven 1993.

[5] T.J. Schmerbeck, R.A. Richetta and L.D. Smith, "A 27MHz Mixed Analog/Digital Magnetic Recording DSP Using Partial Response Signaling with Maximum Likelihood Detection", *Digest of technical papers, ISSCC* 1991,, Feb 1991, pp 136-137.

[6] M. van Heijningen, J. Compiet, P.. Wambacq, S. Donnay, M.G.E. Engels and I. Bolsens, " Analysis and experimental verification of digital substrate noise generation for epi-type substrates", *IEEE Journal of Solid-State Circuits*, vol. 35, pp 1002-1008, July 2000.

[7] D.K. Su, M.J. Loinaz, S. Masui, B.A. Wooley, "Experimental Results and Modeling Techniques for Substrate Noise in Mixed-Signal Integrated Circuits", *IEEE Journal of Solid-State Circuits*, Vol. 28, No. 4, April 1993, pp420-430.

[8] B. Nauta and G. Hoogzaad. " Substrate bounce in Mixed-Mode CMOS IC's" *AACD'98* Denmark.

[9] K.E. Kuijk, "A reference voltage source" *IEEE Journal of Solid-State Circuits*, vol SC-8, pp 222-226, June 1973.

Chapter 13

REDUCING SUBSTRATE BOUNCE IN CMOS RF-CIRCUITRY

On the use of guard rings

Domine M.W. Leenaerts
Philips Research Laboratories

Abstract: We will discuss the use of guard rings as a mean to reduce the effects of substrate bounce in a mixed-signal IC. Measurements have been performed on lightly and heavily doped substrates in several CMOS technologies. Furthermore, we will show some of the problems of substrate bounce in RF applications where the substrate bounce is caused by digital circuitry.

1. INTRODUCTION

Given a particular technology, the designer sometimes has the freedom to choose the substrate (or bulk). Although this option is sometimes provided in bipolar technologies, the type of substrate is mainly a point of discussion in CMOS technologies. Low-ohmic substrates together with a high-ohmic epitaxial layer were used in the past; mainly to increase the immunity to the latch-up problem. Low-ohmic substrate has a resistivity of several milli ohm cm (Ω·cm). Due to the process scaling and the related lower voltage supplies, high-ohmic substrates can now be used without latch-up problems. High-ohmic substrate is available in several gradations, from a few ohm cm up to several 100 Ω·cm. However, most foundries use a 10-15 Ω·cm substrate for their CMOS technologies.

As the back gate of a MOS device is connected to the substrate, it is clear that the type of substrate plays a major role in the RF performance of a CMOS circuit.

The substrate is the connecting layer between all the circuits on a single die. If one circuit generates small (often spiky) signals, and hence the name noise, on the substrate, other circuits will be influenced by this noise. The amount of generated noise in the substrate is therefore a main concern in the design, especially when large parts of digital circuitry are laid out on the same die. There are several paths to inject noise into the substrate as already mentioned in previous chapters. A nice overview is provided in [1].

The major source of noise injection is normally due to the bond wires. The MOS logic is switching and produces current spikes through both the positive and ground supply lines, and hence through the supply pins. In digital standard cells, substrate contacts are normally present for latch-up reasons. The substrate is very well connected to the digital ground on-chip as a result of the multiple substrate contacts. The self inductance of a bond wire causes a voltage drop v_{spike} proportional to the current spikes i_{spike},

$$v_{spike} = L_{bondwire} \cdot di_{spike}/dt$$

This has been visualized in *Figure 13-1*. The actual ground is defined on the PCB, which via the bond wire is connected to the digital ground on-chip. Because of the multiple substrate contacts, the substrate potential is therefore equal to v_{spike}.

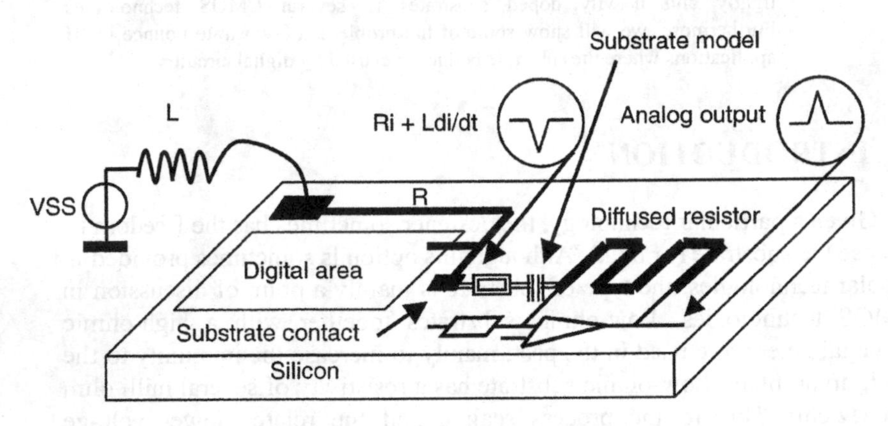

Figure 13-1. Illustration of the cause of substrate bounce. Current spikes from digital switching activities are injected into the substrate via bond wires.

Substrate noise cannot be prevented, because any type of package has bond wires or bond bumps. However, by using some design tricks, we can minimize the noise or, to a certain extent, make the circuits immune to the noise. Consider the simple current mirror in *Figure 13-2*. If the analog ground is separated from the substrate, the substrate noise on the back gate of the MOS transistors will directly modulate the drain current. This is due to the parasitic capacitance of the MOS device. The supply of the analog circuits should therefore be referred to the substrate. The substrate noise is then present at the gate, source and back gate of the transistor and will therefore not (or to a lesser degree) modulate the drain current [2-3][see also Chapter 12]. For noise and matching reasons, the dimensions of both transistors are normally far from the minimum dimensions, still resulting in a large drain-bulk capacitance. Cascode transistors may also be used to overcome this problem.

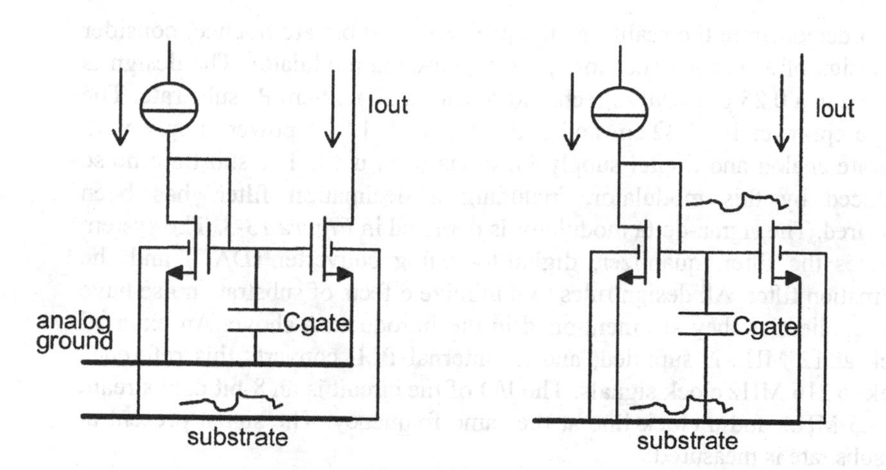

Figure 13-2. Illustration of the effect of substrate noise on analog circuits. On the left side is a current mirror with a clean analog ground. On the right side the mirror is noise immune because the analog ground is connected to the substrate.

A few design rules can be derived from the example:

- Make analog circuits fully differential with a possible clean common-mode signal. The accuracy of balancing is mainly determined by matching properties.
- Use NMOS transistors only as DC current sources, and refer them only to the substrate and not to a clean ground supply. Use PMOS transistors as differential pairs and other signal handling blocks. These transistors have an N-well which can be used to shield the transistor from the substrate. In RF applications, this rule is difficult to fulfill.
- Use different supply lines for analog and digital.

Most designers apply these commonly accepted rules to their designs. The question remains if these rules are enough to bring the substrate bounce down to an acceptable level and if they also apply for RF designs.

Section 2 will discuss the first question and in section 3 until section 5 we will focus on the second question. We will give some conclusions in section 6.

2. SUBSTRATE BOUNCE DUE TO A SIGMA-DELTA MODULATOR

To demonstrate the reality of the problem of substrate bounce, consider the design of a second-order low-pass sigma-delta modulator. The design is realized in a 0.25 μm CMOS technology with a 10 mΩ·cm P^+-substrate. The P-type epi-layer is 11 Ω·cm and 3 μm thick. A 1.8 V power supply with separate analog and digital supply lines has been used. The substrate noise produced by this modulator, including a decimation filter, has been measured. The sigma-delta modulator is depicted in *Figure 13-3.* The system includes the filter, quantizer, digital-to-analog converter (DAC) and the decimation filter. All design rules to minimize effects of substrate noise have been applied, as they are mentioned in the introduction above. An external clock at 12 MHz is supplied, and an internal PLL converts this reference clock to 216 MHz clock signals. The I/O of the circuit is an 8-bit data stream at 13.5 MHz, and a clock line at the same frequency. The signal present at the substrate is measured.

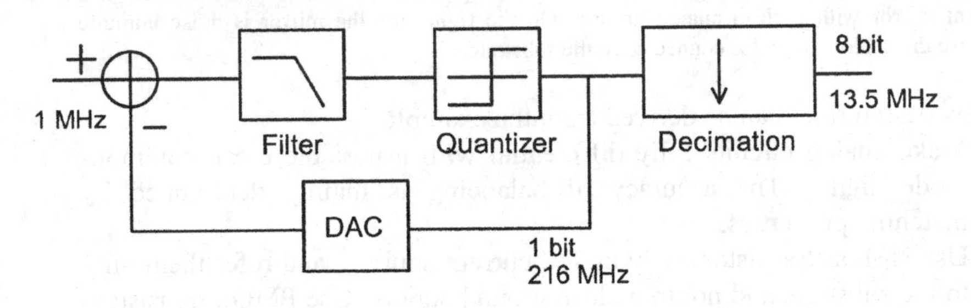

Figure 13-3. Second-order low-pass sigma-delta modulator including the decimation filter.

Figure 13-4 shows the spectrum of this signal when the power supply of the chip and the clock are turned off. The spectrum is flat above 1 GHz (except that we measure some interference from the GSM/DCS band). The substrate bounce has power levels below –90 dBm.

Figure 13-5 shows the spectrum of the substrate signal when the system is active. Odd multiples of 13.5 MHz are visible within the whole band up to the RF frequencies. The substrate noise has been increased by 20 dB in the RF band, and even more at lower frequencies. According to these measure-

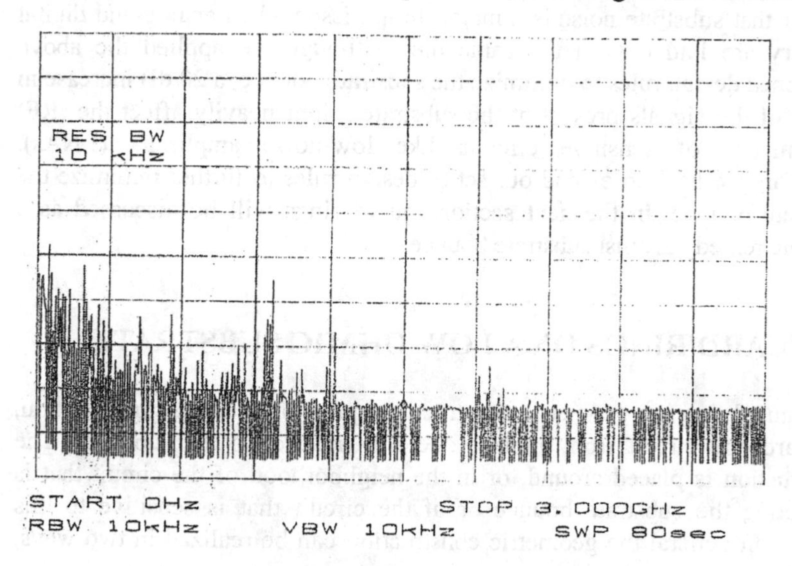

Figure 13-4. Spectrum of the signal present at the substrate when the system is turned off. (RL: -10 dBm, 10 dB/div).

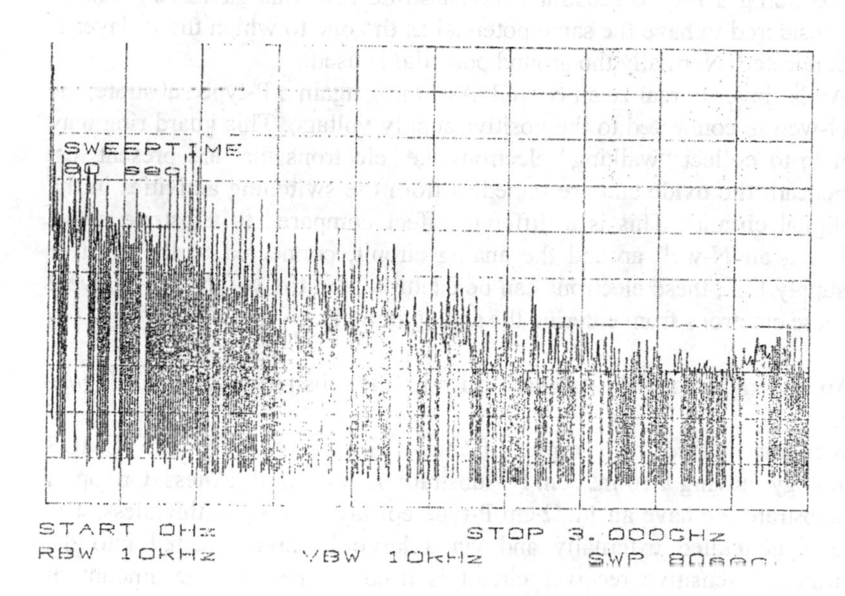

Figure 13-5. Spectrum of the signal present at the substrate when the system is active. (RL: -10 dBm, 10 dB/div).

ments, the substrate noise is concentrated at multiples of the digital clock frequency, also observed in other measurements. From the measurements, it is clear that substrate noise is a major design issue when analog and digital circuitry are laid out on the same die. Although we applied the above mentioned design rules to minimize the substrate bounce, a 20 dB increase in power of the signals present at the substrate might heavily affect the (RF) performance of sensitive circuits like low-noise amplifiers (LNAs). Therefore we have to extend our set of design rules to further minimize the substrate bounce. In the next sections guard rings will be discussed as a possible remedy against substrate bounce.

3. GUARD RINGS ON A LOW-OHMIC SUBSTRATE

A guard ring is a geometric construction, quite often a ring configuration, to guard one circuit from another one with respect to the substrate. The construction is placed around, or in the neighborhood of the circuit that is generating the substrate bounce or of the circuit that is sensitive to this bounce. In general the geometric construction can be realized in two ways, regardless the type of substrate:

1. A P^+ layer can be used with many contact points to the substrate. Assuming a P-type substrate, the substrate near this guard ring can be considered to have the same potential as the one to which the P^+ layer is connected. Normally the ground potential is used.
2. An N^+ layer to realize an N-well. Assuming again a P-type substrate, the N-well is connected to the positive supply voltage. This guard ring may help to collect "walking" electrons, i.e. electrons that are present just beneath the oxide and are remedies from the switching activities in the digital circuits. This is a different effect compared to substrate noise. Using an N-well around the analog circuit, connected to the positive supply line, these electrons can be captured and the N-well will prevent these electrons from entering the analog domain.

An illustration of the two guard rings in a P-substrate is given in *Figure 13-6.*

We have tested both types of guard rings in a 0.18 μm CMOS technology, having a 10 mΩ·cm P^+-substrate of 200 μm thickness. On top of this substrate we have an 11 Ω·cm P-type epi layer of 3 μm thickness. The noise is generated externally and via a large P^+ area injected into the substrate. A sensitive receiver circuit is used to measure the amount of substrate bounce. The measured noise at the receiver side in case of no guard rings is defined as the reference level. The described set up is similar to the one in [6].

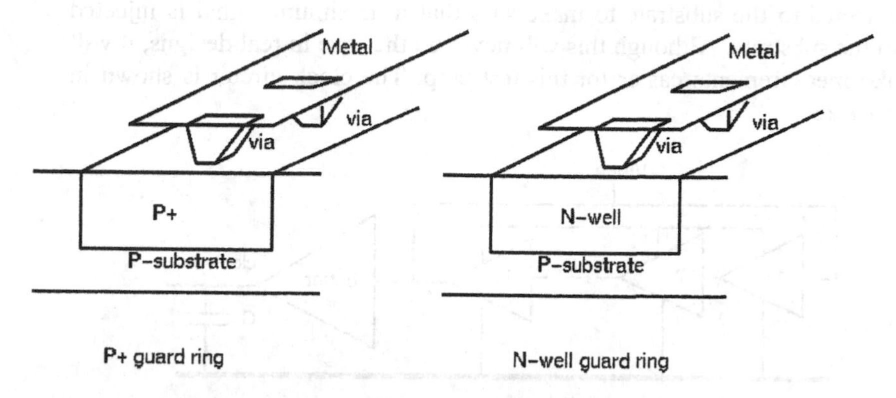

Figure 13-6. Illustration of a P^+ guard ring and a N-well guard ring in a P-substrate.

When a grounded P^+ guard ring is placed around the receiver, we obtain a 13 dB attenuation of the substrate bounce inside the ring. So, indeed such a ring might help although the improvement is not that much. The small positive effect can be explained by the fact that the epi layer near the receiver has been grounded. Therefore one noise path has been eliminated, i.e. the epi layer. Placing only an N-well ring around the receiver does not bring any improvement. The N-well, acting as a reversely biased diode, will not eliminate the noise path through the epi-layer. If around the P^+ guard ring also an N-well guard ring is placed, the attenuation is only 4 dB, indicating that the positive effect of the P^+ guard ring is deteriorated by the N-well.

The problem of having a low-ohmic substrate is the fact that the substrate can be considered as a single node. Therefore the entire disturbance on this node will be seen by all components connected to this node. Guard rings will not improve the situation too much. Preferably for RF applications this type of substrate is not used. We will now give an RF application example to demonstrate the catastrophic influence of substrate bounce on the RF performance while using a CMOS technology with a low-ohmic substrate.

As example, consider the following test chip, where a digital clock oscillator is injecting noise onto the substrate and the RF circuit is a dedicated LNA. The test chip is realized in a 0.25 μm CMOS technology with five metal layers, and a 10 mΩ·cm P^+-substrate beneath a 11 Ω·cm P-type epi layer of 3 μm thickness. The clock circuit consists of an 11-stage ring oscillator based on simple inverters. Each inverter has the following dimensions: $(W/L)_{NMOS}$=1.6μm/0.5μm and $(W/L)_{PMOS}$=2.4μm/0.5μm. The frequency can be tuned using the supply voltage. The frequency f_{cl} is set to 772 MHz during the measurements. A five-stage buffer that is loaded by a MOS capacitor of 160 fF follows the clock circuit. The capacitor is tightly connected to the substrate to make sure that a maximum signal is injected into the substrate. Although this will never be the case in real designs, it will make measurements easier for this test chip. The clock circuit is shown in *Figure 13-7*.

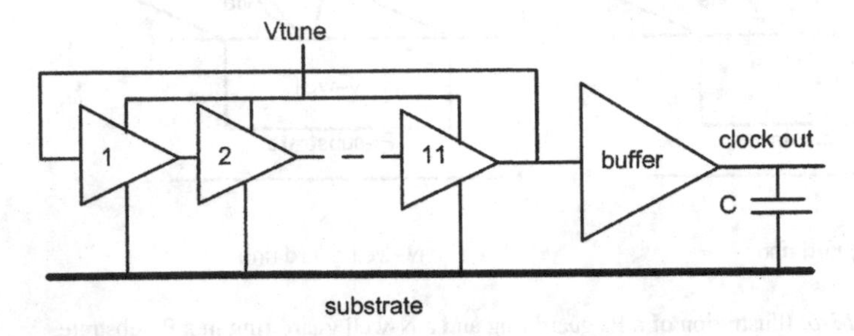

Figure 13-7. Clock circuit.

The die microphotograph of the test chip is shown in *Figure 13-8*. There is also a P^+ guard ring on the chip, laid out around the LNA, to see if it is possible to prevent interference of the substrate noise generated near the clock circuit by the LNA. This guard ring is connected to the ground via a DC probe. There is also a test point available in the guard ring to measure substrate bounce.

As the measurements have been performed on wafers, no bond wires have been used and the measured noise in the substrate is due to the p-n junctions [4-5].

There are several locations on the test chip to measure the substrate noise. One such measurement point is near the clock output. The measured spectrum of the substrate noise is shown in *Figure 13-9.* The fundamental tone and several of its harmonics are clearly visible. Even the 4th harmonic is still at -80 dBm, which is above the sensitivity level of many mobile and wireless telecommunication standards.

The measured spectrum of the substrate noise within the guard ring is identical or almost identical to the spectrum measured near the clock, outside the guard ring. Even measuring the substrate noise at a distance of 1 cm from the clock circuit gave the same spectral results. We can therefore conclude that guard rings will not guard the circuit against substrate bounce on low-ohmic substrates. The substrate can be considered to be a very good conducting plate, and it can therefore be modeled in the circuit as a single node.

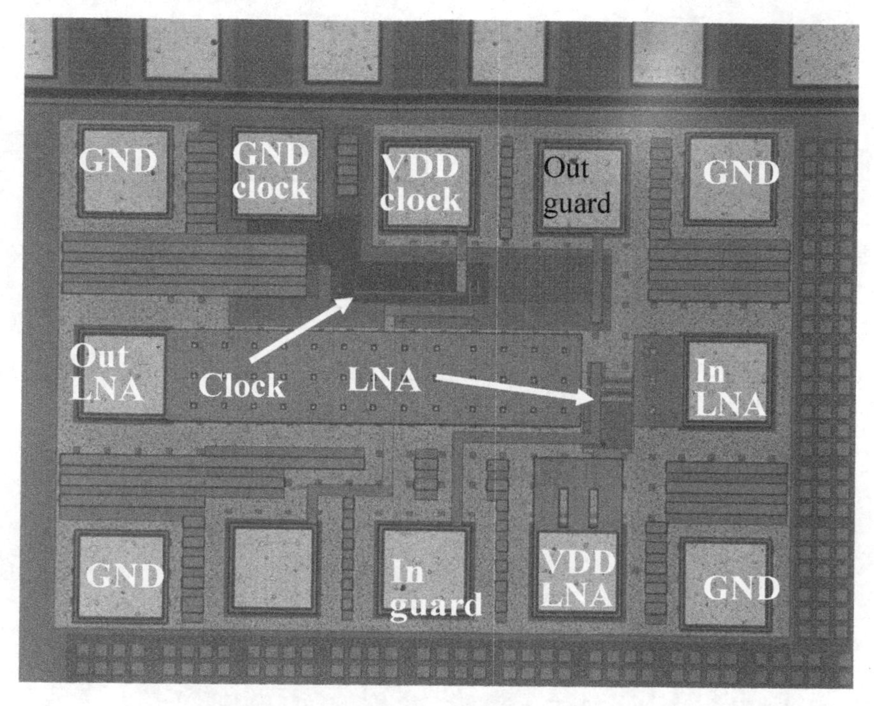

Figure 13-8. Test chip to measure substrate noise. The clock circuit is on the top, the LNA is on the right. Connections to the substrate in and outside the guard ring have also been added.

As can be seen from the measurement results in *Figure 13-9*, the clock frequency and multiples of this frequency are the dominant presence in the substrate. These signals will therefore be injected into the input and output nodes of an analog or RF circuit. As an example, consider the test chip of *Figure 13-8* once again. The measured spectrum available at the output of the LNA is shown in *Figure 13-10*. The LNA has an input signal f_{LNA} at 2.3 GHz, which is three times the clock frequency. The fundamental frequency of the clock and its harmonics are visible in the spectrum. It is clear from the measurement results that it is not only the amplified information signal that is present at the output of the LNA, but also the unwanted third harmonic of the clock. In fact, any substrate noise, with spectral components in the same frequency range as that in which the LNA is operating, will be picked up by the LNA. The information signal will therefore be distorted by this substrate noise. The gain of the LNA is not influenced by the substrate noise, simply because the injected substrate signals are too weak to shift the bias operating point of the LNA.

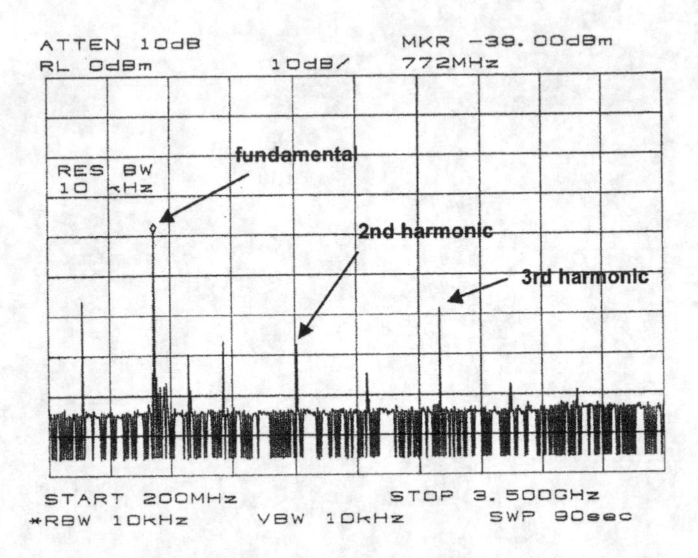

Figure 13-9. Measured spectrum of the substrate inside the guard ring.

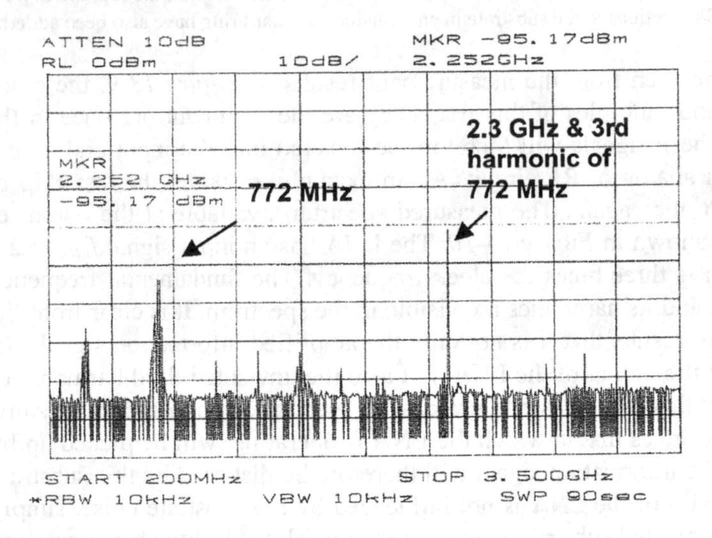

Figure 13-10. Measured spectrum at the output of the LNA when both the LNA and clock are operational.

The s-parameters were measured both when the clock was active and when it was turned off. These measurements gave the same results. Hence, the performance (gain) of the LNA is not affected by the substrate noise. This is to be expected since the RF performance is mainly determined by the bias operating point and the matching network. The 3rd harmonic of the clock is too weak to disturb this operating point. These measurements show us that clock planning is essential in a system with RF and digital circuitry on the same die.

The above experiments show that the digital circuitry injects switching noise into the substrate. In case of a low-ohmic substrate you cannot prevent that this noise is then injected into the signal paths of the analog circuitry, regardless of any guard rings laid out around the analog circuitry. The question then arises if guard rings have effect if a high-ohmic substrate is used. We will consider this option in the next section.

4. GUARD RINGS ON A HIGH-OHMIC SUBSTRATE

We have repeated the experiment of the noise injection and the sensitive receiver of the previous section in the same 0.18 μm CMOS technology, but now with a 10 Ω·cm P-substrate of 200 μm thickness. We will treat this type of substrate as high-ohmic substrate, although substrates with a much higher resistance exist.

Again we observe that only an N-well guard ring does not give any improvement. The noise path beneath the N-well is not eliminated; the situation is similar to the one where we had an epi-layer. However, a single P^+-guard ring now gives a 40 dB attenuation of the received noise compared to the situation when using no guard ring. In this case it also does matter at which distance the noise source is located from the sensitive receiver. At a 100 μm distance we have 40 dB attenuation, at 50 μm distance this reduced to 34 dB and at 25 μm distance we only have 24 dB attenuation. To explain this behavior, consider *Figure 13-11*. The resistances R2 and R4 are representing the substrate resistance and are a function of the distance between the two nodes of the resistor. The P^+ guard ring connects R3 to ground. Resistance R3 is small because of the P^+-P connection. Therefore, the voltage across R3 is a strongly attenuated copy of the injected noise Vnoise. Consequently, the noise picked up at node Vcircuit is also small. Increasing the distance between the source and the guard ring, increases the resistances R2 and R4 and thereby also the attenuation factor.

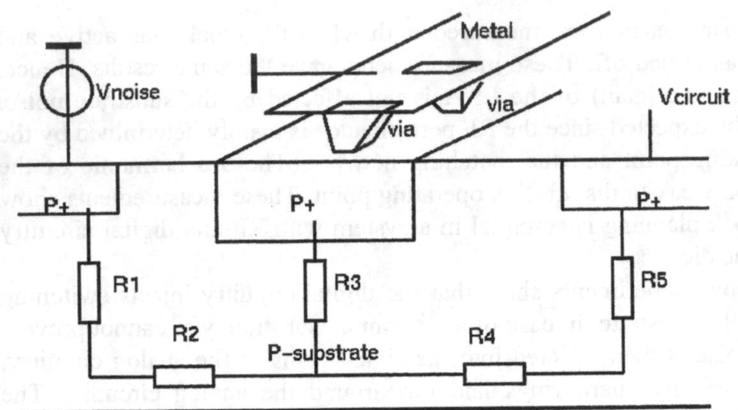

Figure 13-11. Illustration of the P^+ guard ring shielding mechanism in a high-ohmic substrate.

We have observed the above described effect up to 1 GHz, but we believe that it will remain for higher frequencies because of the resistive nature of the effect, rather than a capacitive nature.

The positive effect of using lightly-doped substrate with respect to substrate bounce was already noted in [7]. In their experiment, the authors did not use guard rings and achieved only attenuation by a factor of three in the noise levels. The proper use of guard rings is evident from our experiments.

5. SUBSTRATE BOUNCE IN AN RF SYSTEM

From the above experiments and measurements it is clear that clock planning is of major concern when realizing a mixed-signal system on a single die. Harmonics of the clock signal may give spurious tones in the band of operation, leading to distortion of the wanted signals.

Suppose we want to realize a RF front-end system for the Bluetooth communication standard. Bluetooth has 78 channels of 1 MHz spacing each, resulting in a bandwidth of 78 MHz starting at 2.402 GHz. Digital circuitry on the system is needed to perform (de-) modulation of the signals and to perform analog-to-digital and digital-to-analog conversion of the signals. A central clock frequency is needed to control this digital and mixed-signal circuitry.

It would be preferable to have a clock frequency such that odd multiples of the frequency will not fall in the Bluetooth bandwidth. For instance, the 41th harmonic of 58 MHz is at 2.37 GHz and the 43th harmonic at 2.49 GHz, both outside the Bluetooth band, making this frequency a suitable clock frequency. Obviously, the actual clock frequency is also a result of available crystals and the quality of the synthesizer. We will now demonstrate that the digital clock can have a serious impact on the performance of the RF circuitry, even if all kinds of precautions have been taken into account.

The RF front-end has been realized in a 0.18 μm CMOS technology with a 10 Ω·cm P-substrate of 200 μm thickness. The bond pad ring consists of two separate parts; one part for the analog and RF circuits and one part for the digital circuitry. In this way crosstalk over the bond pad ring is minimized. This setup also implies that the digital and RF circuitry have separate supply domains. The system has been packaged in a standard LQFP package, where the bond wire length is approximately 1 mm.

We will measure the influence of the digital part on the performance of a voltage-controlled oscillator (VCO). This VCO delivers a -5 dBm output power in the load. For measurement purposes the VCO is tuned at 2.379 GHz. The digital circuitry can be considered as 250000 gates, which will be clocked synchronously. This situation can be considered as worst case because it will generate tones at multiples of the clock frequency rather than a spread spectrum when it was asynchronously clocked. We will vary the clock frequency.

In order to reduce substrate bounce from the digital part to the analog and RF part, an N-well guard ring is placed around the digital part. As mentioned earlier, the N-well collects the "walking" electrons of the digital circuitry and prevents them to penetrate into the analog environment. The N-well / P-substrate configuration acts as a diode. Furthermore, each RF circuit has its own P^+ guard ring, connected to a ground line. This guard ring attenuates the substrate bounce.

The measured output spectrum of the VCO is shown in *Figure 13-12* in the situation that the digital part is powered down and no clock frequency is provided. We can observe a rather clean spectrum with no spurious tones. We will now activate the digital clock frequency and power on the digital circuitry. The output spectrum changes dramatically, as can be observed in *Figure 13-13* until *Figure 13-15*, showing the situation for a 13 MHz, a 40 MHz and a 64 MHz clock frequency, respectively. Spurious tones at multiples of the clock frequencies are visible in the output spectrum of the VCO. Clearly frequency modulation takes place between the fundamental frequency of the VCO and the clock frequency. These spurious tones are approximately 50 dBc down. Because the digital circuit has no connections with the VCO, these spurious tones are the result of substrate bounce. It is also interesting to observe that the spectrum at the left-hand side of the VCO frequency is different from that at the right-hand side. This means that the injection mechanism is nonlinear. At the moment there is no good explanation for this behavior.

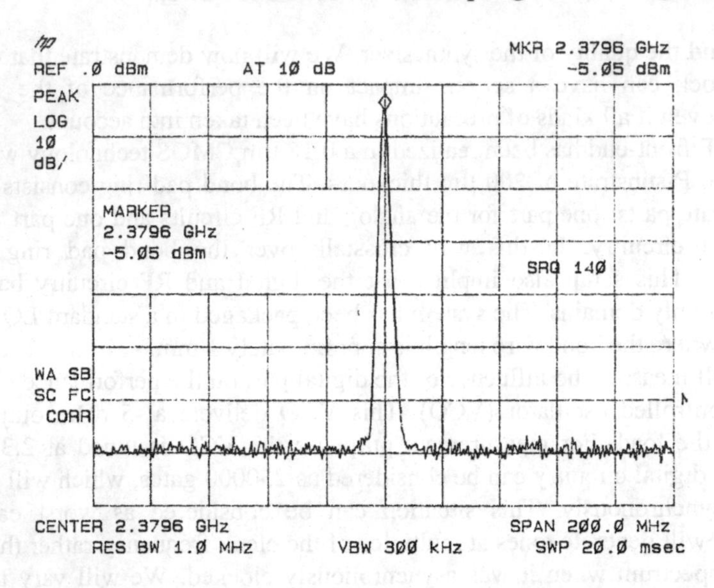

Figure 13-12. Measured output spectrum of the VCO when the digital clock is not active.

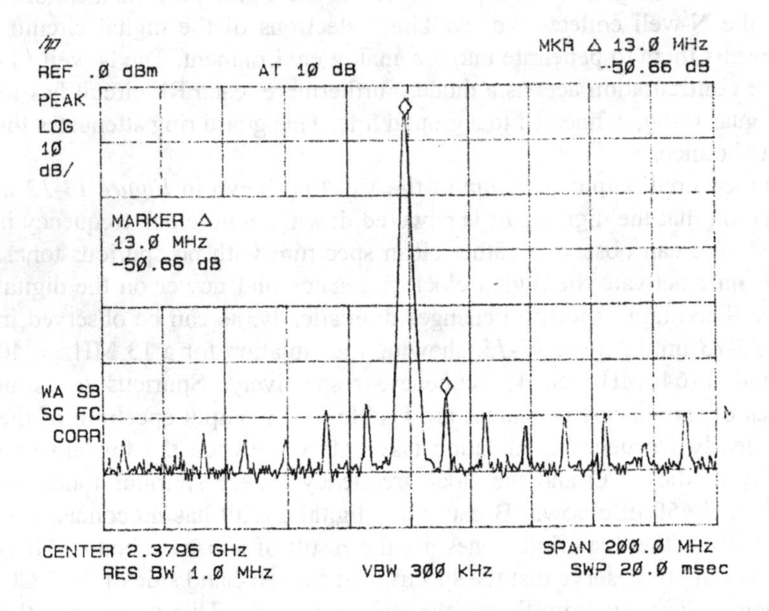

Figure13-13. Measured output spectrum for a 13 MHz clock frequency.

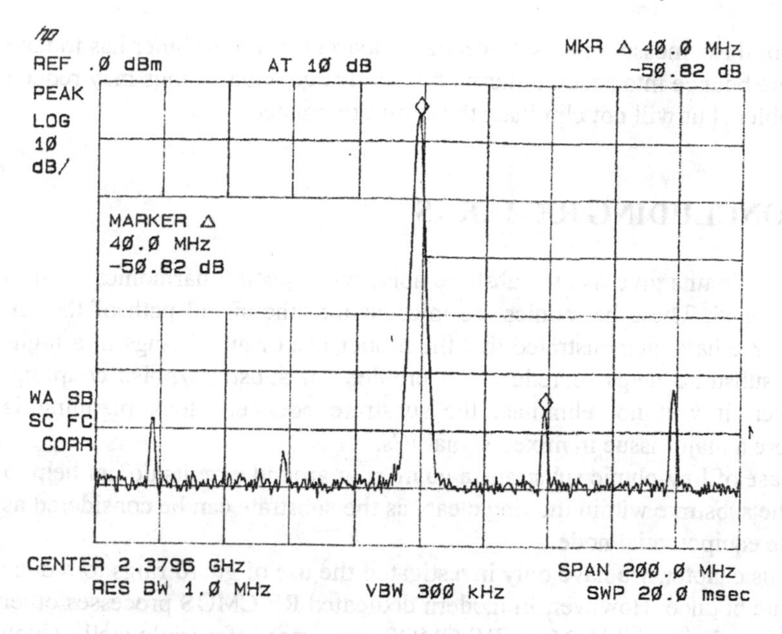

Figure 13-14. Measured output spectrum for a 40 MHz clock frequency.

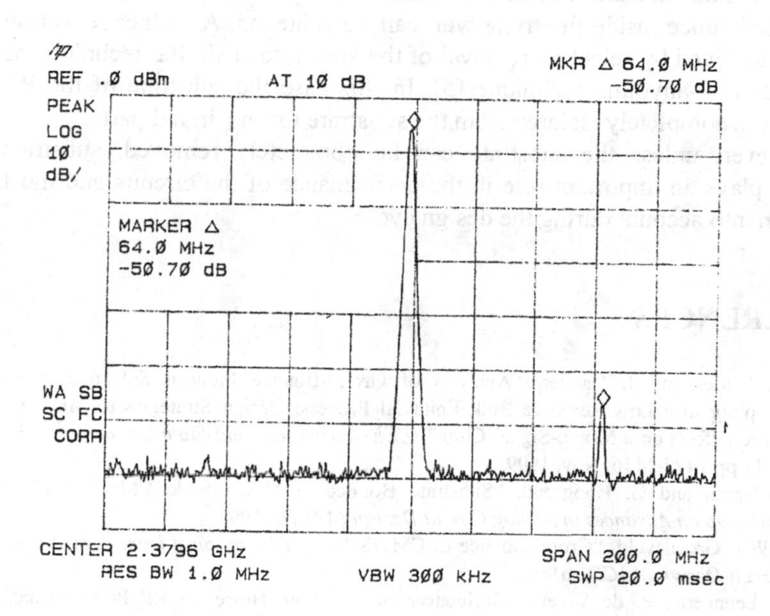

Figure 13-15. Measured output spectrum for a 64 MHz clock frequency.

From these measurements it becomes apparent that a designer has to take substrate bounce into account during his/her design. Guard rings may reduce the problem but will not eliminate the substrate bounce.

6. CONCLUDING REMARKS

Clock circuits give rise to substrate noise with spectral harmonics far into the RF band. These harmonics are injected into the signal path of the RF circuit. We have demonstrated that the proper use of guard rings in a high-ohmic substrate helps to reduce the problem of substrate noise coupling. However, it will not eliminate the substrate bounce. Clock planning is therefore a major issue in mixed-signal ICs.

In case of low-ohmic substrate, a guard ring around circuits do not help to keep the substrate within the ring clean, as the substrate can be considered as a single equipotential node.

In this chapter, we have only investigated the use of guard rings to reduce substrate bounce. However, in modern dedicated RF CMOS processes other means are also possible. Many RF CMOS processes offer triple well, which allows the designer to isolate the NMOS device from the substrate. Measurements indicate that an additional 40 to 50 dB attenuation of the substrate bounce inside the triple well can be achieved. A further reduction can be achieved by selective removal of the substrate, a similar technique as the silicon-on-anything technique [5]. In this case the substrate of the RF circuitry is completely isolated from the substrate for the digital part.

However, unless the substrate can be completely removed, substrate bounce plays an important role in the performance of the circuits and must be taken into account during the design cycle.

REFERENCES

[1] M. Felder and J. Ganger, "Analysis of Groud-Bounce Induced Substrate Noise Coupling in a Low Resistive Bulk Epitaxial Process: Design Strategies to Minimize Noise Effects on a Mixed-Signal Chip," *IEEE Journal of Solid-State Circuits*, vol. 34, no.11 pp.1427-1436, Nov. 1999.

[2] B. Nauta and G. Hoogzaad, "Substrate Bounce in Mixed-mode CMOS ICs," in *workshop on Advances in Analog Circuit Design (AACD)*, 1998.

[3] D.W.J. Groeneveld, "Groundbounce in CMOS," in *workshop on Advances in Analog Circuit Design (AACD)*, 1993.

[4] D. Leenaerts, P. de Vreede, "Influences of Substrate Noise on RF Performance," *European Solid-State Circuits Conf. (ESSCIRC)*, 2000, pp. 300-303.

[5] D.K. Su, et.al. 'Experimental results and modeling techniques for substrate noise in mixed-signal integrated circuits', *IEEE Journal of Solid-State Circuits*, vol. 28, no.4 pp.420-430, April. 1993.

[6] D. Leenaerts, J. van der Tang, C. Vaucher, *Circuit Design for RF Transceivers*, Kluwer Academic Publishers, Dordrecht, 2001.

[7] M. van Heijningen, et.al. 'Modeling of Digital Substrate Noise Generation and Experimental Verification Using a Novel substrate Noise Sensor', proc. ESSCIRC'99, pp.186-189, 1999 Duisburg.

[8] X. Aragonès, A. Rubio, "Experimental Comparison of Substrate Noise Coupling Using Different Wafer Types," *IEEE Journal of Solid-State Circuits*, vol. 34, no.10 pp.1405-1409, Oct. 1999.